Integration of General Relativity and Quantum Theory: Octonion Cosmology, GiFT, Creation/Annihilation Spaces CASe, Reduction of Spaces to a Few Fermions and Symmetries in Fundamental Frames

Stephen Blaha Ph. D.
Blaha Research

Pingree-Hill Publishing
MMXXI

ISBN: 978-1-7372640-8-8

Rev. 00/00/01 December 12, 2021

To Margaret

Some Other Books by Stephen Blaha

All the Megaverse! Starships Exploring the Endless Universes of the Cosmos using the Baryonic Force (Blaha Research, Auburn, NH, 2014)

SuperCivilizations: Civilizations as Superorganisms (McMann-Fisher Publishing, Auburn, NH, 2010)

All the Universe! Faster Than Light Tachyon Quark Starships & Particle Accelerators with the LHC as a Prototype Starship Drive Scientific Edition (Pingree-Hill Publishing, Auburn, NH, 2011).

Unification of God Theory and Unified SuperStandard Model THIRD EDITION (Pingree Hill Publishing, Auburn, NH, 2018).

The Exact QED Calculation of the Fine Structure Constant Implies ALL 4D Universes have the Same Physics/Life Prospects (Pingree Hill Publishing, Auburn, NH, 2019).

Unified SuperStandard Theory and the SuperUniverse Model: The Foundation of Science (Pingree Hill Publishing, Auburn, NH, 2018).

Quaternion Unified SuperStandard Theory (The QUeST) and Megaverse Octonion SuperStandard Theory (MOST) (Pingree Hill Publishing, Auburn, NH, 2020).

Unified SuperStandard Theories for Quaternion Universes & The Octonion Megaverse (Pingree Hill Publishing, Auburn, NH, 2020).

The Essence of Eternity: Quaternion & Octonion SuperStandard Theories (Pingree Hill Publishing, Auburn, NH, 2020).

A Very Conscious Universe (Pingree Hill Publishing, Auburn, NH, 2020).

From Octonion Cosmology to the Unified SuperStandard Theory of Particles (Pingree Hill Publishing, Auburn, NH, 2020).

Beyond Octonion Cosmology (Pingree Hill Publishing, Auburn, NH, 2021).

Available on Amazon.com, bn.com Amazon.co.uk and other international web sites as well as at better bookstores (through Ingram Distributors).

CONTENTS

FIGURES and TABLES

Introduction

The particles that we see generally depend on the coordinate system. The spatial/time measurements that we make depend on particles - an intimate connection that is often ignored in physical theories. This book explores the relation between particles and coordinate systems. It shows that a transformation to a non-static coordinate system from a "normal" static coordinate system can change the nature of the particle spectrum decisively. (Since almost all coordinates systems are non-static it is a weighty issue.)

The book begins by describing the changes undergone by creation/annihilation operators in quantum fields due to a change to non-static coordinates. It then proceeds to develop a new concept of space based on the Generalized Field Theory (GiFT) called Creation/Annihilation Space (CASe). CASe transformations to a non-static reference frame, which we call a Fundamental Frame, can reduce complex quantum field theories such as QUeST in our universe to a field theory with only 16 fermions and three internal symmetries – a one generation Standard Model – in a non-static reference frame. In the other universes of Octonion Cosmology it is also possible to reduce the complexity of quantum field theories to much fewer fermions and internal symmetries.

This book explores the Fundamental Frame of our universe and of the other higher dimension universes of Octonion Cosmology in detail. It shows the contents of the universes of Octonion Cosmology can be contracted to smaller size. The formalism is strongly based on Cayley-Dickson hypercomplex numbers: octonions, sedenions, …, n-ions, … where an n-ion is a hypercomplex number of Cayley-Dickson number n.

An analysis of the nature of fermions shows that the four types of fermions: e-type, ν-type, quark-up-type, and quark-down-type, can be viewed as generated by Complex Lorenz group boosts. The author's 2007 book *The Origin of the Standard Model ...* , describing this in detail is reprinted for the reader's convenience. The fermions of Fundamental Frames are of these types.

Gravitation joins the other gauge theories of interactions in GiFT. It has a finite perturbation theory due to GiFT's Two-Tier Theory and PseudoQuantum Theory. Due to its long range nature (like QED) Gravitation has interesting solutions such as Black Holes–just as Electrodynamics has interesting solutions that are used in Solid State Physics.

1. Unification of General Relativity and Quantum Theory

The unification of General Relativity and Quantum Theory has been a goal of Physics for approximately one hundred years. The meaning, and process, of unification has been the subject of discussion for almost as long. This book formulates a unification procedure and then proceeds to examine the ingredients and form of unification. We will see that Octonion Cosmology contains the essential ingredients for unity: a combination of space-times with internal symmetries in each of its universes.

The form of General Relativity, in relation to the form of other particle interactions, reveals a unity based on GiFT. A careful analysis of elementary particle field theory in differing coordinate systems leads to the Generalized Field Theory's (GiFT) Creation/Annihilation Space (CASe), which furnishes a further basis for unification with General Relativity.

1.1 The Form of General Relativity

General Relativity was first formulated by Einstein based on the Classical Physics (differential equations) of coordinate systems. It appeared different from the gauge field formulation of internal symmetries. As a result there seemed to be a gap between General Relativity and Elementary Particle Theory. Attempts were made to bridge the gap by quantizing General Relativity directly. These attempts seem to have not changed the view that General Relativity and Elementary particle theory were inherently different.

A different approach within a *vierbein* framework appeared due to the work of T. Kibble and others. This approach led to a gauge field formulation of General Relativity Thus this type of formulation of General Relativity places it on the same basis as the formulation of internal symmetry gauge fields.

A significant issue for unification is the appearance of infinities in gravitation perturbation theory. These infinities do not appear when Gravitation perturbation theory is done within the framework of the Two-Tier Theory within GiFT.

1.2 Octonion Cosmology Framework

The similarity between the *vierbein* formulation of General Relativity and elementary particle internal symmetry gauge theory raises the question of origin: Is there a common origin?

Octonion Cosmology provides a common origin in each of its ten universes (spaces). Part of each space has coordinates for each fundamental irreducible representation. Part of each space is for space-time coordinates. The transformation properties of the space-times lead to a General Relativity for each universe. Both the General Relativity and the internal symmetries can be formulated as gauge theories. Are there relations between coupling constants? Yes, but they differ in principle from universe to universe. Is there an ultimate unification value for all coupling constants in a universe? There may be such a unification value. But one must also consider the partition of internal symmetries in each universe due to the original form of spinor spaces, and in hypercomplex number multiplication tables, as we described in previous work. Partitioning may obviate the value of a common coupling constant value.

1.3 Quantization Process

The process of quantization begins with the issue of particle quantization vs. wave quantization: Should matter and energy be quantized as particles or as waves? The answer is particle quantization is simple if one wishes creation and annihilation events. Particles can interact to produce other particles in a "billiard ball" picture. Waves require complex differential equations at best.

Given a particle formulation it makes sense to define particle creation, a, and annihilation operators, $a^\dagger$, that can embody interactions that can change a set of incoming particles to a set of outgoing particles. At this point, if we introduce a "momentum" α for each particle, the question arises: Can particles have the same momentum? The Pauli Exclusion Principle states that fermion particles cannot. So we can set $(a^\dagger)^2 = 0$ for fermions. Bosons are not constrained by the Pauli Exclusion Principle so $(a^\dagger)^2 \neq 0$. One concludes the particle definition concepts by defining field operators for particles with Fourier expansions in α.

Having defined fields for particles in a coordinate system, then we note that a particle in that coordinate system becomes a superposition of particles when viewed in another coordinate in general. Einstein encountered this problem of exploring dynamic equations in one coordinate system in comparison to their equivalent in other coordinate systems. He resolved the problem with General Relativity.

1.4 Quantization in Differing Coordinate Systems

In our earlier work we showed that transformations between coordinate systems, where one or both coordinate systems do not have a Killing vector, are questionable in ordinary quantum field theory. We then showed quantization via GiFT provides a clear definition of particle states and their relation in the respective coordinate systems. (We pursue the GiFT approach further for bosons and fermions later in this book.)

The GiFT formulation of field theory is based on transformations between coordinate systems as is General Relativity. Thus both General Relativity and GiFT support the unity of General Relativity and Quantum Theory. General Relativity performs for coordinate systems what GiFT does for quantization in coordinate systems. Fig. 1.1 shows the correspondence between General Relativistic coordinate transformations and GiFT transformations in the space of creation/annihilation operators. A General Relativistic transformation between coordinate systems with Killing vectors does not induce a GiFT transformation. A General Relativistic transformation between coordinate systems where one coordinate system has a Killing vector, and the other does not, does induce a non-trivial GiFT transformation.

As we pointed out in earlier work GiFT embodies Quantum Theory in its entirety: PseudoQuantum Theory, Two-Tier Theory, Quantum Field Theory, and

Quantum Mechanics. GiFT supports CQ Mechanics which contains Quantum Mechanics and Classical Mechanics. Thus we see the unity in General Relativity and Quantum Theory.

1.5 GiFT Creation/Annihilation Spaces

GiFT has creation/annihilation spaces created with CASe transformations. In section 3.2 and in chapter 4 we show that creation/annihilation operators can form a basis for a 16-dimension space, a CASe space, which has a metric that is invariant under su(4,4) transformations for fermion particles, and su(n,m) in general for fermion particles in higher dimensions where n and m depend on the space-time dimension.

The structure of CASe spaces and their group transformations parallels that of Special Relativity and General Relativity in space-time.

Thus General Relativity and GiFT Quantum Theory, which contains Quantum Field Theory and Quantum Mechanics, closely parallel each other in the features of transformations between coordinate systems, and in the gauge field format of interactions.

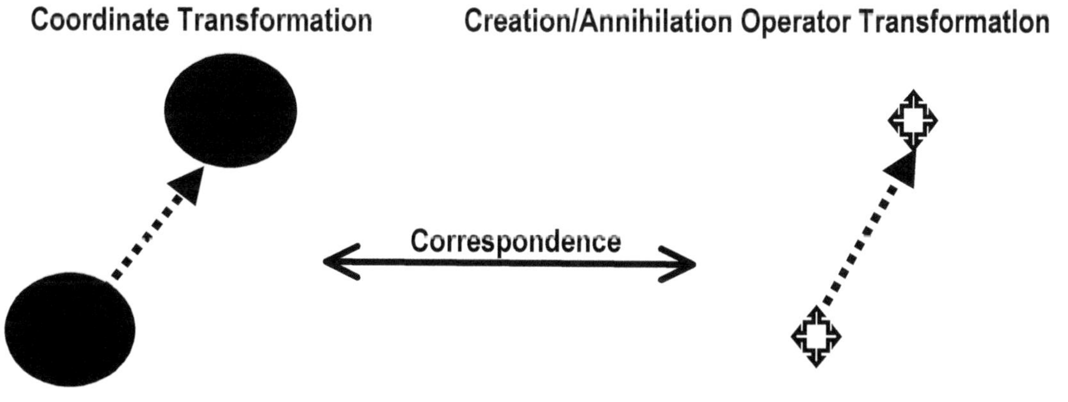

Figure 1.1. General Coordinate Transformations parallel Creation/Annihilation Operator Transformations.

1.6 Unification

The above considerations show the unity of General Relativity and Quantum Theory in our formulation. We find a unifying basis for General Relativity and Quantum Theory in considerations of transformations between coordinate systems, and in the existence of the relation between space-time transformations and CASe transformations. Vierbein gravity and the local gauge theory of internal symmetries exhibit a similar formulation. The structure of an overall fully unified theory of General Relativity and GiFT is evident.

2a. Pioneer's Progress in Octonion Cosmology

Recent years have witnessed a new effort to develop a comprehensive, deeper theory of Elementary Particles, Cosmology, and Quantum Theory. In this chapter we epitomize the developments of the Unified SuperStandard Theory and the Octonion Cosmology that embodies it. Octonion Cosmology requires a deeper quantum formulation that we have provided in GiFT,[1] which contains our PeseudoQuantum Theory and our Two-Tier Theory. Together these efforts open a much wider framework beyond the four space-time dimensions of the Standard Model and the Cosmological Standard Model. The developments are based on features of Spinor spaces that provide a new arena for the exploration of dynamical and group features of Elementary Particles and the Cosmos.

This chapter, and the following chapter 2b, outline the theory of Octonion Cosmology and its connection to the Unified SuperStandard Theory (NEWUST) through NEWQUeST. It is based on the author's books: *Quantum Space Theory With Application to Octonion Cosmology & Possibly To Fermionic Condensed Matter* and *Beyond Octonion Cosmology*, and *Beyond Octonion Cosmology II* as well as other previous books by the author. The discussion will be in a narrative form to make it accessible to a wider range of readers. Readers familiar with these books may proceed directly to chapter 3.

2a.1 Fundamental Required Features

Certain features are necessary for a satisfactory modern theory of fundamental Physics. These features include:

1. Space-times within which dynamic processes can take place.
2. A satisfactory Quantum formalism – GiFT.
3. Matter.
4. Energy.
5. A space of internal symmetries for SU(3) and so on.

In addition to these necessary requirements are a viable Quantum Field Theory, and a satisfactory Quantum Space Theory[2] of fermion and boson quantum fields containing subspaces, and describing the generation and features of these spaces.

2a.2 The Primary Space of Octonion Cosmology

Octonion Cosmology begins with the definition of a primary space,[3] which we will call space 0, from which the other spaces evolve. Space 0 has a twenty dimension

---

[1] Blaha (2021i).
[2] See Blaha (2021g).
[3] The space corresponds to Cayley number 2048. See Fig. 1.1 or Blaha (2021c) for details

space-time (nineteen spatial dimensions and one time dimension.) It contains $2^{22} - 20 =$ 4,194,284 additional dimensions for internal symmetries such as SU(2)⊗U(1) and SU(3) symmetries.

A space is a universe. Each universe, after space 0, has a space-time with dimensions ranging from zero to eighteen. In addition it has dimensions for irreducible representations of the internal symmetry groups that it supports. Examples of groups are SU(3) – the Strong interactions and SU(2)⊗U(1) – the ElectroWeak interactions. We will use the terms space and universe interchangeably.

We view space 0 as initially having a certain size in both space and time. In that space-time a fermion has a *spinor array* with row and column widths of 1024×1024 and thus 1,048,576 entries. We assume a fermion(s) exists in the 20 space-time dimension space 0.

We now contract the space-time to a 20 dimension point. The fermion spinor array size (rows and columns) remains the same and generates a spinor array – a *spinor space* with $1024^2 = 1,048,576$ dimensions.[4] This space has Cayley-Dickson number n = 10 in Fig. 2a.1. This space is a universe.

THE TEN OCTONION SPACES SPECTRUM

Octonion Space Number O_s	Cayley-Dickson Number n	Cayley Number d_c	Dimension Array Total d_d	Space-time-Dimension r	Fermion Spinor Array Total d_s
0	10	1024	1024 × 1024	18	512 × 512
1	9	512	512 × 512	16	256 × 256
2	8	256	256 × 256	14	128 × 128
3	7	128	128 × 128	12	64 × 64
4	6	64	64 × 64	10	32 × 32
5	5	32	32 × 32	8	16 × 16
6	**4**	**16**	**16 × 16**	**6 → 4**[5]	**8 × 8**
7	**3**	8	8 × 8	4	4 × 4
8	**2**	4	4 × 4	2	2 × 2
9	**1**	2	2 × 2	0	1 × 1

Figure 2a.1. The Octonion Cosmology ten space spectrum. The space for our universe, number 6, (Cayley-Dickson number 4) is in bold type.

2a.3 Space 10

A fermion in the 20 space-time dimension space 0 contains a spinor space (space 10) populated with matter and energy due to its mass. The n = 10 spinor space, with its elementary particles and energies, is the space[6] of a sub-universe. It has an 18

[4] The transformation of a fermion spinor array to a dimension array is described in detail in the case of fermion annihilation in chapter 6 of Blaha (2021g).

[5] The space-time dimensions become 4 as in the Unified SuperStandard Theory (UST) through either transfer of dimensions to internal symmetries or by the compactification of two dimensions. As a result the pattern of fermion-antifermion annihilations producing spaces n = 7, 8, and 9 is disrupted.

[6] See Fig. 1.1.

dimension space-time. The choice of 18 dimensions is motivated by the goal of having a Cayley-Dickson spectrum of spaces. Later we will find a deeper reason for this choice. See chapters 5 and 6.

In space n = 10 a fermion-antifermion annihilation can take place. It can be visualized with the following Feynman-like diagram:[7]

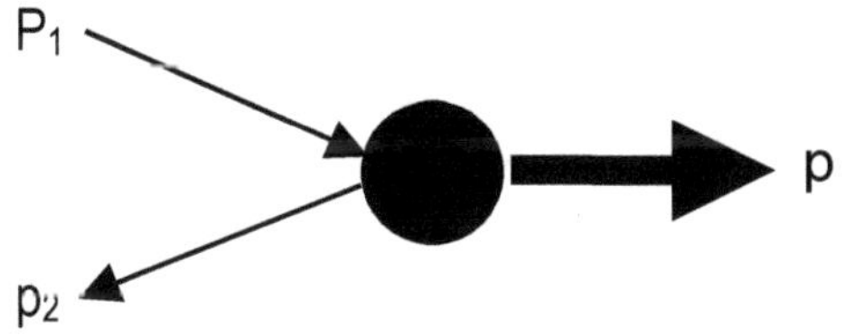

where p_1 and p_2 are the momentum of the fermion and antifermion respectively and p is the momentum of the produced space-containing particle. The produced (scalar) space particle, which exists in space 10,[8] contains a spinor array/space that is the space of another universe of the type of space n = 9 in Fig. 2a.1.

Space n = 9, with its mass and energy, is created using the mass and energy from the annihilation in space 10. It has 512^2 = 262,144 dimensions, of which 16 are space-time dimensions.

This space-time is the first of the series of nine spaces (universes) generated through fermion-antifermion annihilations. They produce a Cayley-Dickson number based spectrum of spaces. See Figs. 2a.1 and 2a.2 below.

2a.4 The Further Generation of Spaces

The spectrum of spaces (universes) is generated through a series of fermion-antifermion annihilations in space after space. The Cayley-Dickson pattern of spaces emerges from the choice of the number of space-time dimensions in each space. We descend from space to space by decreasing the space-time dimension by two at each space.[9] The below Fig. 2a.3 shows the general pattern of nesting of space instances in a nested form.

The result of the sequence of "annihilations" is the spectrum of Fig. 2a.1, which corresponds to the Cayley-Dickson numbered spaces from 10 through 1 with space-time dimensions from 18 through 0. See Fig. 2a.5.

In this iterative process there can be any number of subspace instances generated from a space instance by multiple fermion-antifermion annihilations. For example see Fig. 2a.4.

---

[7] We assume off shell fermions.

[8] Note the produced particle has a mass and momentum in space 10.

[9] This choice is motivated by deeper reasons tied to General Relativity. See chapters 5 and 6.

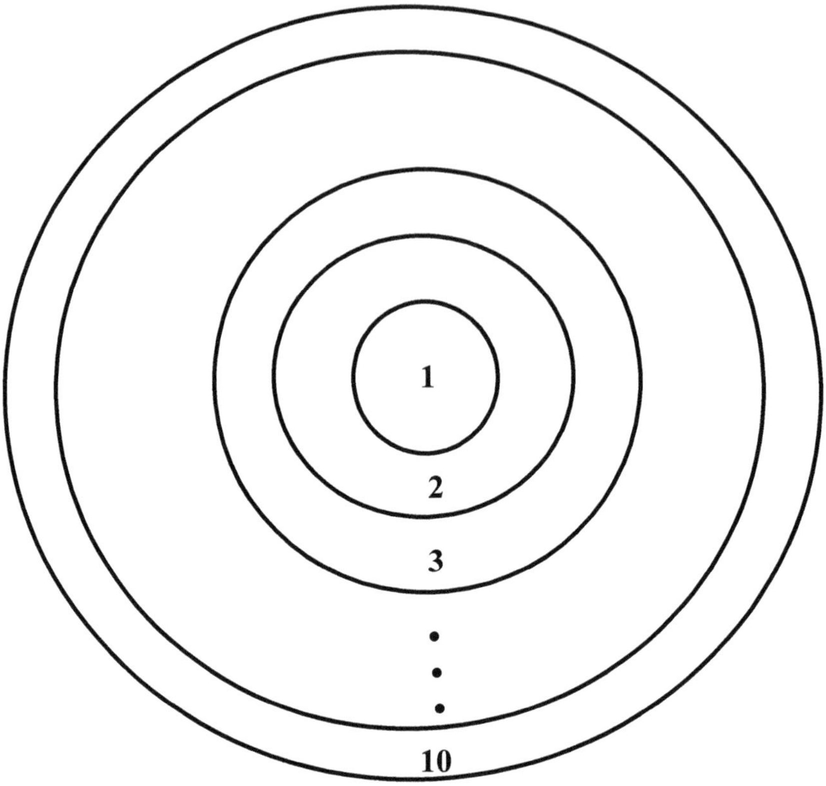

Figure 2a.2. Numbered form of the ten space instances of Octonion Cosmology. They are generated by fermion-antifermion annhilations. The outermost space instance (Cayley-Dickson space 10) is generated first from the 20 space-time dimension space. The Space 9 instance is generated within space 10 and so on.

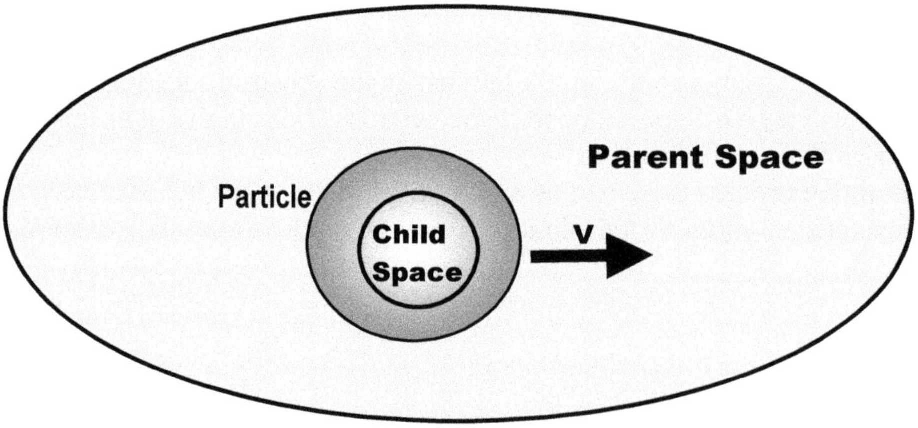

Figure 2a.3. The general pattern of a (parent) space containing a particle of velocity **v** that has an internal space instance (child) of lower dimension and

lower Cayley-Dickson number. The space is itself within an instance of a higher space. The spectrum's spaces form a nested sequence. Each space contains the next lower space within it. The instances of spaces also form a nested sequence as shown in Fig. 1.2 below.

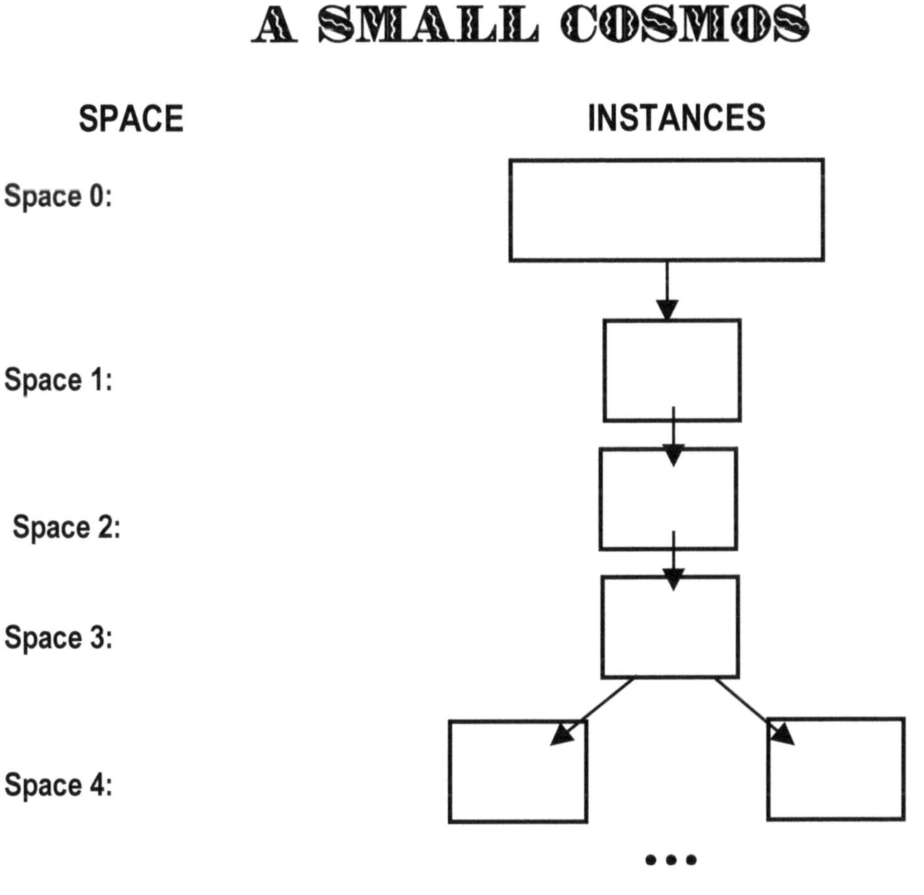

Figure 2a.4. A hierarchy of space instances one "sibling" and three "cousin" universes. *The entire hierarchy resides in space 0.*

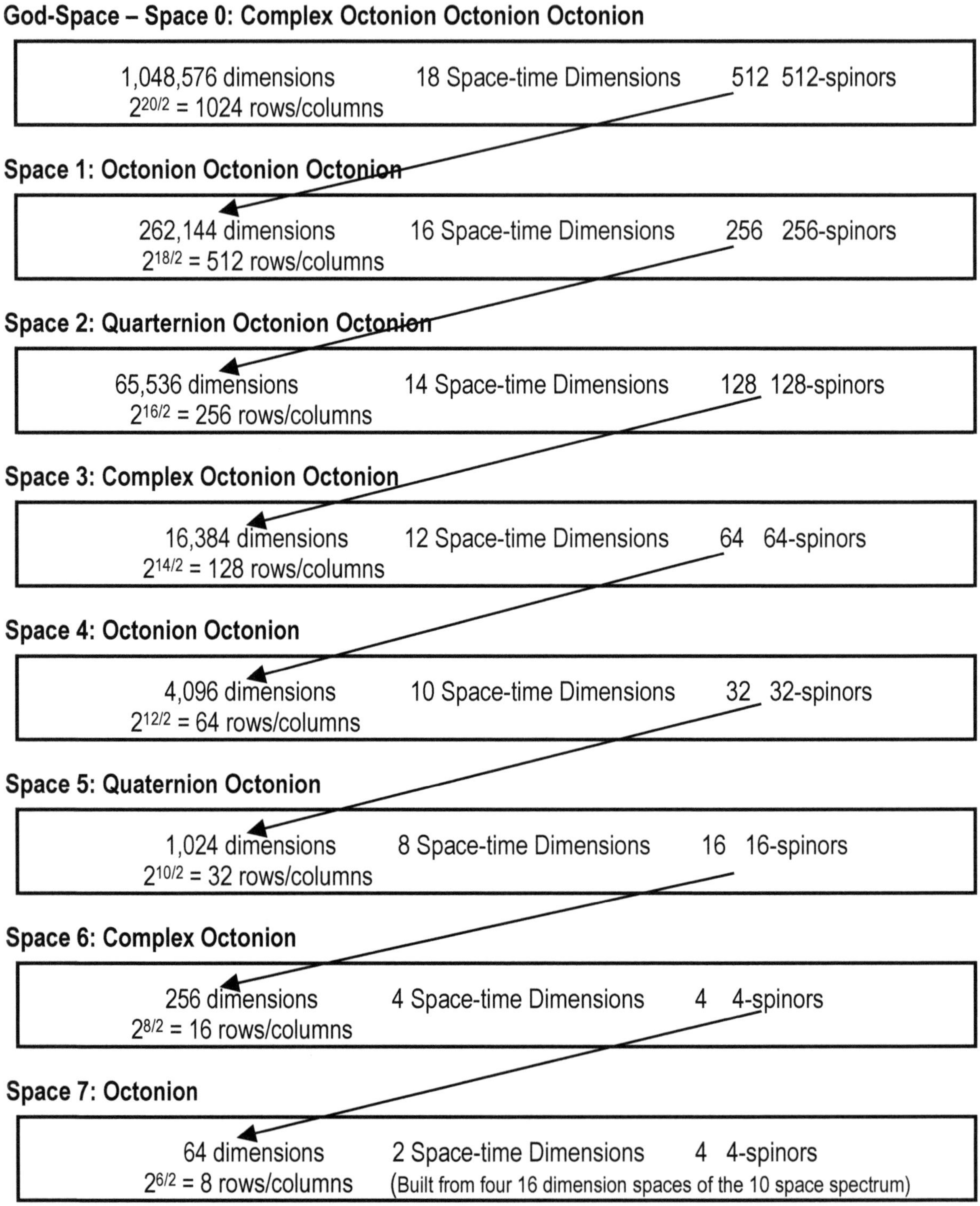

Figure 2a.5. A sequence of fermion-antifermion annihilations generating instances in the set of Octonion Cosmology spaces.

2a.5 Comments

One might expect that generating the sequence of spaces would take an extraordinarily long time. However time, as we know it, only exists in our space 4 universe. Time is not a problem.

The octonion spaces have many dimensions beyond their space-time dimensions. These additional dimensions furnish the fundamental representations of the internal symmetry groups in the space. We show this for space 4 in chapter 2b in Fig. 2b.2. Fig. 2b.5 shows the *form* of the fermion part of the instance of universe (described by our NEWQUeST theory.) The fermions of our universe have this form.

An interesting feature: The creation of each space instances always leads to a Big Bang —justifying the Big Bang for our universe for, perhaps, the first time.[10]

Point of Clarification: When we say a space (universe) is generated by a fermion-antifermion annihilation we mean that an *instance* of the space is generated. A space is defined independently of its contents. A space plus its contents is called an *instance* of the space. Annihilations generate instances. The source of the matter and energy of an instance is the mass-energy of the annihilating particles. We sometimes say space to refer to a space instance to avoid cumbersome text.

2a.6 The Golden Triangle of Mass, Energy and Space

Examining the Octonion Cosmology spectrum and its origin in fermion-antifermion annihilations, we see a "golden" triangle of transformations between mass, energy, and space. Mass can be transformed into energy and vice versa.

Mass and/or energy can also be transformed into a space and vice versa. The creation of a space (instance) is the result of a mass and energy transformation. A space instance generates mass and energy transformations as it evolves. A space instance also generates instances of particles and of subspaces through fermion-antifermion annihilations within it.

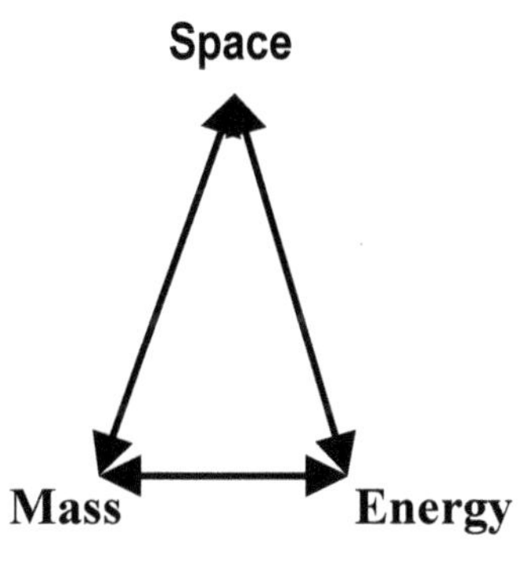

Figure 2a.5. The golden triangle of mass, energy, and space.

[1010] See Blaha (2021d) for the author's model for universe expansion and the Hubble Constant.

A remarkable reflection of the connection between mass and space is the equality of the number of fundamental fermion types in each of the ten spaces and the dimension of its space. In Cayley-Dickson number 4 space, which is that of our universe, we have found 256 fundamental fermion types matching the $16^2 = 256$ dimensions of the space.

2a.7 Octonion Cosmology Connection to Elementary Particle Theory

The Octonion Cosmology with its ten spaces originated in the author's Unified SuperStandard Theory (NEWUST) that was developed over a number of years, and described in:

Blaha, 2018e, *Unification of God Theory and Unified SuperStandard Model THIRD EDITION*

Blaha, 2020a, *Quaternion Unified SuperStandard Theory (The QUeST) and Megaverse Octonion SuperStandard Theory (MOST)*

Blaha, 2020c, *Unified SuperStandard Theories for Quaternion Universes & The Octonion Megaverse*

Blaha, 2021c, *Beyond Octonion Cosmology*

Blaha, 2021d, *Universes are Particles*

Blaha, 2021e, *Octonion-like dna-based life, Universe expansion is decay, Emerging New Physics*

Blaha, 2021f, *The Science of Creation New Quantum Field Theory of Spaces*

and other books by the author.

In the Fall of 2019 the author considered the possibility that the UST was based on a deeper theory and proceeded to consider theories based on hypercomplex numbers such as quaternions and octonions. *Thus Octonion Cosmology did not "emerge out of thin air."* It provided a deeper base for the UST.

Remarkably the author found that the hypercomplex theories, QUeST and NEWQUeST, had similar internal symmetry groups, and fundamental fermions, as UST. In addition a hypercomplex theory NEWUTMOST furnished an acceptable theory of the Megaverse (Multiverse). Over the following two years 2020 and 2021 Octonion Cosmology emerged with its ten spaces.

The history of this effort, which has some relevance to Natural Philosophy, appears below. It is extracted from *Quantum Space Theory With Application to Octonion Cosmology Beyond Octonion Cosmology*.

2b. Octonion Cosmology Connection to the Unified SuperStandard Theory (UST and NEWUST)

The author's UST and NEWUST theories originated in the past twenty years from the Standard Model of Particles with SU(2)⊗U(1)⊗SU(3) internal symmetries combined with Two-Tier Quantum and PseudoQuantum Field Theory (GiFT).

Noting the presence of conserved particle numbers, and the presence of at least three fermion generations, we introduced the U(4) Generation Groups and the U(4) Layer Groups together with four layers of four generation of "Normal" fermions necessitated by Generation Group and Layer group symmetries at the point of the Big Bang prior to symmetry breaking. The Dark sector had a corresponding set of four layers of four generations of "Dark" fermions. The result was the Unified SuperStandard Theory (UST) with the symmetry:

$$\{[SU(2)\otimes U(1)\otimes SU(3)]^2\otimes U(4)^4\}^4$$

supplemented with an additional Strong Interaction U(1) group. Space-time had four dimensions.

In the Fall, 2019 the author discovered that an octonion-based theory that he constructed and named QUeST had the same internal symmetries as UST with the addition of $U(1)^8$. The addition of $U(1)^8$ indicates that the Strong Interactions in the theory are broken Strong SU(4).

QUeST's internal symmetries, which could be based on a 16 × 16 dimension array of dimensions, was

$$[SU(2)\otimes U(1)\otimes SU(3)\otimes U(1)]^8\otimes U(4)^{16}$$

During 2020 the author developed an octonion space spectrum for both universes and Megaverses, and other spaces. The spectrum was shown to arise from a generation mechanism whereby fermion-antifermion annihilation in a higher space produced an instance of a lower space. A critical part of the derivation of the octonion spectrum was the realization that even space-time dimension spinor arrays are composed of Cayley number rows and columns. Spinor arrays of annihilating fermion-antifermion pairs were shown to generate the arrays of dimensions of subspace instances.

Analyzing the spinor arrays the author noted that the dimension array could be viewed as composed of 64 dimension subblocks, which were further subdivided into 16 dimension subblocks.

The subblock structuring, using the known contents of the Standard Model plus Generation and Layer groups for guidance, gave the dimension array structure containing 4×4 subblocks in Figs. 2b..1 and 2b.2.

Thus there was a *most* satisfactory match between UST and QUeST with the only significant difference being the space-time: four octonion (complex quaternion) coordinates for QUeST and four real space-time coordinates for UST. This difference was resolved in NEWUST and NEWQUeST. See the Connection groups in Fig. 2b.3.

Figs. 2b.4 and 2b.5 show the detailed group and fermion structures.

The form of the square spinor arrays generated by fermion-antifermion annihilation gives 64 dimension blocks and 16 dimension blocks as well as 32 dimension composite blocks that are evidenced in the NEWQUeST fermion spectrums and internal symmetry group structure.

2b.1 Dimensions of Symmetries and Coordinates

Since we see only real dimensions in Reality, we transferred 28 QUeST dimensions from space-time to $U(2)^7$ internal symmetry dimensions. The set of internal symmetries was increased by $U(2)^7$, which we call Connection Groups. Each Connection group specifies interactions between corresponding fermions (e with e, q with q, and so on) in separate layers and between Normal and Dark fermions. The connections between the various blocks of fermions are shown in Fig. 2b.3. *We implement the very practical rule that all blocks must be connected by interactions or they would not be of physical interest. A totally isolated block effectively does not exist physically (except possibly for gravitation effects).*

The interactions of the Connection groups must be very weak and/or their gauge bosons must be very massive.

The addition of the Connection Groups and the reduction of space-time dimensions accordingly results in NEWQUeST and NEWUST as summarized below.

Note: the Generation, Layer, and Connection groups are all badly broken. Their vector bosons must be very massive since they have not been detected in experiments.

2b.1.1 Internal Symmetries

The groups are ElectroWeak SU(2)⊗U(1), Strong SU(3), Generation Group U(4), Layer Group U(4), and U(2) and U(4) Connection groups obtained by transfer from space-time coordinates (See Blaha 2012c). The SU(3)⊗U(1) symmetries may be a broken SU(4) symmetries. The internal symmetries for the theories are:

UST

$$[SU(2)\otimes U(1)\otimes SU(3)]^8\otimes U(4)^{16} \qquad (2b.1)$$

QUeST

$$[SU(2)\otimes U(1)\otimes SU(3)\otimes U(1)]^8\otimes U(4)^{16} \qquad (2b.2)$$

NEWQUeST

$$[SU(2)\otimes U(1)\otimes SU(3)\otimes U(1)]^8\otimes U(4)^{16}\otimes U(2)^7 \qquad (2b.3)$$

The only change is in internal symmetries: Twenty-eight real dimensions transferred from space-time coordinates to Connection group $U(2)^7$ internal symmetry.

NEWUST

$$[SU(2)\otimes U(1)\otimes SU(3)\otimes U(1)]^8 \otimes U(4)^{16} \otimes U(2)^7 \quad (2b.4)$$

The only change in internal symmetries: Twenty-eight real dimensions added for $U(2)^7$ Connection group internal symmetry.

2b.1.2 Space-Time Coordinates

UST

Four real space-time coordinates.

QUeST

Four octonion (complex quaternion) coordinates.

NEWUST

Four real space-time coordinates. No change from UST space-time.

NEWQUeST

Four real space-time coordinates. The six coordinates in the n = 4 octonion space were lowered to four space-time coordinates with two coordinates transferred to Connection groups.

The only change is in space-time coordinates: Fourteen dimensions transferred from QUeST space-time coordinates to Connection group $U(2)^7$ internal symmetry.

2b.2 Fundamental Fermion Spectrum

There are 256 fundamental fermions in NEWQUeST and NEWUST. Conceptually their structure can be viewed as an extrapolation of the known three generations of The Standard Model. For good reason (U(4) Generation groups) a fourth generation was indicated and a corresponding Dark sector of similar structure was added. In addition, because of the need for Layer groups, the overall structure consisted of four copies of this layer (due to U(4) Layer groups).

Correspondingly, each layer also has its own set of internal symmetry gauge groups to limit mixing between the layers to Layer group interactions and Connection group interactions.

Fig. 2b.1 shows the structure of the NEWQUeST/NEWUST fermions. The blocks are 4 × 4 reflecting the origin of the NEWQUeST/NEWUST space (universe) instance from Megaverse fermion-antifermion annihilation. The spinor analysis of their spinor arrays yields a 16 dimension block structure. The 64 dimension fermion layers reflect the 64 dimension structuring of the Megaverse obtained from its creation by fermion-antifermion creation in the Maxiverse.

2b.3 Total Dimensions

The total of internal symmetry and space-time dimensions is 256 in all four theories listed above. It is based on the 16 × 16 dimension array of the Cayley number n = 4 space of the Octonion spectrum.

2b.4 Pattern of Internal Symmetries

The NEWQUeST dimension array for internal symmetries is subdivided into four layers of 56 dimensions—just as in NEWUST (and UST). Fig. 2b.2 displays the layers using SU(4) in place of SU(3)⊗U(1). Each layer has a block of 28 dimensions for Normal and 28 dimensions for Dark sectors. There are also seven U(2) Connection groups plus four real-valued space-time coordinates bringing the NEWQUeST total to 256 dimensions.= 4*56 + 28 + 4 = 256 dimensions. The Connection groups are shown in Fig. 2b.3.

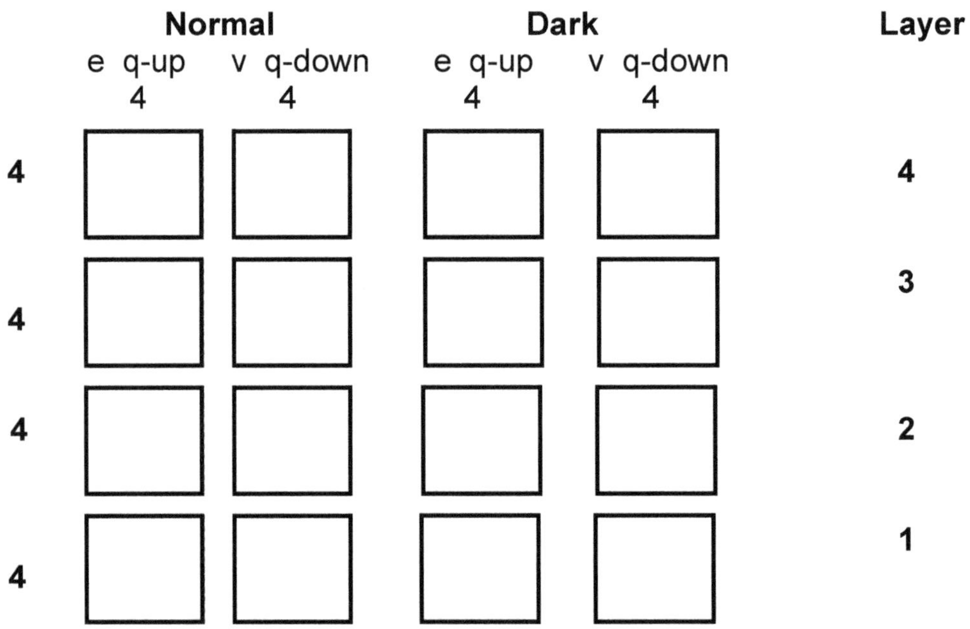

Figure 2b.1. Block form of a 16 × 16 NEWQUeST/NEWUST fermion array with each block row corresponding to one layer. Each block contains four generations of fermions. The result is 4 × 4 blocks. The label e q-up indicates a charged lepton – up-type quark pair, v q-down indicates a neutral lepton – down-type quark pair, and so on. The blocks can be viewed as SU(3)⊗U(1) or broken SU(4) blocks.

Layers	NORMAL 4	4	DARK 4	4
4	$SU(2)\otimes U(1)\otimes SU(3)\otimes U(1)$ 4 Space-time Dimensions	Generation + Layer Groups	$SU(2)\otimes U(1)\otimes SU(3)\otimes U(1)$ 4 Space-time Dimensions	Generation + Layer Groups
4	$SU(2)\otimes U(1)\otimes SU(3)\otimes U(1)$ 4 Space-time Dimensions	Generation + Layer Groups	$SU(2)\otimes U(1)\otimes SU(3)\otimes U(1)$ 4 Space-time Dimensions	Generation + Layer Groups
4	$SU(2)\otimes U(1)\otimes SU(3)\otimes U(1)$ 4 Space-time Dimensions	Generation + Layer Groups	$SU(2)\otimes U(1)\otimes SU(3)\otimes U(1)$ 4 Space-time Dimensions	Generation + Layer Groups
4	$SU(2)\otimes U(1)\otimes SU(3)\otimes U(1)$ 4 Space-time Dimensions	Generation + Layer Groups	$SU(2)\otimes U(1)\otimes SU(3)\otimes U(1)$ 4 Space-time Dimensions	Generation + Layer Groups

Figure 2b.2.. Four layers of Internal Symmetry groups in NEWQUeST and NEWUST (omitting Connection Groups) showing the 4 by 4 sublocks. The groups in each layer are independent of those in other layers. The groups in each subblock of each layer are independent of those in the other subblocks. Each subblock contains 16 dimensions. The block dimensions furnish fundamental representations for the groups listed. The entire set of blocks contains 256 dimensions. Each layer contains 56 internal symmetry dimensions. The first two columns are for the "Normal" sector. The last two columns are for the "Dark" sector (although most of the Normal sector is Dark observationally at present.) This figure also holds for UST with the addition of U(1) groups. The eight sets of 4 real dimension space-times combine to give a 4 real dimension space-time and seven U(2) Connection groups.

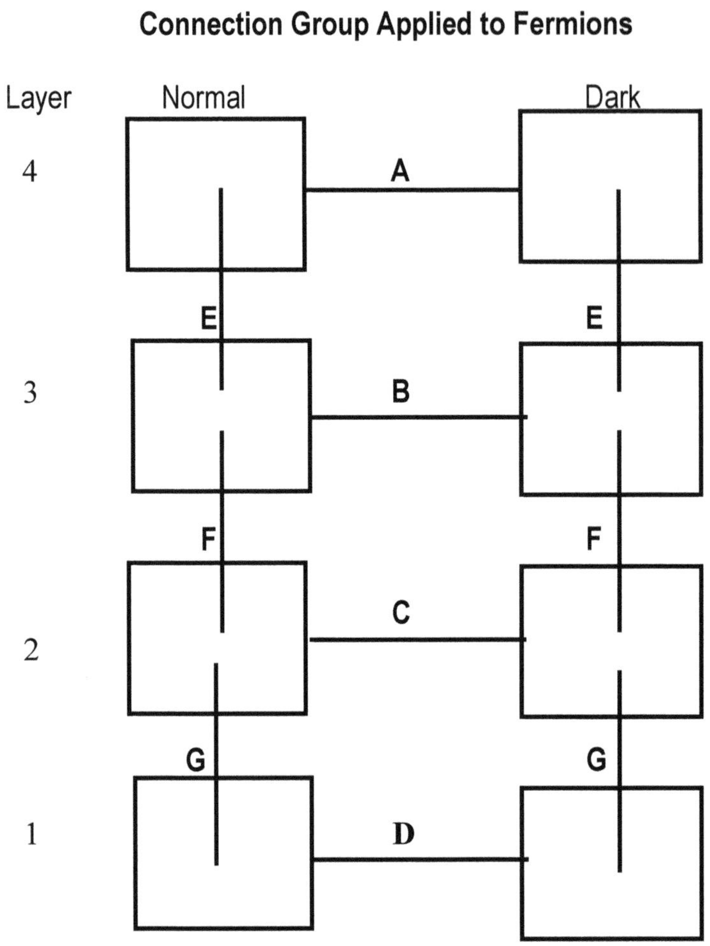

Figure 2b.3. The seven U(2) Connection groups (shown as 10 lines) between the eight NEWQUeST/NEWUST blocks. Connection groups are obtained by transfering 28 dimensions from QUeST space-time to internal symmetries with the consequent reduction of the space-time from four octonion (complex quaternion) coordinates to four real coordinates. The Connection groups generate rotations and interactions between corresponding fermions and vector bosons of each pair of blocks. The Normal and Dark sector U(2) vertical connections above (E, F, G) represent the same U(2) groups.

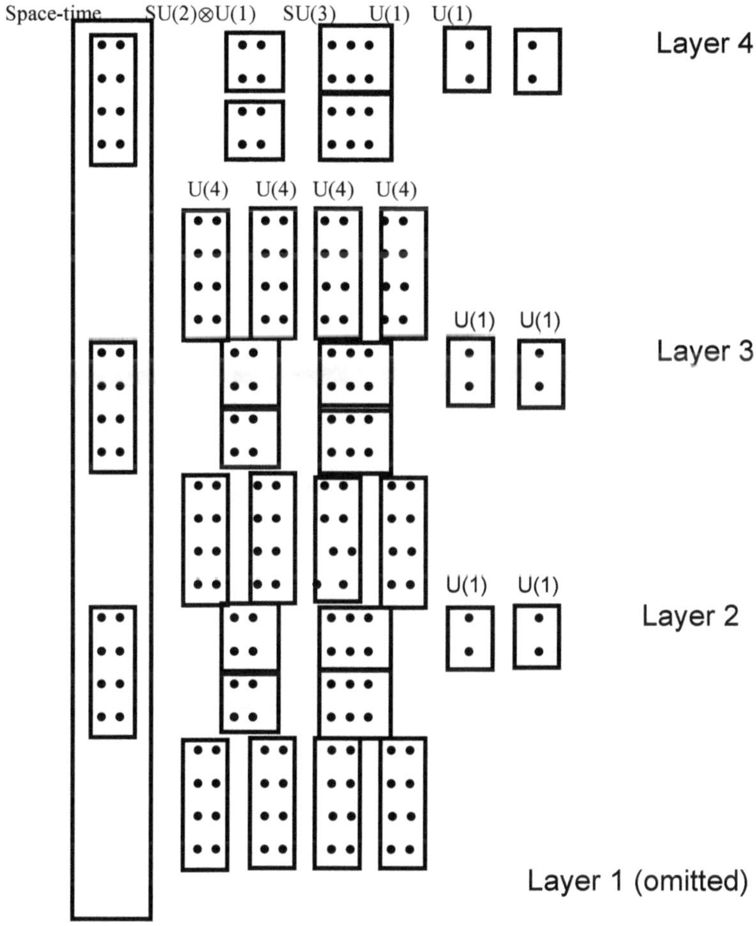

Figure 2b.4. Three of the four layers of QUeST internal symmetry groups (and space-time) for Cayley-Dickson space 4. Layer 1 which has an identical form was omitted due to "page space" limitations. Note the left column of blocks combine to specify a 4 dimension octonion space-time. Note each layer has 64 dimensions.

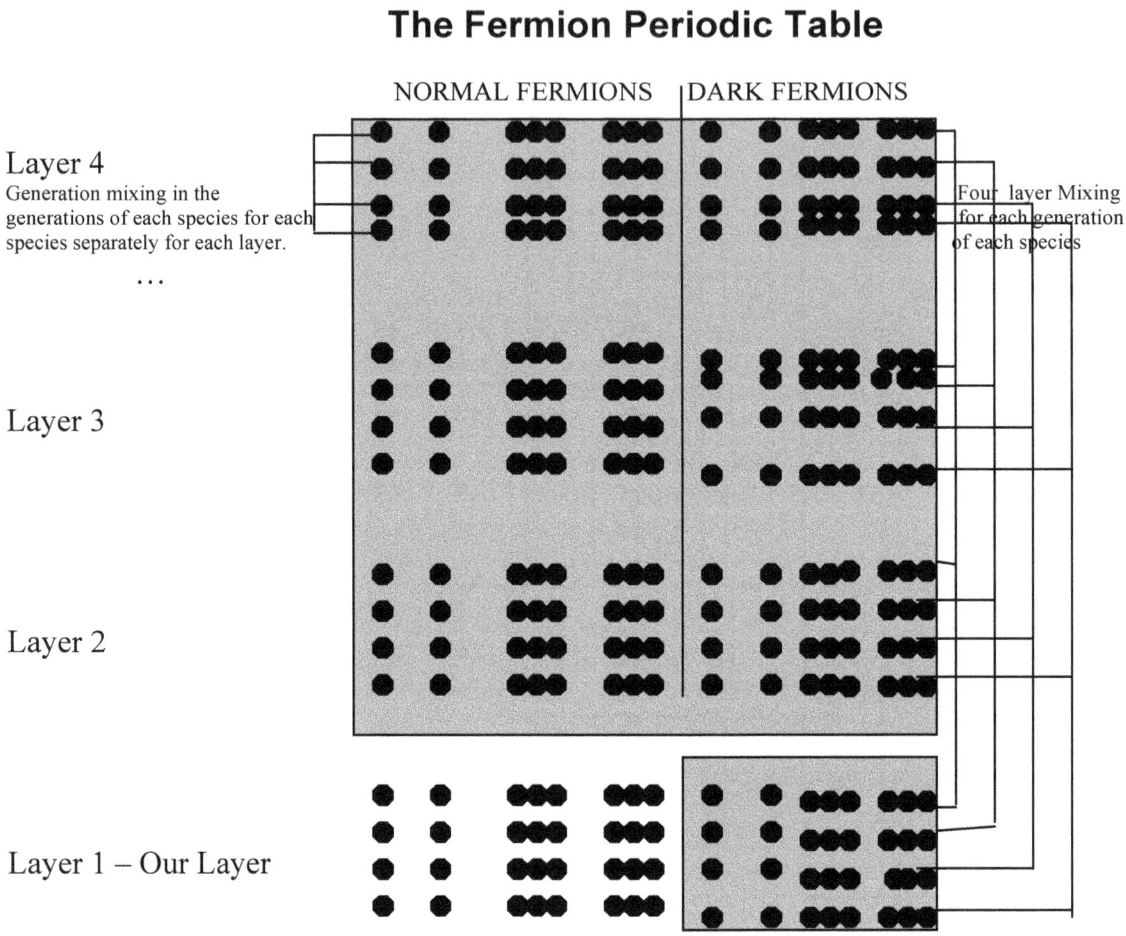

Figure 2b.5. Fermion particle spectrum and partial examples of the pattern of mass mixing of the Generation group and of the Layer group. Unshaded parts are the known fermions with an additional, as yet not found, 4th generation. The lines on the left side (only shown for one layer) display the Generation mixing within each layer. The Generation mixing occurs within each layer using a separate Generation group for each layer. The lines on the right side show Layer group mixing (for Dark matter) with the mixing among all four layers for each of the four generations individually. There are four Layer groups for Normal matter and four Layer groups for Dark matter.. There are 256 fundamental fermions. QUeST and UST have the same fermion spectrum.

3. GiFT for *Scalar* Particle Creation/Annihilation Operator Transformations

In Blaha (2021i) we developed the features of Generalized Field Theory (GiFT). In this chapter we apply the formalism to the case of fermions. We show spin is naturally incorporated within GiFT. We begin by considering two 4-dimension coordinate systems denoted α and β and develop a relation between the quantizations in these coordinate systems in the case where the spins of the fermions in the respective coordinate systems are the same. Then we consider the case where the spins of states in one coordinate system correspond to superpositions of spin states in the other coordinate system.

Since higher dimension space-times appear in Octonion Cosmology and other theories we develop the relation between quantizations of coordinate systems in higher dimensions. We then consider the relations of these results to other spin-based theories such as Twistor Theory.

3.1 GiFT for *Scalar* Creation/Annihilation Operator Transformations

We illustrate the procedure by considering the case of a scalar particle in four dimensions[11] and expand on the development in section 4 of S. Blaha, "The Local Definition of Asymptotic Particle States", Il Nuovo Cimento **49A**, 35 (1979), which we call I. We associate the scalar particle with two scalar fields: $\varphi_1(x)$ and $\varphi_2(x)$. We choose $\varphi_1(x)$ to have a zero equal time commutator with $d\varphi_1(x)/dx^0$ and $\varphi_2(x)$ to have a conventional equal time commutator with $d\varphi_2(x)/dx^0$. Conceptually $\varphi_1(x)$ is a "classical" field and $\varphi_2(x)$ is a quantum field. A Lagrangian that implements these choices of commutation relations is:

$$\mathcal{L} = \partial^\mu \varphi_1(x)\partial_\mu\varphi_2(x) - \tfrac{1}{2}\,\partial^\mu \varphi_1(x)\partial_\mu\varphi_1(x) - m^2\,\varphi_1(x)\varphi_2(x) + \tfrac{1}{2}\,m^2\,\varphi_1(x)^2 \tag{3.1}$$

$$(\Box + m^2)\varphi_1(x) = 0 \tag{3.2}$$

$$(\Box + m^2)\varphi_2(x) - (\Box + m^2)\varphi_1(x) \quad = 0 \tag{3.3}$$

The canonical momenta are

$$\pi_1 = d\varphi_2(x)/dt - d\varphi_1(x)/dt \tag{3.4}$$
$$\pi_2 = d\varphi_1(x)/dt \tag{3.5}$$

and the equal time commutation relations are

---

[11] The case of a space with a higher dimension spacetime is completely analogous.

$$[\varphi_i(x), \pi_j(y)] = i\delta_{ij}\delta(\mathbf{x} - \mathbf{y}) \tag{3.6}$$
$$[\varphi_i(x), \varphi_j(y)] = [\pi_i(x), \pi_j(y)] = 0 \tag{3.7}$$

for i, j = 1, 2 implying

$$[\varphi_1(x), d\varphi_1(y)/dt] = 0 \tag{3.8}$$
$$[\varphi_2(x), d\varphi_2(y)/dt] = i\delta^3(\mathbf{x} - \mathbf{y}) \tag{3.9}$$
$$[\varphi_1(x), d\varphi_2(y)/dt] = i\delta^3(\mathbf{x} - \mathbf{y}) \tag{3.10}$$

The *most* general mode expansion of the fields in the coordinate system A with Fourier momentum α is

$$\varphi_{1A}(x) = \Sigma_\alpha\, [a_{1\alpha}\, f_\alpha(x) + a^\dagger_{1\alpha}\, f_\alpha{}^*(x)] \tag{3.11}$$
$$\varphi_{2A}(y) = \Sigma_\alpha\, [a_{2\alpha} f_\alpha(x) + a^\dagger_{2\alpha}\, f_\alpha{}^*(x)] \tag{3.12}$$

and in coordinate system B with Fourier momentum β is

$$\varphi_{1B}(x) = \Sigma_\beta\, [a_{1\beta}\, f_\beta(x) + a^\dagger_{1\beta}\, f_\beta{}^*(x)] \tag{3.13}$$
$$\varphi_{2B}(y) = \Sigma_\beta\, [a_{2\beta}\, f_\beta(x) + a^\dagger_{2\beta}\, f_\beta{}^*(x)] \tag{3.14}$$

The creation/annihilation operators of A have the commutation relations:

$$[a_{i\alpha'}, a_\alpha] = 0 \tag{3.15}$$
$$[a^\dagger_{i\alpha'}, a^\dagger_{j\alpha}] = 0$$

$$[a_{i\alpha}, a^\dagger_{j\alpha'}] = (1 - \delta_{ij})\delta^3(\boldsymbol{\alpha} - \alpha') \tag{3.16}$$

with i, j = 1, 2. Two related vacuums are defined: $|0>_1$ and $|0>_2$ by

$$a_{1\alpha}|0>_{2\alpha} = a^\dagger_{1\alpha}|0>_{2\alpha} = 0 \tag{3.17}$$
$$a_{2\alpha}|0>_{1\alpha} = a^\dagger_{2\alpha}|0>_{1\alpha} = 0$$

and

$$a_{i\alpha}|0>_{i\alpha} \neq 0 \qquad a^\dagger_{i\alpha}|0>_{i\alpha} \neq 0 \tag{3.18}$$

for i = 1, 2 and all α and α′. And similarly for the creation/annihilation operators of B with α replaced with β.

The general coordinate system B Fourier expansion in terms of A creation/annihilation operators is

$$\varphi_{1B}(x) = \Sigma_\beta\, \Sigma_\alpha\, [(c_{11}a_{1\alpha} + c_{12}a_{2\alpha} + C_{11}a^\dagger_{1\alpha} + C_{12}a^\dagger_{2\alpha})f_\beta(x) + \\ + (c'_{11}a^\dagger_{1\alpha} + c'_{12}a^\dagger_{2\alpha} + C'_{11}a_{1\alpha} + C'_{12}a_{2\alpha})f_\beta{}^*(x)] \tag{3.19}$$

$$\varphi_{2B}(y) = \Sigma_\beta\, \Sigma_\alpha\, [(c_{21}a_{1\alpha} + c_{22}a_{2\alpha} + C_{21}a^\dagger_{1\alpha} + C_{22}a^\dagger_{2\alpha})\, f_\beta(x) + \\ + (c'_{21}a^\dagger_{1\alpha} + c'_{22}a^\dagger_{2\alpha} + C'_{21}a_{1\alpha} + C'_{22}a_{2\alpha})f_\beta{}^*(x)] \tag{3.20}$$

where the c_{ij} and C_{ij}, and c'_{ij} and C'_{ij} are all functions of α.

We define a sets of Bogoliubov transformations B_1 and B_2 that give

$$a_{i\beta} = \Sigma_\alpha (f_\beta, f_\alpha) B_{2\alpha} B_{1\alpha} a_{i\alpha} (B_{1\alpha} B_{2\alpha})^{-1} = \Sigma_\alpha (f_\beta, f_\alpha)[c_{i1} a_{1\alpha} + c_{i2} a_{2\alpha} + C_{i1} a^\dagger_{1\alpha} + C_{i2} a^\dagger_{2\alpha}] \tag{3.21}$$

$$a^\dagger_{i\beta} = \Sigma_\alpha (f^*_\beta, f^*_\alpha) B_{2\alpha} B_{1\alpha} a^\dagger_{i\alpha} (B_{1\alpha} B_{2\alpha})^{-1} = \Sigma_\alpha (f^*_\beta, f^*_\alpha)[c'_{i1} a^\dagger_{1\alpha} + c'_{i2} a^\dagger_{2\alpha} + C'_{i1} a_{1\alpha} + C'_{i2} a_{2\alpha}] \tag{3.22}$$

3.1.1 The B_1 Transformation

The B_1 transformation has the form:

$$B_{1\alpha}(x_1, x_2) = \exp[x_1 \Gamma^1{}_{3\alpha}] \exp[x_2 \Gamma^1{}_{2\alpha}] \tag{3.23}$$

where the Hermitian operators $\Gamma^1{}_{1i\alpha}$ are

$$\Gamma^1{}_{3\alpha} = (a^\dagger_{2\alpha} a_{1\alpha} + a_{2\alpha} a^\dagger_{1\alpha})/2 \tag{3.24}$$
$$\Gamma^1{}_{2\alpha} = i(a^\dagger_{2\alpha} a^\dagger_{1\alpha} - a_{2\alpha} a_{1\alpha})/2 \tag{3.25}$$

We also define

$$\Gamma^1{}_{1\alpha} = -(a^\dagger_{2\alpha} a^\dagger_{1\alpha} + a_{2\alpha} a_{1\alpha})/2 \tag{3.26}$$

and

$$\Gamma^1{}_{4\alpha} = (a^\dagger_{2\alpha} a_{1\alpha} - a_{2\alpha} a^\dagger_{1\alpha})/2 \tag{3.27}$$

Together the three operators satisfy su(1,1) algebra commutation relations:

$$[\Gamma^1{}_{1\alpha}, \Gamma^1{}_{2\alpha'}] = -i\delta_{\alpha\alpha'} \Gamma^1{}_{3\alpha} \qquad [\Gamma^1{}_{2\alpha}, \Gamma^1{}_{3\alpha'}] = i\delta_{\alpha\alpha'} \Gamma^1{}_{1\alpha} \qquad [\Gamma^1{}_{3\alpha}, \Gamma^1{}_{1\alpha'}] = i\delta_{\alpha\alpha'} \Gamma^1{}_{2\alpha} \tag{3.28}$$

They transform $a_{i\alpha}$ and $a^\dagger_{i\alpha}$ to terms of the form $c_{i\alpha} a_{i\alpha} + c_{i\alpha}' a^\dagger_{i\alpha}$ where $c_{i\alpha}$ and $c_{i\alpha}'$ are constants.

Some properties of the su(1,1) algebra are detailed in S. Blaha, "The Local Definition of Asymptotic Particle States", IL Nuovo Cimento **49A**, 35 (1979).

3.1.1 The B_2 Transformation

The B_2 transformation has the form:

$$B_{2\alpha}(x_1, x_2) = \exp[x_1 \Gamma^2{}_{3\alpha}] \exp[x_2 \Gamma^2{}_{2\alpha}] \tag{3.29}$$

where the Hermitian operators $\Gamma^2{}_{1i\alpha}$ are

$$\Gamma^2{}_{3\alpha} = (a^\dagger_{1\alpha} a_{1\alpha} + a_{1\alpha} a^\dagger_{1\alpha})/2 \tag{3.30}$$
$$\Gamma^2{}_{2\alpha} = (a^\dagger_{2\alpha} a_{2\alpha} + a_{2\alpha} a^\dagger_{2\alpha})/2 \qquad (3.30)$$
$$\Gamma^2{}_{1\alpha} = (a^\dagger_{2\alpha} a^\dagger_{2\alpha} + a_{2\alpha} a_{2\alpha})/2 \tag{3.31}$$

We also define the Hermitian operators

$$\Gamma^2_{4\alpha} = (a^\dagger_{1\alpha}a^\dagger_{1\alpha} + a_{1\alpha}a_{1\alpha})/2 \quad (3.32)$$

and

$$\Gamma^2_{5\alpha} = i(a^\dagger_{2\alpha}a^\dagger_{2\alpha} - a_{2\alpha}a_{2\alpha})/2 \quad (3.33)$$
$$\Gamma^2_{6\alpha} = i(a^\dagger_{1\alpha}a^\dagger_{1\alpha} - a_{1\alpha}a_{1\alpha})/2 \quad (3.34)$$

They transform $a_{1\alpha}$ and $a^\dagger_{1\alpha}$ to have added terms of the form $c_\alpha a_{2\alpha} + c_\alpha' a^\dagger_{2\alpha}$ where c_α and c_α' are constants; and they transform $a_{2\alpha}$ and $a^\dagger_{2\alpha}$ to have added terms of the form $c_\alpha a_{1\alpha} + c_\alpha' a^\dagger_{1\alpha}$.

The complete set of Γ operators are part of an su(2,2) group. This group appears in Twistor Theory. A relation/unification of the group of these Bogoliubov transformations and Twistor Theory may eventually emerge.

The transformation $B_{2\alpha}B_{1\alpha}$ generates the maps of eqs. 3.21 and 3.22. As we will see next the key terms for the transition between annihilation/creation operators of two coordinate systems, A and B, are in

$$[c'_{i1}a^\dagger_{1\alpha} + c'_{i2}a^\dagger_{2\alpha} + C'_{i1}a_{1\alpha} + C'_{i2}a_{2\alpha}] \quad (3.35)$$

if we construct states based on $a^\dagger_{2\beta}$ and $a^\dagger_{2\alpha}$ with the vacuum $|0>_2$. Note that there are four constants, which are functions of α, in eq. 3.35. These constants are determined by the four parameters in $B_{2\alpha}B_{1\alpha}$. Thus the transformation between coordinate systems embodied in $B_{2\alpha}B_{1\alpha}$ determines the constants in eq. 3.35.

3.1.2 States in the Respective Coordinate Systems

The B states are superpositions of the A states as eq. 3.21 and 3.22 demonstrate. The one particle state in the B coordinate system is

$$a^\dagger_{2\beta}\,|0>_2 = \Sigma_\alpha\,(f^*_\beta, f^*_\alpha)[c'_{21}a^\dagger_{1\alpha} + c'_{22}a^\dagger_{2\alpha} + C'_{21}a_{1\alpha} + C'_{22}a_{2\alpha}]\,|0>_2$$

$$= \Sigma_\alpha\,(f^*_\beta, f^*_\alpha)[c'_{22}a^\dagger_{2\alpha} + C'_{22}a_{2\alpha}]\,|0>_2 \quad (3.36)$$

where the $a^\dagger_{2\alpha}$ term generates a one particle a state, and the $a_{2\alpha}$ term generates an A particle negative energy state. The two particle B coordinate system state

$$a^\dagger_{2\beta}\,a^\dagger_{2\beta}\,|0>_2$$

generates a superposition in momenta α) of two A particles, two A negative energy particles and an A vacuum state term.

Later, when we consider fermions, we will see that parton momentum distributions, xF(x), in deep inelastic e-p scattering may partially originate in the transition from a rest frame coordinate system to a different internal coordinate system.

3.2 Scalar Particle Creation/Annihilation Space (CASe)

The set of creation and annihilation operators a_1, a_2, $a^\dagger_1$, $a^\dagger_2$ form an operator basis for spaces, for each value α. In the scalar particle case in four space-time dimensions we define a four dimension space that we call (2,2) Creation/Annihilation Space or CASe(2,2) that supports an su(2,2) representation for each α, which we call a GiFT transformation. Defining basis "ortho" vectors $a_\alpha = (a^\dagger_{1\alpha}, a^\dagger_{2\alpha}, a_{1\alpha}, a_{2\alpha})$ we specify complex coordinates for an irreducible su(2,2):

$$x_\alpha = (x_{1\alpha}{}^0, x_{2\alpha}{}^0, x_\alpha{}^1, x_{2\alpha}{}^1) \tag{3.37}$$

Defining the metric

$$ds^2 = |dx_{1\alpha}{}^0|^2 + |dx_{2\alpha}{}^0|^2 - |dx_{1\alpha}{}^1|^2 - |dx_{2\alpha}{}^1|^2 \tag{3.38}$$

we note that there is an su(2,2) group invariance.

Thus for scalar particles an su(2,2) symmetry exists for each Fourier component α in a space-time of any number of space-time dimensions.

The vectors in the space have the form

$$a_\alpha \cdot x_\alpha = x_{1\alpha}{}^0 a^\dagger_{1\alpha} + x_{2\alpha}{}^0 a^\dagger_{2\alpha} + x_\alpha{}^1 a_{1\alpha} + x_{2\alpha}{}^1 a_{2\alpha} \tag{3.39}$$

The "ortho" vectors a_α form an algebra reminiscent of hypercomplex quaternions. The lack of complete commutativity is not a source for disquiet when one considers the anti-commutativity of quaternion, octonion and sedenion multiplication. It can be viewed positively as a step in the direction of a larger unified theory such as

$$\text{Spinor Space+CASe(n,m)} \tag{3.40}$$

CASe is enlarged for fermions in the various space-time dimensions. See chapter 4.

Note: A General Relativistic transformation between coordinate systems with the same Killing vector does not induce a GiFT transformation. A General Relativistic transformation between coordinate systems with differing sets of Killing vectors (or to a coordinate system with no Killing vector) does induce a GiFT CASe transformation.

4. GiFT for *Fermion* Creation/Annihilation Operator Transformations

The GiFT formalism developed in Blaha (2021i) and in section 4 of S. Blaha, "The Local Definition of Asymptotic Particle States", IL Nuovo Cimento **49A**, 35 (1979), which we call I, is directly extendable to fermions. We begin by considering 4-dimension space-time coordinate systems. Following the framework developed in section 4 of I we define fermion fields $\psi_1(x)$ and $\psi_2(x)$ in a coordinate system A labeled with α as

$$\psi_{1A}(x) = \Sigma_{\alpha,s}[b_{1\alpha s}u_{\alpha s}f_\alpha(x) + d^\dagger{}_{1\alpha s}v_{\alpha s}f_\alpha{}^*(x)] \tag{4.1}$$

$$\psi_{2A}(x) = \Sigma_{\alpha,s}[b_{2\alpha s}u_{\alpha s}f_\alpha(x) + d^\dagger{}_{2\alpha s}v_{\alpha s}f_\alpha{}^*(x)] \tag{4.2}$$

and in another coordinate system B labeled with β as

$$\psi_{1B}(x) = \Sigma_{\beta,s}[b_{1\beta s}u_{\beta s}\, g_\beta(x) + d^\dagger{}_{1\beta s}v_{\beta s}\, g_\beta{}^*(x)] \tag{4.3}$$

$$\psi_{2B}(x) = \Sigma_{\beta,s}[b_{2\beta s}u_{\beta s}\, g_\beta\, (x) + d^\dagger{}_{2\beta s}v_{\beta s}\, g_\beta{}^*(x)] \tag{4.4}$$

where $f_\alpha(x)$ and $g_\beta(x)$ are Fourier components.

To begin we assume the spin sectors transform independently We also anticipate the Fourier expansions are related in a manner similar to the scalar case of chapter 3. Then for spin value s we have

$$b_{1\beta s} = \Sigma_{\alpha,x}\,(g_\beta,\, f_\alpha)\; u^\dagger{}_{\beta s}u_{\alpha s}\,(c_{11s}b_{1\alpha s} + c_{12s}b_{2\alpha s} + C_{11s}b^\dagger{}_{1\alpha s} + C_{12s}b^\dagger{}_{2\alpha s}) \tag{4.6a}$$

$$b_{2\beta s} = \Sigma_{\alpha,x}\,(g_\beta,\, f_\alpha)\; u^\dagger{}_{\beta s}u_{\alpha s}\,(c_{21s}b_{1\alpha s} + c_{22s}\, b_{2\alpha s} + C_{21s}b^\dagger{}_{1\alpha s} + C_{22s}b^\dagger{}_{2\alpha s}) \tag{4.6b}$$

$$d^\dagger{}_{1\beta s} = \Sigma_{\alpha,x}\,(g^*{}_\beta\, f^*{}_\alpha)\; v^\dagger{}_{\beta s}v_{\alpha s}\,(c'_{11s}d^\dagger{}_{1\alpha s} + c'_{12s}d^\dagger{}_{2\alpha s} + C'_{11s}d^\dagger{}_{1\alpha s} + C'_{12s}d^\dagger{}_{2\alpha s}) \tag{4.6c}$$

$$d^\dagger{}_{2\beta s} = \Sigma_{\alpha,x}\,(g^*{}_\beta\, f^*{}_\alpha)\; v^\dagger{}_{\beta s}v_{\alpha s}\,(c'_{21s}d^\dagger{}_{1\alpha s} + c'_{22s}d^\dagger{}_{2\alpha s} + C'_{21s}d^\dagger{}_{1\alpha s} + C'_{22s}d^\dagger{}_{2\alpha s}) \tag{4.6d}$$

generalizing eq. 118 of I where the c_{ij} are all constants and where x is a space-like surface. The inner products have the form

$$\Sigma_x\,(g_\beta,\, f_\alpha) = i\; \Sigma_x\, g^*{}_\beta \overleftrightarrow{\partial_0} f_\alpha \tag{4.6e}$$

where "0" indicates the time-like coordinate. For example, in rectilinear coordinates

$$\Sigma_x (g_{k'}, f_k) = i\, \Sigma_x\, g^*{}_k \overleftrightarrow{\partial_0} f_k = \delta^3(\mathbf{k'} - \mathbf{k}) \tag{4.6e}$$

for momenta k and k′.

The non-zero anti-commutation relations of the creation/annihilation operators (with those of the rest being zero) are:

$$\begin{aligned}\{b_{i\alpha s}(\alpha), b^\dagger{}_{j\alpha s'}(\alpha')\} &= \{d_{i\alpha s}(\alpha), d^\dagger{}_{j\alpha s'}(\alpha')\} = (1 - \delta_{ij})\delta_{ss'}\, \delta^3(\boldsymbol{\alpha} - \boldsymbol{\alpha}') \\ \{b_{i\beta s}(\beta), b^\dagger{}_{j\beta s'}(\beta')\} &= \{d_{i\beta s}(\beta), d^\dagger{}_{j\beta s'}(\beta')\} = (1 - \delta_{ij})\delta_{ss'}\, \delta^3(\boldsymbol{\beta} - \boldsymbol{\beta}')\end{aligned} \tag{4.7}$$

Note the transformations of eq. 4.6 superimpose "momenta" in relating the coordinate system quantizations. But they do not superimpose spins.[12] As a result the transformations can be factored into two sets: one set for spin up and one set for spin down.

4.1 Transformations of the b and $b^\dagger$ Operators

The sets of Bogoliubov transformations B_1 and B_2 for spin up or down b creation/annihilation operators are defined by

$$B_{1\alpha}(x_1, x_2) = \mathrm{ex}[x_1\Gamma^1{}_{3\alpha}]\, \exp[x_2\Gamma^1{}_{2\,\alpha}] \tag{4.8}$$

where the Hermitian operators $\Gamma^1{}_{1i\alpha}$ are

$$\Gamma^1{}_{3\alpha}(b_{1\alpha s}, b_{2\alpha s}) = (b^\dagger{}_{2\alpha s} b_{1\alpha s} + b_{2\alpha s} b^\dagger{}_{1\alpha s})/2 \tag{4.9}$$

$$\Gamma^1{}_{2\alpha}(b_{1\alpha s}, b_{2\alpha s}) = i(b^\dagger{}_{2\alpha s} b^\dagger{}_{1\alpha s} - b_{2\alpha s} b_{1\alpha s})/2 \tag{4.10}$$

We also define

$$\Gamma^1{}_{1\alpha}(b_{1\alpha s}, b_{2\alpha s}) = -\,(b^\dagger{}_{2\alpha s} b^\dagger{}_{1\alpha s} + b_{2\alpha s} b_{1\alpha s})/2 \tag{4.11}$$

and

$$\Gamma^1{}_{4\alpha}(b_{1\alpha s}, b_{2\alpha s}) = -\,(b^\dagger{}_{2\alpha s} b_{1\alpha s} + b_{2\alpha s} b^\dagger{}_{1\alpha s})/2 \tag{4.12}$$

The B_2 transformation has the form:

$$B_{2\alpha}(x_1, x_2) = \exp[x_1\Gamma^2{}_{3\alpha}]\, \exp[x_2\Gamma^2{}_{2\,\alpha}] \tag{4.13}$$

where the Hermitian operators $\Gamma^2{}_{1i\alpha}$ are

$$\Gamma^2{}_{3\alpha} = \Gamma^1{}_{3\alpha}(b_{1\alpha s}, b_{1\alpha s}) \tag{4.14}$$

$$\Gamma^2{}_{2\alpha} = \Gamma^1{}_{3\alpha}(b_{2\alpha s}, b_{2\alpha s}) \tag{4.15}$$

$$\Gamma^2{}_{1\alpha} = \Gamma^1{}_{1\alpha}(b_{2\alpha s}, b_{2\alpha s}) \tag{4.16}$$

We also define the Hermitian operators

---

[12] We consider spin mixing later.

$$\Gamma^2{}_{4\alpha} = \Gamma^1{}_{1\alpha}(b_{1\alpha s}, b_{1\alpha s}) \tag{4.17}$$
$$\Gamma^2{}_{5\alpha} = \Gamma^1{}_{2\alpha}(b_{2\alpha s}, b_{2\alpha s}) \tag{4.18}$$
$$\Gamma^2{}_{6\alpha} = \Gamma^1{}_{2\alpha}(b_{1\alpha s}, b_{1\alpha s}) \tag{4.19}$$

The above set of operators generate the su(1,1) transformations to the forms of eqs. 4.6a and 4.6b. The su(1,1) representation has four real components for b_i, and $b_i^\dagger$ for i = 1, 2.

4.2 Transformations of the d and d† Operators

These operators for "holes" have an analogous form to the b operators. They have the same form as those in eqs. 4.8 – 4.19 with each "b" replaced with a "d". These operators generate the transformations to the forms of eqs. 4.6c and 4.6d. These operators are part of the operators of another su(1, 1) in a manner similar to those of chapter 3.

These operators, plus those of section 4.1, are part of the operators of an su(2, 2), which contains operators mixing b and d operators. The su(2,2) representation has eight real components for b_i, and $b_i^\dagger$ and for d_i, and $d_i^\dagger$ for i = 1, 2 for each spin.

$$su(1,1) \otimes su(1,1) \rightarrow su(2,2) \tag{4.20}$$

4.3 Operators of Types b and d *with* Spin Mixing

We can anticipate that certain General Relativistic transformations may cause spin mixing. Some typical (2,2) Creation/Annihilation Space (CASe) transformations are (in the notation of section 4.1)

$$B_{1\alpha}(x_1, x_2) = \exp[x_1\Gamma^1{}_{3\alpha}]\exp[x_2\Gamma^1{}_{2\alpha}] \tag{4.21}$$

with

$$\Gamma^1{}_{3\alpha}(b_{1\alpha s}, b_{2\alpha s'})$$
$$\Gamma^1{}_{2\alpha}(b_{1\alpha s}, b_{2\alpha s'})$$
$$\Gamma^1{}_{1\alpha}(b_{1\alpha s}, b_{2\alpha s'})$$
$$\Gamma^1{}_{4\alpha}(b_{1\alpha s}, b_{2\alpha s'})$$

where $s \neq s'$.

The B_2 transformation has the form:

$$B_{2\alpha}(x_1, x_2) = \exp[x_1\Gamma^2{}_{3\alpha}]\exp[x_2\Gamma^2{}_{2\alpha}] \tag{4.22}$$

where

$$\Gamma^2{}_{3\alpha} = \Gamma^1{}_{3\alpha}(b_{1\alpha s}, b_{1\alpha s'})$$
$$\Gamma^2{}_{2\alpha} = \Gamma^1{}_{3\alpha}(b_{2\alpha s}, b_{2\alpha s'})$$
$$\Gamma^2{}_{1\alpha} = \Gamma^1{}_{1\alpha}(b_{2\alpha s}, b_{2\alpha s'})$$
$$\Gamma^2{}_{4\alpha} = \Gamma^1{}_{1\alpha}(b_{1\alpha s}, b_{1\alpha s'})$$
$$\Gamma^2{}_{5\alpha} = \Gamma^1{}_{2\alpha}(b_{2\alpha s}, b_{2\alpha s'})$$
$$\Gamma^2{}_{6\alpha} = \Gamma^1{}_{2\alpha}(b_{1\alpha s}, b_{1\alpha s'})$$

and $s \neq s'$.

4.3.1 Four Dimension Case

In four dimensions there are two spin values, up and down. Consequently the combination of spin mixed transformations gives su(4,4) as the CASe group, which contains the combined spin up and spin down groups. The su(4,4) representation has sixteen real components for b_i, and $b_i^\dagger$, and d_i, and $d_i^\dagger$, for i = 1, 2 for both spins.

$$\text{su}(4,4) \tag{4.23}$$

4.3.2 Higher Dimension Case

For space-time dimension r there are $2^{r/2-1}$ spin values. In this case there is an $\text{su}(2^{r/2}, 2^{r/2})$ CASe group containing the direct product subgroup:

$$\bigotimes_{k=1}^{2^{r/2-1}} \text{su}(2,2)_k \tag{4.24}$$

where k labels the spin value. The combined spin mixing gives the CASe group

$$\text{su}(2^{r/2}, 2^{\,r/2}) \tag{4.25}$$

with $2^{r/2+2}$ real components for b_i, and $b_i^\dagger$, and d_i, and $d_i^\dagger$, for i = 1, 2 for all spins.

4.4 Operators of Types b and d *with* Spin and b – d Mixing

If we consider transformations with spin and b – d mixing then the form of the transformed creation/annihilation eq. 4.6a operator becomes

$$\begin{aligned} b_{1\beta s} = \Sigma_{\alpha,x,s}\,(g_\beta,\, f_\alpha)\, u^\dagger_{\beta s}u_{\alpha s}\,(&c_{11s}b_{1\alpha s} + c_{12s}b_{2\alpha s} + C_{11s}b^\dagger{}_{1\alpha s} + C_{12}b^\dagger{}_{2\alpha s} + \\ &+ c'_{11s}d_{1\alpha s} + c'_{12s}d_{2\alpha s} + C'_{11s}d^\dagger{}_{1\alpha} + C'_{12s}d^\dagger{}_{2\alpha}) \end{aligned} \tag{4.26}$$

with similar forms for the other transformed operators.

The transformations of this type form require an extension of su(4, 4) to su(4, 4) over sedenion hypercharge numbers. (Sedenions have 16 components. They are the next Cayley-Dickson number after octonions.)

In the present case we therefore express each component creation/annihilation operator in eqs. 4.1 and 4.2 as sedenions, and then perform transformations with sedenion array multiplication to generate the terms of the form of eq. 4.26.

4.5 The CASe Groups

The CASe groups of fermions of the universes of Octonion Cosmology are determined above in eq. 4.25. Fig. 4.1 presents them for b and d creation/annihilation operators separately. They have spin mixing and generate transformations of the type of eq. 4.6.

A General Relativistic transformation to a non-static coordinate system without a Killing vector causes fermion wave function creation/annihilation operators to be CASe transformed in every Octonion Cosmology universe (space-time). The correspondence between General Relativistic transformations and the CASe transformations of GiFT reflects the unity of General Relativity and Quantum Theory.

CASe Groups of Eight of the Ten Octonion Universes

Octonion Space Number o_s	Cayley-Dickson Number n	Cayley Number d_c	Dimension Array Total d_d	Space-time-Dimension r	Fermion Spinor Array Total d_s	CASe Group $su(2^{r/2},2^{r/2})$ CASe
0	10	1024	1024×1024	18	512×512	su(512,512)
1	9	512	512×512	16	256×256	su(256,256)
2	8	256	256×256	14	128×128	su(128,128)
3	7	128	128×128	12	64×64	su(64,64)
4	6	64	64×64	10	32×32	su(32,32)
5	5	32	32×32	8	16×16	su(16,16)
6	**4**	**16**	$\mathbf{16 \times 16}$	$\mathbf{6 \to 4}$	$\mathbf{8 \times 8}$	su(4,4)
7	**3**	8	8×8	4	4×4	su(4,4)

Figure 4.1. The CASe groups of the eight Octonion Cosmology spaces. Space n = 4 has 6 – 2 = 4 space-time dimensions due to compacted dimensions.

5. A General Relativistic Basis for Octonion Cosmology

The spaces of Octonion Cosmology have been derived from a recursive generation of subspaces through fermion-antifermion annihilation of space fermion particles. In this chapter we consider an alternate, not necessarily unrelated, approach based on General Relativistic considerations. We begin by considering a universe with a 4-dimension space-time. Then we go on to consider the case of spaces with higher dimension space-times in chapters 6 and 7.

5.1 Universe with Four Dimension Space-Time

We begin by considering a space with a four dimension space-time[13] such as our universe. In that space we can define a GiFT formulation for fermions as we did in chapter 4. The fermion wave function has an su(4,4) (2,2) Creation/Annihilation Space (CASe) group for transformations of the b and d creation/annihilation operators from a coordinate system to a non-static coordinate system without a Killing vector. These transformations parallel the transformations of General Relativity (GR).

It is worth noting that almost all coordinate systems are non-static.[14] Consequently the set of GR transformations is dominated by transformations to non-static coordinate systems. A CASe transformation to a non-static coordinate system has the effect of changing the Fermi sea such that some fermions are raised above or below the sea. Thus GR transformations cause the set of physical fermion states to be modified resulting in fermion creation/annihilation.

5.1.1 Matrix Representations of su(1,1)

The matrix representations of the su(4,4) group form a 16 × 16 fundamental representation in a 16 real dimension space. The space vectors have a form analogous to the set of creation/annihilation operators:

$$(b_{1\uparrow}, b_{2\uparrow}, b^{\dagger}{}_{1\uparrow}, b^{\dagger}{}_{2\uparrow}, \ b_{1\downarrow}, b_{2\downarrow}, b^{\dagger}{}_{1\downarrow}, b^{\dagger}{}_{2\downarrow}, \ d_{1\uparrow}, d_{2\uparrow}, d^{\dagger}{}_{1\uparrow}, d^{\dagger}{}_{2\uparrow}, \ d_{1\downarrow}, d_{2\downarrow}, d^{\dagger}{}_{1\downarrow}, d^{\dagger}{}_{2\downarrow}) \qquad (5.1)$$

Note that the space has four quartets of vectors.[15] This arrangement will be found to persist to the level of the dimension and fermion arrays of NEWQUeST, the octonion spectrum space of our universe. The subdivisions appear in the known fermion and internal symmetry fundamental representations as seen in Figs. 2b.1 and 2b.2.

5.1.2 Matrix Form of CASe Transformations

The form a CASe transformation can be symbolized by

[13] This space has space-time dimension 4 due to the compaction of two space-time dimensions.
[14] Some examples are the Kerr solution, space-times with gravity waves, and and spherically symmetric space-times.
[15] Note the form of eq. 5.1 appears to be complex-valued. But it is equivalent to a real-valued set.

$$x' = Tx \tag{5.2}$$

where T is a 16 × 16 su(4, 4) transformation, and x and x′ are 16-vectors in the fundamental representation of su(4, 4). In CASe the vectors are creation/annihilation operators that are transformed due to a GR transformation.

5.1.3 CASe Transformations for su(4, 4) over Sedenion Hypercomplex Numbers

CASe transformations take sixteen creation/annihilation operator terms (eq. 5.1) of eqs. 4.1 and 4.2, and generate 16 terms of the form of 4.26. We now extend CASe transformations to su(4, 4) over sedenion numbers. Thus each of the 16 terms in eqs. 4.1 and 4.2 becomes a sedenion number.

The use of sedenions, with their 16 parts, enables us to treat sets of 16 creation/annihilation operators as a unit as eqs. 5.1, 4.1, 4.2 and 4.26 suggest.

The vectors e_i are the 16 unit sedenions. They are numbered from 0 through 15. A sedenion S has the form

$$\begin{aligned} S &= a_0e_0 + a_1e_1 + a_2e_2 + \ldots + a_{15}e_{15} \\ &\equiv (a_0, a_1, \ldots, a_{15}) \end{aligned} \tag{5.3}$$

We now define the 16 sedenion terms as

$$S_i = (0, 0, \ldots \overset{i^{th}}{x_i}, 0, 0, \ldots) \tag{5.4}$$

for i = 0, 1, 2, …, 15 where x_i i s the ith entry in eq. 5.1 above. We form a 16-vector

$$x = (S_0, S_1, \ldots, S_{15}) \tag{5.5}$$

We define an su(4, 4) transformation over sedenions with

$$T = [a_{ij}] \tag{5.6}$$

where

$$a_{ij} = s_{ij}e_ie_j = s_{ij}f_{ijk}e_k$$

where the s_{ij} are coefficients. The sedenion multiplication table appears in Fig. 5.1. Sedenions have an identity

$$e_ie_j = f_{ijk}e_k \tag{5.7}$$

for integers f_{ijk} which are implicit in the table.

An su(4, 4) sedenion transformation has the form

$$x' = Tx \tag{5.8}$$

and generates a sedenion 16-vector with each term of the form of eq. 4.26.

e_0	e_1	e_2	e_3	e_4	e_5	e_6	e_7	e_8	e_9	e_{10}	e_{11}	e_{12}	e_{13}	e_{14}	e_{15}
e_0	e_1	e_2	e_3	e_4	e_5	e_6	e_7	e_8	e_9	e_{10}	e_{11}	e_{12}	e_{13}	e_{14}	e_{15}
e_1	$-e_0$	e_3	$-e_2$	e_5	$-e_4$	$-e_7$	e_6	e_9	$-e_8$	$-e_{11}$	e_{10}	$-e_{13}$	e_{12}	e_{15}	$-e_{14}$
e_2	$-e_3$	$-e_0$	e_1	e_6	e_7	$-e_4$	$-e_5$	e_{10}	e_{11}	$-e_8$	$-e_9$	$-e_{14}$	$-e_{15}$	e_{12}	e_{13}
e_3	e_2	$-e_1$	$-e_0$	e_7	$-e_6$	e_5	$-e_4$	e_{11}	$-e_{10}$	e_9	$-e_8$	$-e_{15}$	e_{14}	$-e_{13}$	e_{12}
e_4	$-e_5$	$-e_6$	$-e_7$	$-e_0$	e_1	e_2	e_3	e_{12}	e_{13}	e_{14}	e_{15}	$-e_8$	$-e_9$	$-e_{10}$	$-e_{11}$
e_5	e_4	$-e_7$	e_6	$-e_1$	$-e_0$	$-e_3$	e_2	e_{13}	$-e_{12}$	e_{15}	$-e_{14}$	e_9	$-e_8$	e_{11}	$-e_{10}$
e_6	e_7	e_4	$-e_5$	$-e_2$	e_3	$-e_0$	$-e_1$	e_{14}	$-e_{15}$	$-e_{12}$	e_{13}	e_{10}	$-e_{11}$	$-e_8$	e_9
e_7	$-e_6$	e_5	e_4	$-e_3$	$-e_2$	e_1	$-e_0$	e_{15}	e_{14}	$-e_{13}$	$-e_{12}$	e_{11}	e_{10}	$-e_9$	$-e_8$
e_8	$-e_9$	$-e_{10}$	$-e_{11}$	$-e_{12}$	$-e_{13}$	$-e_{14}$	$-e_{15}$	$-e_0$	e_1	e_2	e_3	e_4	e_5	e_6	e_7
e_9	e_8	$-e_{11}$	e_{10}	$-e_{13}$	e_{12}	e_{15}	$-e_{14}$	$-e_1$	$-e_0$	$-e_3$	e_2	$-e_5$	e_4	e_7	$-e_6$
e_{10}	e_{11}	e_8	$-e_9$	$-e_{14}$	$-e_{15}$	e_{12}	e_{13}	$-e_2$	e_3	$-e_0$	$-e_1$	$-e_6$	$-e_7$	e_4	e_5
e_{11}	$-e_{10}$	e_9	e_8	$-e_{15}$	e_{14}	$-e_{13}$	e_{12}	$-e_3$	$-e_2$	e_1	$-e_0$	$-e_7$	e_6	$-e_5$	e_4
e_{12}	e_{13}	e_{14}	e_{15}	e_8	$-e_9$	$-e_{10}$	$-e_{11}$	$-e_4$	e_5	e_6	e_7	$-e_0$	$-e_1$	$-e_2$	$-e_3$
e_{13}	$-e_{12}$	e_{15}	$-e_{14}$	e_9	e_8	e_{11}	$-e_{10}$	$-e_5$	$-e_4$	e_7	$-e_6$	e_1	$-e_0$	e_3	$-e_2$
e_{14}	$-e_{15}$	$-e_{12}$	e_{13}	e_{10}	$-e_{11}$	e_8	e_9	$-e_6$	$-e_7$	$-e_4$	e_5	e_2	$-e_3$	$-e_0$	e_1
e_{15}	e_{14}	$-e_{13}$	$-e_{12}$	e_{11}	e_{10}	$-e_9$	e_8	$-e_7$	e_6	$-e_5$	$-e_4$	e_3	e_2	$-e_1$	$-e_0$

Figure 5.1. The sedenion multiplication table. We have divided it into 4× 4 parts reflecting the divisions of the fundamental fermions and the subgroups of Cayley-Dickson number n = 4 space (our universe).

The sedenion x′ is the result of a CASe transformation instigated by a GR transformation. It has 256 parts. We can display it as an array:

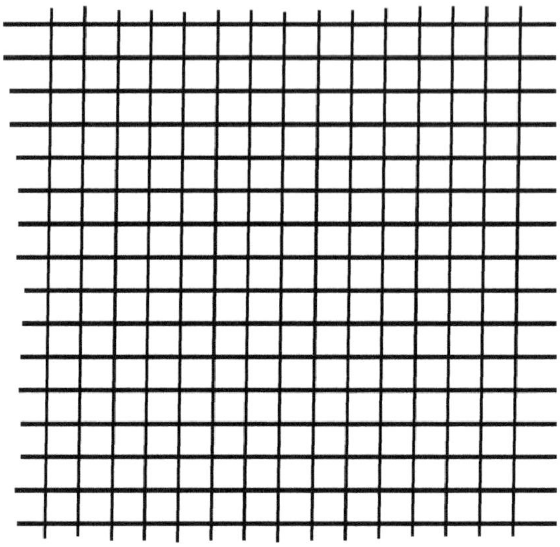

Figure 5.2. The sixteen sedenions of x′ expanded to form a 16 × 16 array of entries. Each row has the form of eq. 4.26.

In chapter 7 we will see that higher spaces also have a hypercomplex number aspect that emerges when GR in their space-time diimensions causes GiFT transformations of creation/annihilation operators.

6. Generation of Fermions and Internal Symmetries from a GR CASe Transformation

This chapter uses the form of GiFT transformations of creation/annihilation operators as a paradigm for transformations of fundamental fermions and internal symmetries to a non-static reference frame where the number of fermions is reduced from 256 to 16 and the internal symmetries are reduced to SU(3) ⊗U(1)⊗SU(2)⊗U(1) (or SU(4)⊗SU(2)⊗U(1)) forming a one generation Standard Model.

6.1 Fundamental Frame

A common feature of calculations in Physics is to choose a reference frame which facilitates computation. Common choices are the rest frame and the center of mass frame. In the preceding chapter we found that GR transformations to non-static reference frames induced creation/annihilation operator transformations. We further found that the transformations can be formulated using sedenion numbers. The result was transformations producing 16 × 16 arrays of numbers with a form like Fig. 5.2.

The appearance of sedenions and 16 × 16 arrays is analogous to the sedenions and dimension array of NEWQUeST in Octonion Cosmology. The 4 × 4 subblocks of 16 dimensions in NEWQUeST mirror the subblocks in Fig. 5.1 and the subblocking appearing in the creation/annihilation vector of eq. 5.1. Fig. 6.1 shows the similarity of the array structures in NEWQUeST and (2,2) Creation/Annihilation Space (CASe).

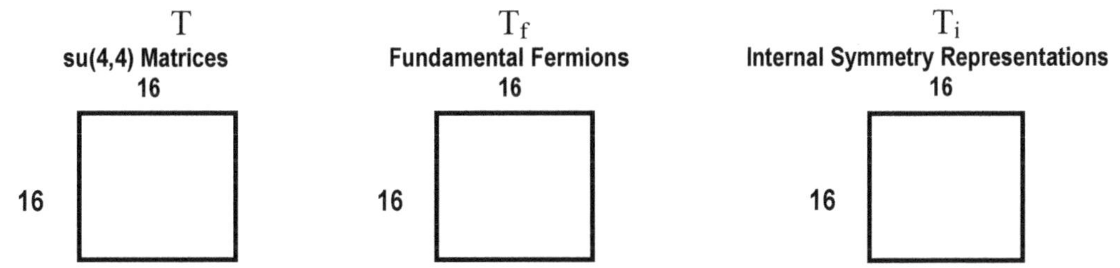

Figure 6.1 Comparison of arrays in CASe and NEWQUeST.

The analogous nature of these arrays raises the possibility that the fundamental fermion array and the Internal Symmetry array both are generated from a non-static reference frame via an su(4, 4) CASe-like transformation.

In chapter 5 we found

$$x' = Tx \tag{5.8}$$

We now propose analogous transformations, T′ and T″, that generate the fundamental fermion array and the Internal Symmetry array from a non-static coordinate system fermions and symmetries.

$$F_{ermions} = T'F_{fundamental} \tag{6.1}$$
$$I_{QUeST} = T''I_{fundamental} \tag{6.2}$$

Ideally $T' = T''$. In this approach $F_{fundamental}$ would be a sedenion vector composed of 16 sedenions of the form of eq. 5.4 where each sedenion has one primitive fermion. Thus it would represent 16 primitive fermions. Similarly $I_{fundamental}$ would be a sedenion vector composed of sedenions of the form of eq. 5.4 where each sedenion has one dimension. It would thus contain 16 real dimensions.

We call the reference frame for these reduced sets of fermions and dimensions the *Fundamental Frame*. It has a non-static coordinate system. (Almost all reference frames are non-static.)

The results of the transformations from the non-static Fundamental Frame are the NEWQUeST fermion and dimension arrays. Consequently the plethora of fermions and dimensions would be much reduced.

6.2 Contents of the Reduced Fermion and Internal Symmetry Arrays of the Fundamental Frame

The fermions and symmetries of the Fundamental Frame appear to mirror the Standard Model in part. On that basis we suggest the fermions appear in a 16-vector sedenion in analogy with eqs. 5.4 and 5.5. Each of the 16 sedenions contains one of the 16 fermions with each represented by an integer such as 1. This sedenion vector is mapped by a CASe transformation, eq. 6.1, to the 256 fermions of NEWQUeST and NEWUST.

Conceptually we can group the set of Fundamental Frame fermions into four fermion subblocks. Two of the subblocks contain Normal fermions. Two of the subblocks contain Dark fermions.

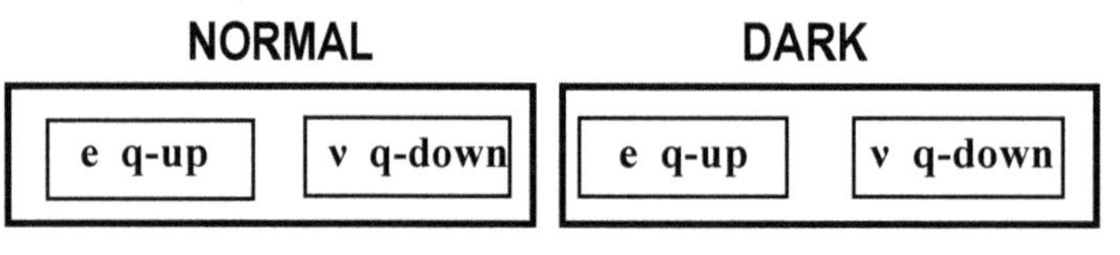

Figure 6.2. The 16 fermions of the Fundamental Frame grouped into Normal and Dark color SU(4) sets. Note each q-up and q-down are color triplets.

6.2.1 Four Types of Fermions

There are four types (species) of fermions: e-type, ν-type, up-quark type, and down-quark type. The species originate in a previous derivation of four fermion species (types) in the early 2000's. See Appendix D for the author's 2007 book *The Origin of the Standard Model* where it is derived in detail based on the four types of Complex Lorentz group boosts.

6.2.2 Transformation of Fundamental Frame Fermions to a "Normal" Frame

The proposed CASe transformation eq. 6.1 generates a 16-vector of sedenions in a "normal" static reference frame. Each sedenion component contains 16 fermions, which can be arranged to have the form of Fig. 6.3. We assume the arrangement of each of the output 16 sedenions, which contain numeric values, can map to the form of Fig. 6.2: Normal: e, q-up_1, q-up_2, q-up_3, ν, q-$down_1$, q-$down_2$, q-$down_3$; Dark: e, q-up_1, q-up_2, q-up_3, ν, q-$down_1$, q-$down_2$, q-$down_3$. Fig. 6.3 shows the 16 output sedenions displaying four layers with each containing four generations as in Fig. 6.4, and in NEWQUeST and NEWUST.

	NORMAL		DARK	
	e q-up	ν q-down	e q-up	ν q-down
	e q-up	ν q-down	e q-up	ν q-down
Layer 1	e q-up	ν q-down	e q-up	ν q-down
	e q-up	ν q-down	e q-up	ν q-down
	e q-up	ν q-down	e q-up	ν q-down
	e q-up	ν q-down	e q-up	ν q-down
Layer 2	e q-up	ν q-down	e q-up	ν q-down
	e q-up	ν q-down	e q-up	ν q-down
	e q-up	ν q-down	e q-up	ν q-down
	e q-up	ν q-down	e q-up	ν q-down
Layer 3	e q-up	ν q-down	e q-up	ν q-down
	e q-up	ν q-down	e q-up	ν q-down
	e q-up	ν q-down	e q-up	ν q-down
	e q-up	ν q-down	e q-up	ν q-down
Layer 4	e q-up	ν q-down	e q-up	ν q-down
	e q-up	ν q-down	e q-up	ν q-down

Figure 6.3. The 16-vector of sedenions with each containing 16 different fermions. Note the appearance of 4 layers and 4 generations of each fermion in each layer.

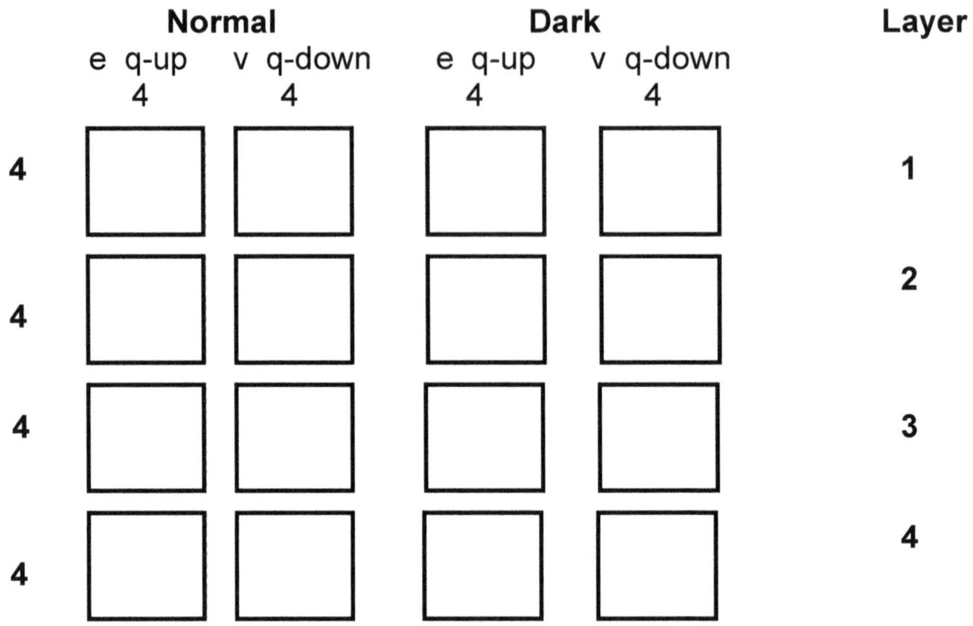

Figure 6.4. Block form of NEWQUeST's and NEWUST's 256 fermions grouped by sixteens. Each block contains 4 generations of each fermion within the block.

6.2.3 Symmetry Groups within the Fundamental Frame – One Generation Standard Model

We provisionally assume that there are 16 symmetry group irreducible representation dimensions in a Fundamental Frame sedenion 16-vector in analogy with eqs. 5.4 and 5.5. Each of the 16 sedenions contains one of the 16 symmetry group dimensions in the form of an integer such as 1. This sedenion vector is mapped by a CASe transformation, eq. 6.1, to the 256 symmetry groups irreducible representations dimensions of NEWQUeST and NEWUST.

The 16 dimensions may be viewed as the dimensions of irreducible representations of U(8) broken to

$$\mathrm{SU}(3)\otimes\mathrm{U}(1) \quad \mathrm{SU}(2)\otimes\mathrm{U}(1)\otimes\mathrm{SL}(2, \mathbf{C})$$ [16]

In this case we can see a SU(3)⊗U(1)⊗SU(2)⊗U(1)⊗SL(2, **C**) set of symmetries arises giving a one generation Standard Model in 4 space-time dimensions. We have considered this type of Standard Model in the author's 2007 book *The Origin of the Standard Model*, which is reprinted in Appendix D.

The Fundamental Frame has the internal symmetry groups, the set of 8 normal fermions and 4-dimension space-time to form a Standard Model. The Standard Model, and its generalization to the Unified SuperStandard Theory, can be viewed as a

[16] We use SL(2, C) to represent $SO^{+}(1, 3)$.

transformation from the Fundamental Frame. The Fundamental Frame is non-static. There is an infinite set of possible Fundamental Frames since there is not a 1:1 correspondence between a General Relativistic transformation and a specific su(4, 4) transformation. A General Relativistic transformation corresponds to an infinite set of possible Fundamental Frames However the fermion and symmetry group contents of the Fundamental Frame is specific.

6.2.4 Transformation of Fundamental Frame Dimensions to a "Normal" Frame

The proposed CASe transformation eq. 6.1 generates a 16-vector of sedenions. Each sedenion component contains 16 irreducible representation dimensions with each in the form of a numeric value. A typical output sedenion has the form:

xxxxxxxxxxxxxxxx

where each x represents has some numeric value. The values may be the same or differ from entry to entry. We interpret these values as markers for irreducible symmetry group representation dimensions.

We arrange the 16 output sedenions in layers of Normal and Dark with the form of Fig. 6.5 with the form of each sedenion initially being 16 real U(4)⊗U(4) irreducible representation dimensions. Some sedenions are further subdivided into 8 real SU(4) irreducible representation dimensions,[17] and 4 real SU(2)⊗U(1) irreducible representation dimensions, and 4 real SL(2, **C**) space-time dimensions. See Figs. 6.5, 6.6. 6.7 and 6.8.

	NORMAL	**DARK**
	U(4)⊗U(4)	
	U(4)⊗U(4)	
Layer 1		U(4)⊗U(4)
		U(4)⊗U(4)
	U(4)⊗U(4)	
	U(4)⊗U(4)	
Layer 2		U(4)⊗U(4)
		U(4)⊗U(4)
	U(4)⊗U(4)	
	U(4)⊗U(4)	
Layer 3		U(4)⊗U(4)
		U(4)⊗U(4)
	U(4)⊗U(4)	
	U(4)⊗U(4)	
Layer 4		U(4)⊗U(4)
		U(4)⊗U(4)

Figure 6.5. The sixteen output dimension sedenions generated by a CASe transformation from the Fundamental Frame arranged in layers and generations.

---

[17] Or SU(3)⊗U(1) dimensions.

NORMAL		DARK	
SU(3)⊗U(1) Generation U(4)	SU(2)⊗U(1)⊗SL(2, C) Layer U(4)	SU(3)⊗U(1) Generation U(4)	SU(2)⊗U(1)⊗SL(2, C) Layer U(4)
SU(3)⊗U(1) Generation U(4)	SU(2)⊗U(1)⊗SL(2, C) Layer U(4)	SU(3)⊗U(1) Generation U(4)	SU(2)⊗U(1)⊗SL(2, C) Layer U(4)
SU(3)⊗U(1) Generation U(4)	SU(2)⊗U(1)⊗SL(2, C) Layer U(4)	SU(3)⊗U(1) Generation U(4)	SU(2)⊗U(1)⊗SL(2, C) Layer U(4)
SU(3)⊗U(1) Generation U(4)	SU(2)⊗U(1)⊗SL(2, C) Layer U(4)	SU(3)⊗U(1) Generation U(4)	SU(2)⊗U(1)⊗SL(2, C) Layer U(4)

Figure 6.6. The rearranged sixteen sedenions of Fig. 6.5 generated by a CASe transformation from the Fundamental Frame. This arrangement is the same as NEWQUeST and NEWUST.

6.3 A Basis for NEWQUeST and NEWUST in CASe Transformations from the Fundamental Frame

The set of internal symmetries of NEWQUeST and NEWUST emerges from Fig. 6.6. GiFT together with its General Relativistic aspect enables a reduction of the symmetries and fermions of NEWQQUeST and NEWUST to 16 symmetry dimensions and 16 fermions.

The CASe transformations of GiFT, that provide a Fundamental Frame formulation of particles and interactions, *require* the PseudoQuantum formulation of GiFT. They implement a unification of Quantum theory and General Relativity since Quantum Theory is based on GiFT, which embodies Quantum Field Theory, which, in turn, embodies Quantum Mechanics.

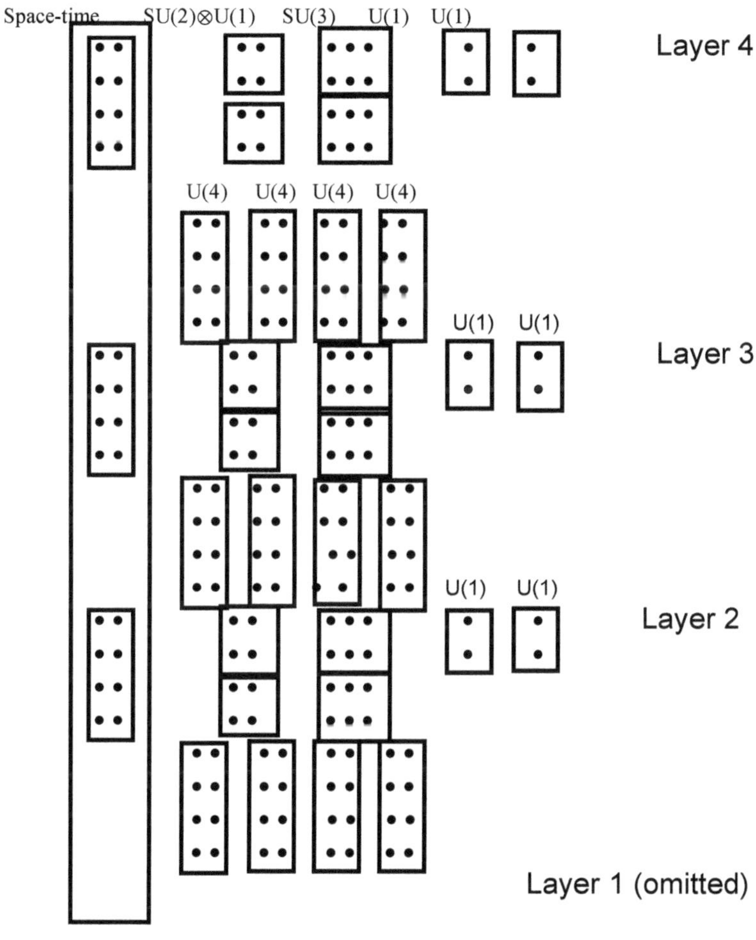

Figure 6.7. Three of the four layers of QUeST internal symmetry groups (and space-time) for Cayley-Dickson space 4. Layer 1 which has an identical form was omitted due to "page space" limitations. Note the left column of blocks combine to specify a 4 dimension octonion space-time. Note each layer has 64 dimensions.

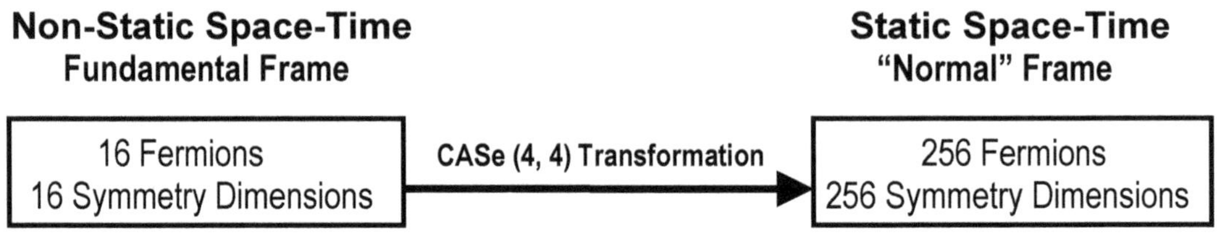

Figure 6.8. The CASe transformation from a Fundamental Frame to a "Normal" static space-time reference frame.

7. Higher Spaces CASe Transformations and Their Fundamental Frames

The higher Octonion Cosmology spaces where the Cayley-Dickson number is > 4 have Fundamental Frames that are analogous to the Fundamental Frame for n = 4 discussed in Chapter 6.

Figure 4.1 below displays the space-time dimension and the CASe group for these spaces. Since there is no name for hypercomplex numbers beyond sedenion, we will define *n-ion* to represent the hypercomplex Cayley-Dickson number n.

Each Octonion Cosmology n-ion space has a General Relativity. We call it n-ion GR. This chapter uses GiFT transformations of creation/annihilation operators as a paradigm for n-ion GR transformations of fermions and internal symmetries in an n-ion Fundamental Frame to a "normal" reference frame.

In an n-ion Fundamental Frame there are 2^n fermions and 2^n irreducible representation dimensions for internal symmetries. They are "boosted" to 2^{2n} fermions and 2^{2n} dimensions in a "normal" n-ion space reference frame by a CASe transformation. Correspondingly an n-ion GR transformation takes a Fundamental Frame to a "normal" reference frame.

The n-ion Fundamental Frame is non-static. There is an infinite set of possible n-ion Fundamental Frames since there is not a 1:1 correspondence between an n-ion General Relativistic transformation and a specific n-ion CASe transformation. An n-ion General Relativistic transformation corresponds to an infinite set of possible n-ion CASe transformations and thus to an infinite set of possible n-ion Fundamental Frames However the fermion and symmetry group contents of all the possible n-ion Fundamental Frames are the same.

CASe Groups of Eight of the Ten Octonion Universes

Octonion Space Number	Cayley-Dickson Number	Cayley Number	Dimension Array Total	Space-time-Dimension	Fermion Spinor Array Total	CASe Group $su(2^{r/2},2^{r/2})$
o_s	n	d_c	d_d	r	d_s	CASe
0	10	1024	1024×1024	18	512×512	su(512,512)
1	9	512	512×512	16	256×256	su(256,256)
2	8	256	256×256	14	128×128	su(128,128)
3	7	128	128×128	12	64×64	su(64,64)
4	6	64	64×64	10	32×32	su(32,32)
5	5	32	32×32	8	16×16	su(16,16)
6	**4**	**16**	$\mathbf{16 \times 16}$	$\mathbf{6 \to 4}$	$\mathbf{8 \times 8}$	su(4,4)
7	**3**	8	8×8	4	4×4	su(4,4)

Figure 4.1. The CASe groups of the ten Octonion Cosmology spaces. Space n = 4 has 6 – 2 = 4 space-time dimensions due to 2 compacted dimensions.

We begin by noting the n-ion CASe group is su(2^{n-1}, 2^{n-1}), the total dimensions of the group representation is $2^n \times 2^n = 2^{2n}$, the space-time dimension is r = 2n -2, and the dimension of the symmetry array of the Octonion Cosmology space is 2^{2n}.

In a manner similar to that of chapter 6 we see the CASe group dimensions equals the symmetry array dimensions 2^{2n}. See Fig. 7.1. On this basis this chapter uses n-ion GiFT CASe transformations of creation/annihilation operators as a paradigm for transformations between fermions and internal symmetries of a "normal" reference frame and an n-ion Fundamental Frame.

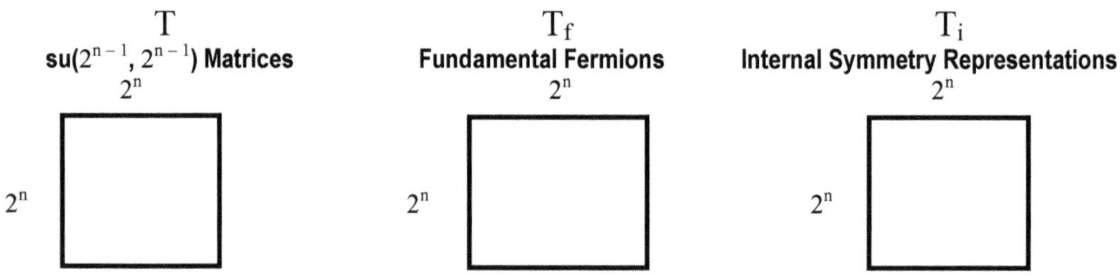

Figure 7.1 Comparison of arrays in CASe and NEWQUeST for the n-ion case.

$$F_{\text{fermions}} = T'F_{\text{fundamental}} \tag{7.1}$$
$$I_{\text{QUeST}} = T''I_{\text{fundamental}} \tag{7.2}$$

We assume T′ = T″. Then $F_{\text{fundamental}}$ would be a n-ion vector composed of 2^n n-ions of the form of eq. 5.4 where each n-ion has one primitive fermion within it. The n-ion vector represents 2^n primitive fermions. Similarly $I_{\text{fundamental}}$ would be a n-ion vector composed of 2^n n-ion components of the form of eq. 5.4 where each n-ion component has one dimension. Thus the vector contains 2^n real-valued dimensions.

We call the reference frame containing these reduced sets of fermions and dimensions the *n-ion Fundamental Frame*. The results of the transformations from the non-static Fundamental Frame are the space n fermion and dimension arrays.

7.1 Contents of the Reduced Fermion and Internal Symmetry Arrays of the Fundamental Frame

We choose the fermions to appear in a 2^n-vector of n-ions in analogy with eqs. 5.4 and 5.5. Each of the 2^n n-ions contains one of the 2^n fermions with each represented by a number such as 1. This n-ion vector is mapped by a CASe transformation, eq. 7.1, to a set of n-ion vectors containing a total of 2^{2n} fermions in space n in a "normal" coordinate system. *Thus 2^n n-ion Fundamental Frame fermions become 2^{2n} fermions in the "normal" reference frame.*

Conceptually we can group the set of n-ion Fundamental Frame fermions into a set of 16 fermion subblocks with each containing the four types (species) of fermions: e-type, ν-type, up-quark type, and down-quark type. There are 2^{n-4} such subblocks.

7.1.1 Transformation of n-ion Fundamental Frame Fermions to a "Normal" Frame

A CASe transformation corresponding to eq. 7.1 generates a 2^n-vector of n-ions. Each n-ion component contains 2^n fermions expressed as numeric values. . The arrangement of each of the 2^n output n-ions can be chosen to have the form of a multiple of 16 fermion subblocks: Normal: e, q-up$_1$, q-up$_2$, q-up$_3$, ν, q-down$_1$, q-down$_2$, q-down$_3$; Dark: e, q-up$_1$, q-up$_2$, q-up$_3$, ν, q-down$_1$, q-down$_2$, q-down$_3$. The 2^{2n} n-ion fermions then consist of 2^{2n-4} subblocks of a form similar to the form of the fermions in NEWQUeST and NEWUTMOST. See Figs. 6.2 – 6.4 for the NEWQUeST case.

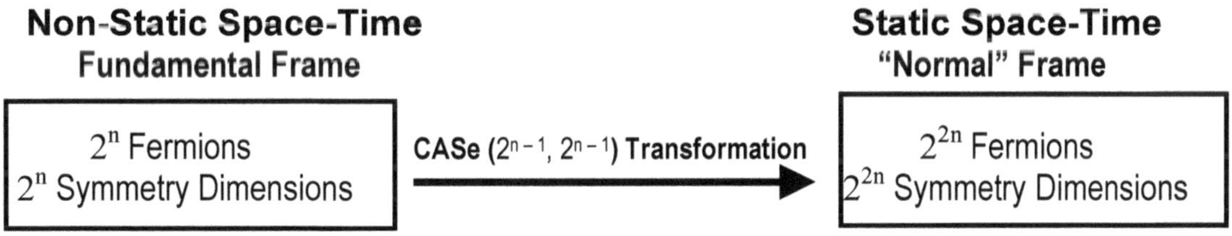

Figure 7.2. The CASe transformation from a non-static Fundamental Frame to a "Normal" static space-time reference frame.

7.1.2 Symmetry Groups within the n-ion Fundamental Frame

There are 2^n symmetry group irreducible representation dimensions in an n-ion Fundamental Frame n-ion-vector in analogy with eqs. 5.4 and 5.5. Each of the 2^n n-ions contains one of the 2^n symmetry group dimensions in the form of an integer such as 1. This n-ion vector is mapped by a CASe transformation, eq. 7.1, to 2^{2n} symmetry groups irreducible representations dimensions of space n.

7.1.3 Transformation of Fundamental Frame Dimensions to a "Normal" Frame

The CASe transformation eq. 7.1 generates a 2^n-vector of 2^n-n-ions from a Fundamental Frame 2^n-vector with one dimension in each n-ion. Each output n-ion component contains 2^n irreducible representation dimensions with each in the form of a numeric value. The 2^{2n} resulting dimensions are the symmetries' irreducible representations dimensions of space n. They represent the symmetries in Figs. 6.5 - 6.7 for the NEWQUeST case.

The n = 5 with su(16, 16) corresponds to NEWUTMOST, our proposed Megaverse space. Its 5-ion Fundamental Frame has 32 fermions and 32 symmetry group dimensions for

$$SU(3)\otimes U(1)\otimes SU(2)\otimes U(1)\otimes SL(2,\mathbf{C})\otimes SU(3)\otimes U(1)\otimes SU(2)\otimes U(1)\otimes SL(2,\mathbf{C})$$

presumably giving interactions to both the Normal and Dark fermion sectors.

The n = 8 CASe group case is interesting because its Fundamental Frame has 256 fermions and 256 symmetry dimensions. NEWQUeST has similar sets of fermions and dimensions. The n = 8 Fundamental Frame may contain an analogue to

NEWQUeST. This case raises the, perhaps remote, possibility that our universe has a non-static reference frame in the large.

The n = 10 CASe group case is interesting because its Fundamental Frame has 1024 fermions and 1024 symmetry dimensions. NEWUTMOST for the Megaverse has similar sets of fermions and dimensions. The n = 10 Fundamental Frame may contain an analogue to the NEWUTMOST Megaverse.

7.2 A Basis for NEWQUeST and NEWUST in CASe Transformations from Fundamental Frames

The set of internal symmetries of space n emerges. Again GiFT together with its General Relativistic aspect enables a reduction of the symmetries and fermions to 2^n symmetry dimensions and 2^n fermions in a non-static Fundamental Frame with a non-static reference frame.

The CASe transformations of GiFT, that provide an n-ion Fundamental Frame formulation of particles and interactions, again require the PseudoQuantum formulation of GiFT. They implement a unification of Quantum theory and General Relativity since Quantum Theory is based on GiFT, which embodies Quantum Field Theory, which, in turn, embodies Quantum Mechanics.

GiFT embodies General Relativity (and Special Relativity) in its PseudoQuantum formalism. This formalism supports a *canonical* formulation of higher derivative gravity with three distance scales that appears to give similar results to the MoND formulation of gravity. The PseudoQuantum formalism also supports higher derivative strong interactions with explicit quark confinement.[18] *NEWUST relates the three scale parameters of Gravitation to the coupling constant of the higher derivative, quark confining strong linear potential.*[19]

Gravitation joins the other gauge theories of interactions. It has a finite perturbation theory due to GiFT's Two-Tier Theory and PseudoQuantum Theory. Due to its long range nature (like QED) Gravitation has interesting solutions such as Black Holes just as Electrodynamics has interesting solutions that are used in Solid State Physics.

---

[18] Blaha (2018c) pages 460 – 465 and chapters 23 and 24.

[19] Chapters 22 – 24 in Blaha (2018e).

8. Future Developments in Fundamental Physics

This work displays the unification of Quantum Theory and General Relativity within the GiFT formulation. The further development of GiFT seems to require a specific example(s) of the GiFT CASe transformation from "normal" space-time with 2^{2n} fermions and 2^{2n} symmetry space dimensions to a non-static coordinate system where the particles and space dimensions are compressed to 2^n fermions and 2^n space dimensions.

Appendix A. Space Particle Wave Functions

In this appendix we list the Two-Tier and PseudoQuantum free field space wave functions and Lagrangians for scalar, fermion, gauge vector boson and graviton *space* particles. Details of their features may be found in Blaha (2021g). The Two-Tier and PseudoQuantum formulations for normal fermion and boson particles are considered in Appendix C.

A.1 Two-Tier and PseudoQuantum Free Fermion Space Particle Wave Function and Lagrangian

Following S. Blaha, Il Nuovo Cimento **49A**, 35 (1979) as we did in earlier we define the Two-Tier, PseudoQuantum Lagrangian for the spin ½ space particle interacting with a scalar space particle to be:

$$\mathcal{L} = \overline{\psi}_2(i\partial_\mu\gamma^\mu - m)\psi_1 + \overline{\psi}_1(i\partial_\mu\gamma^\mu - m)\psi_2 - \overline{\psi}_1(i\partial_\mu\gamma^\mu - m)\psi_1 + \mathcal{L}_\Phi(\psi_2,\Phi_2) \tag{A.1}$$

where Φ_2 is a scalar space field with a pair of spinor indices specified later, where the fields are all functions of $X^\mu = y^\mu + iY^\mu(y)$, and where $\partial_\mu = \partial/\partial X^\mu$.

From the development[20] in S. Blaha, Il Nuovo Cimento **49A**, 35 (1979) we find the free space fermion wave function for an r space-time dimension space fermion (with a q-dimension inner space-time) has the form[21]

$$\Psi_{ia}(r, q, X) = (\text{const}) \int d^{r-1}p(p^0)^{-½}\{e^{-ip\cdot X}\, U_{a\beta}(r, q, p)(c_{i1\beta}\, b_{1\beta}(p) + c_{i2\beta}\, b_{2\beta}(p)) +$$

$$+\ e^{ip\cdot X}\, V_{a\beta}(r, q, p)\,(c'_{i1\beta}\, d^\dagger_{1\beta}(p) + c'_{i2\beta}\, d^\dagger_{2\beta}(p))\,\} \tag{A.2}$$

Its Hermitian conjugate is

$$\Psi^\dagger_{ia}(r, q, X) = (\text{const}) \int d^{r-1}p(p^0)^{-½}\{e^{ip\cdot X}\, U^\dagger_{a\beta}(r, q, p)(c_{i1\beta}\, b^\dagger_{1\beta}(p) + c_{i2\beta}\, b^\dagger_{2\beta}(p)) +$$

$$+\ e^{-ip\cdot X}\, V^\dagger_{a\beta}(r, q, p)\,(c'_{i1\beta}\, d_{1\beta}(p) + c'_{i2\beta}\, d_{2\beta}(p))\,\} \tag{A.3}$$

where a and β are spinor indices.

The spinor arrays U and V have the form:[22]

---

[20] Eqs. 61- 68.

[21] This form makes β into the spinor index. It implements the close connection of the space feature and the spinors. Alternately an internal symmetry index could be introduced (as done in the scalar case later) for the internal symmetry. The sets of b_β and of d_β operators each constitute a fundamental q-number representation of $U(2^{r/2})$. The sums over β embody a sum over all spin states including fermion and antifermion spins.

[22] Blaha(2021g). Eqs. 1.32 and 1.33. The origin and properties of these spinor arrays are discussed in detail there.

$$U(r, q, p, m) = ((m + p^0)/2m)^{1/2}\gamma^0 S(p) = (\not{p} + m)/2m \quad (A.4)$$

$$V(r, q, p, m) = ((m - p^0)/2m)^{1/2}\gamma^0 S'(p) = (m - \not{p})/2m \quad (A.5)$$

One may visualize the form of the spinor matrix U as a matrix with $2^{r/2}$ spinor columns, in which each column is a spinor with $2^{r/2}$ components.[23] (Fig. 1.3) We view U as a matrix in spinor space. See chapter 5 of Blaha (2021g)..

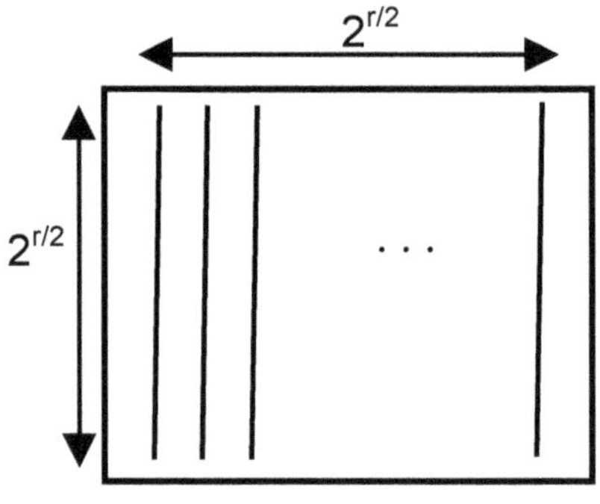

Figure A.1. The U array with $2^{r/2}$ columns and $2^{r/2}$ rows. The V array has the same form.

The creation/annihilation operators anti-commute:

$$\{b_{i\beta}(p), b_{j\beta'}(p')\} = 0 \quad (A.6)$$
$$\{b^\dagger_{i\beta}(p), b^\dagger_{j\beta'}(p')\} = 0$$
$$\{b_{i\beta}(p), b^\dagger_{j\beta'}(p')\} = (1 - \delta_{ij})\delta_{\beta\beta'}\, \delta^{r-1}(\mathbf{p} - \mathbf{p'})$$

$$\{d_{i\beta}(p), d_{j\beta'}(p')\} = 0$$
$$\{d^\dagger_{i\beta}(p), d^\dagger_{j\beta'}(p') = 0$$
$$\{d_{i\beta}(p), d^\dagger_{j\beta'}(p')\} = (1 - \delta_{ij})\delta_{\beta\beta'}\, \delta^{r-1}(\mathbf{p} - \mathbf{p'})$$

for i, j = 1, 2.

The equal-time commutation relations and the self-adjointness of the Hamiltonian H imply the fields in eq. A.2 and A.3 (after some algebra) have the parameters[24]

$$c_{11\beta} = \cos(\theta_{1\beta} - \theta_{2\beta})/\sin\theta_{1\beta} \quad (A.7)$$
$$c_{12\beta} = \sin(\theta_{1\beta} - \theta_{2\beta})/\sin\theta_{1\beta}$$
$$c'_{11\beta} = \cos(\theta_{1\beta} - \theta_{2\beta})/\cos\theta_{1\beta}$$
$$c'_{12\beta} = -\sin(\theta_{1\beta} - \theta_{2\beta})/\cos\theta_{1\beta}$$

---

[23] r is the spacetime dimension.

[24] See eqs. 41 and 42 of S. Blaha, Il Nuovo Cimento **49A**, 35 (1979).

$$c_{21\beta} = \sin\theta_{2\beta}$$
$$c_{22\beta} = \cos\theta_{2\beta}$$
$$c'_{21\beta} = \cos\theta_{2\beta}$$
$$c'_{22\beta} = \sin\theta_{2\beta}$$

where θ_1 and θ_2 are functions of k and the spinor index β. These constants must depend on β since the creation/annihilation operators are indexed by β and form a $U(2^{r/2})$ spinor group representation.

A.2 Two-Tier and PseudoQuantum Free Scalar Space Particle Wave Function and Lagrangian

Using eq. A.1, we define the free scalar space particle Lagrangian terms as:

$$\mathcal{L}_\Phi(\psi_2,\Phi_2) = -g'\overline{\psi}\,\Phi_2\psi + \overline{\Phi}_1(i\partial_\mu\gamma^\mu - m')\,\Phi_2 - \tfrac{1}{2}\,\overline{\Phi}_1(i\partial_\mu\gamma^\mu - m')\,\Phi_1 \qquad \text{(A.8)}$$

In order to have a spinor array within a scalar space particle we require the free Lorentz *scalar* space particle Dirac equation to satisfy:[25]

$$(i\partial_\mu\gamma^\mu - m')\,\Phi(r, q, X) = 0 \qquad \text{(A.9)}$$

with $\partial_\mu = \partial/\partial X^\mu$. It implies the usual scalar Klein-Gordon equation

$$(\partial_\mu\partial^\mu + m'^2)\,\Phi(r, q, X) = 0 \qquad \text{(A.10)}$$

We choose the Two-Tier, PseudoQuantum, free scalar space particle wave function with second rank spinor indices for an r space-time dimension space boson (with a q-dimension inner space-time) to be

$$\Phi_{iab}(r, q, X) = \int d^{r-1}p(2\pi)^{-(r-1)}(m'/p^0)^{1/2}\,\{\exp^{-ip\cdot X}\,U_{a\beta}(r, q, p, m')a'_{i\beta b}(p) +$$
$$+ \exp^{ip\cdot X}\,V_{a\beta}(r, q, p, m')c''^{\dagger}_{i\beta b}(p)\} \qquad \text{(A.11)}$$

where, due to a PseudoQuantum transformation, have the form

$$a'_{i\beta b}(p) = c_{i1\beta}\,a_{1\beta}(p) + c_{i2\beta}\,a_{2\beta}(p) \qquad \text{(A.12a)}$$
$$c''^{\dagger}_{i\beta b}(p) = c'_{i1\beta}\,c^{\dagger}_{1\beta}(p) + c'_{i2\beta}\,c^{\dagger}_{2\beta}(p)$$

where a, b, and β are spinor internal symmetry indices. Its Hermitian conjugate is

$$\Phi^{\dagger}_{iab}(r, q, X) = \int d^{r-1}p(2\pi)^{-(r-1)}(m'/p^0)^{1/2}\,\{\exp^{ip\cdot X}U^{\dagger}_{a\beta}(r, q, p, m')\,a'^{\dagger}_{i\beta b}(p) +$$

---

[25] The Dirac equation appears necessary for scalar space particles. Together with the requirement that the scalar wave function have a pair of spino indices the form of the wave function is determined.

$$+ \ \exp^{-ip\cdot X} \, V^{\dagger}{}_{a\beta}(r, q, p, m') \, c''_{i\beta b}(p) \} \quad \text{(A.12)}$$

The transformation constants of the creation/annihilation operators in eq. A.12a are

$$\begin{aligned} c_{11\beta} &= \cos(\theta_{1\beta} - \theta_{2\beta})/\sin \theta_{1\beta} \\ c_{12\beta} &= \sin(\theta_{1\beta} - \theta_{2\beta})/\sin \theta_{1\beta} \\ c'_{11\beta} &= \cos(\theta_{1\beta} - \theta_{2\beta})/\cos \theta_{1\beta} \\ c'_{12\beta} &= - \sin(\theta_{1\beta} - \theta_{2\beta})/\cos \theta_{1\beta} \end{aligned} \quad \text{(A.13)}$$

$$\begin{aligned} c_{21\beta} &= \cos \theta_{2\beta} \\ c_{22\beta} &= \sin \theta_{2\beta} \\ c'_{21\beta} &= \sin \theta_{2\beta} \\ c'_{22\beta} &= \cos \theta_{2\beta} \end{aligned}$$

The constant m' is the scalar particle mass as well as the energy of the created space, and U and V are the same as in fermion case. The spinor arrays, U and V, have $2^{r/2}$ columns and $2^{r/2}$ rows. The spinor arrays U and V have their size determined by r, which we call the *parent* space space-time dimension. The space-time dimension q is the *child* space space-time dimension. Note Φ is a second rank spinor. The sets of $a_{\beta b}$ and of $c_{\beta b}$ operators each constitute a q-number representation of $U(2^{r/2})$.

The creation/annihilation operators satisfy the commutation relations:

$$\begin{aligned} [a_{j\beta b}(p), a^{\dagger}{}_{j'\beta'b'}(p')] &= (1 - \delta_{jj'})\delta_{\beta\beta'}\, \delta_{'b\, b'}\, \delta^{r-1}(\mathbf{p} - \mathbf{p'}) \quad \text{A.14)} \\ [a_{j\beta b}(p), a_{j'\beta'b'}(p')] &= 0 \\ [a^{\dagger}{}_{j\beta b}(p), a^{\dagger}{}_{j'\beta'b'}(p')] &= 0 \\ [c_{j\beta b}(p), c^{\dagger}{}_{j'\beta'b'}(p')] &= (1 - \delta_{jj'})\delta_{\beta\beta'}\, \delta_{'b\, b'}\, \delta^{r-1}(\mathbf{p} - \mathbf{p'}) \quad \text{(A.15)} \\ [c_{j\beta b}(p), c_{j'\beta'b'}(p')] &= 0 \\ [c^{\dagger}{}_{j\beta b}(p), c^{\dagger}{}_{j'\beta'b'}(p')] &= 0 \\ [c_{j\beta b}(p), a^{\dagger}{}_{j'\beta'b'}(p')] &= 0 \\ [c_{j\beta b}(p), a_{j'\beta'b'}(p')] &= 0 \\ [c^{\dagger}{}_{j\beta b}(p), a^{\dagger}{}_{j'\beta'b'}(p')] &= 0 \end{aligned}$$

The $a_{jk}(p)$ and $a^{\dagger}{}_{j'k'}(p')$ commute with $c_{jk}(p)$ and $c^{\dagger}{}_{j'k'}(p')$. There are two sets: 2^r operators $a_{jk}(p)$ and 2^r operators c_{jk}. Each set furnishes a representation of $U(2^{r/2})$.

A.3 Two-Tier and PseudoQuantum Free Vector Space Particle Wave Function and Lagrangian

In this section we define the wave functions of a free, vector space particle. We use T matrices as the generators of the $U(2^{r/2})$ Dirac-Good group,[26] with $A_{k\mu}$ as the

[26] See Blaha (2021g).

gauge vector fields of that group, and g' is its coupling constant. The $U(2^{r/2})$ group will be seen to be a local Yang-mills gauge group. The group acts on spinors. Therefore T has spinor indices. T also consists of 2^r generators: the r Dirac matrices and $2^r - 1$ other generators. T has the form $T^i{}_{ab}$ for $i = 1, \ldots, 2^r$ and for $a, b = 1, \ldots, 2^{r/2}$.

The vector field in the radiation gauge satisfies

$$(\partial_\mu\partial^\mu + \xi^2)\, A^i{}_k(r, q, X) = 0 \qquad (A.16)$$

for k = 1, 2, for spatial coordinates i = 1, … , r – 1, where ξ is a small mass that we take to zero at the end of calculations (as done in QED infrared studies.) In order to introduce the U and V spinor arrays seen earlier we require

$$(i\partial_\mu\gamma^\mu - \xi)\, A_k{}^i(r, q, X) = 0 \qquad (A.17)$$

Note eq. A.16 follows from eq. A.17.

Emulating the scalar wave function of section A.2, we choose the vector space particle wave function for an r space-time dimension space gauge boson (with a q-dimension inner space-time) in the radiation gauge to be[27]

$$\mathbf{A}^i{}_{kab}(r, q, X) = \int d^{r-1}p(2\pi)^{-(r-1)}(2p^0)^{-1/2}\, \Sigma_{\lambda\alpha\beta}\varepsilon(p,\lambda)T^i{}_{a\alpha}\{\exp^{-ip\cdot X}\, U_{\alpha\beta}(r, q, p, \xi)\, w'_{k\beta b}(p, \lambda) +$$

$$+ \exp^{ip\cdot X}\, V_{\alpha\beta}(r, q, p, \xi)\, z'_{k\beta b}{}^\dagger(p,\lambda)\} \qquad (A.18)$$

for k = 1, 2 where ξ is a mass that is set to zero after perturbation theory calculations; where **ε** is the polarization unit vector; w and z are creation/annihilation operators with spinor indices—each taking $2^{r/2}$ values, λ takes r – 2 values, and U and V are similar to U and V in the fermion case. $\mathbf{A}^i{}_{ab}(r, q, X)$ is a vector of the spatial components of the vector field in the radiation gauge. The creation/annihilation operators under a PseudoQuantum transformation have the form:

$$\begin{aligned} w'_{k\beta b}(p, \lambda) &= c_{k1\beta}\, w_{1\beta}(p, \lambda) + c_{k2\beta}\, w_{2\beta}(p, \lambda) \\ z'_{k\beta b}{}^\dagger(p, \lambda)) &= c'_{k1\beta}\, z^\dagger{}_{1\beta}(p, \lambda) + c'_{k2\beta}\, z^\dagger{}_{2\beta}(p, \lambda) \end{aligned} \qquad (A.19)$$

where

$$\begin{aligned} c_{11\beta} &= \cos(\theta_{1\beta} - \theta_{2\beta})/\sin\theta_{1\beta} \\ c_{12\beta} &= \sin(\theta_{1\beta} - \theta_{2\beta})/\sin\theta_{1\beta} \\ c'_{11\beta} &= \cos(\theta_{1\beta} - \theta_{2\beta})/\cos\theta_{1\beta} \\ c'_{12\beta} &= -\sin(\theta_{1\beta} - \theta_{2\beta})/\cos\theta_{1\beta} \\ c_{21\beta} &= \cos\theta_{2\beta} \\ c_{22\beta} &= \sin\theta_{2\beta} \\ c'_{21\beta} &= \sin\theta_{2\beta} \\ c'_{22\beta} &= \cos\theta_{2\beta} \end{aligned} \qquad (A.20)$$

---

[27] With repeated indices summed.

The commutation relations of the creation/annihilation operators are

$$[w_{j\beta b}(p, \lambda), w^{\dagger}_{j'\beta' b'}(p', \lambda')] = (1 - \delta_{jj'})\delta_{\beta\beta'}\delta_{b\,b'}\delta_{\lambda\lambda'}\delta^{r-1}(\mathbf{p} - \mathbf{p'}) \quad (A.21)$$
$$[w_{j\beta b}(p, \lambda), w_{j'\beta' b'}(p'\, \lambda')] = 0$$
$$[w^{\dagger}_{j\beta b}(p, \lambda), w^{\dagger}_{j'\beta' b'}(p'\, \lambda')] = 0$$
$$[z_{j\beta b}(p, \lambda), z^{\dagger}_{j'\beta' b'}(p'\, \lambda')] = (1 - \delta_{jj'})\delta_{\beta\beta'}\delta_{b\,b'}\delta_{\lambda\lambda'}\delta^{r-1}(\mathbf{p} - \mathbf{p'})$$
$$[z_{j\beta b}(p, \lambda), z_{j'\beta' b'}(p'\, \lambda')] = 0$$
$$[z^{\dagger}_{j\beta b}(p, \lambda), z^{\dagger}_{j'\beta' b'}(p'\, \lambda')] = 0$$
$$[z_{j\beta b}(p, \lambda), w^{\dagger}_{j'\beta' b'}(p'\, \lambda')] = 0$$
$$[z_{j\beta b}(p, \lambda), w_{j'\beta' b'}(p'\, \lambda')] = 0$$
$$[z^{\dagger}_{j\beta b}(p, \lambda), w^{\dagger}_{j'\beta' b'}(p'\, \lambda')] = 0$$

for j, j' = 1, 2.

The spinor arrays, U and V, have $2^{r/2}$ columns and $2^{r/2}$ rows. The spinor arrays U and V size is determined by r, the *parent* space space-time dimension. The space-time dimension q is the *child* space space-time dimension. Since U and V are not orthogonal for zero mass, we have introduced a "fictitious" mass ξ which is taken to zero at the end of calculations.

The Hermitian conjugate is

$$\mathbf{A}^{i}_{kab}(r, q, X)^{\dagger} = \int d^{r-1}p(2\pi)^{-(r-1)}(2p^0)^{-\frac{1}{2}}\Sigma_{\lambda\alpha\beta}\varepsilon(p,\lambda)T^{i}_{a\alpha}\{\exp^{ip\cdot X}U_{\alpha\beta}(r, q, p, \xi)^{\dagger}w'_{k\beta b}{}^{\dagger}(p, \lambda) +$$
$$+ \exp^{-ip\cdot X}V_{\alpha\beta}(r, q, p, \xi)^{\dagger}z'_{k\beta b}(p,\lambda)\} \quad (A.22)$$

A.4 Two-Tier and PseudoQuantum Free Graviton Space Particle Wave Function and Lagrangian

Our consideration of the fermion, scalar boson, and vector boson space fields leads to the possibility of graviton space fields. In a weak field approximation we define a metric tensor with two spinor indices a and b:

$$g_{ab}{}^{\mu\nu}(r, q, X) \cong \eta_{ab}{}^{\mu\nu}(r, q, X) + h_{ab}{}^{\mu\nu}(r, q, X) \quad (A.23)$$

Following the conventional approach we arrive at the dynamic equation:

$$-\tfrac{1}{2}\,\square(h_{ab}{}^{\mu\nu} - \tfrac{1}{2}h\,\eta_{ab}{}^{\mu\nu}) = -16\pi G T_{ab}{}^{\mu\nu} \quad (A.24)$$

based on the

$$R_{\mu\nu ab} - \tfrac{1}{2}R_{a\beta}g_{\mu\nu\beta b} = 8\pi G T_{\mu\nu ab}$$

The spin 2 $h_{ab}{}^{\mu\nu}(r, q, x)$ field can be PseudoQuantum expanded to

$$h^{\mu\nu}{}_{kab}(r, q, X) = \int d^{r-1}p(2\pi)^{-(r-1)}(2p^0)^{-½} \Sigma_{\lambda,\beta}\, \varepsilon^{\mu\nu}(p,\lambda)\{\exp^{-ip\cdot X}\, U_{a\beta}(r, q, p, \xi)\, w'_{k\beta b}(p, \lambda) +$$

$$+ \exp^{ip\cdot X}\, V_{a\beta}(r, q, p, \xi)\, z'_{k\beta b}{}^{\dagger}(p,\lambda)\} \qquad (A.25)$$

with quantities defined similarly to the previous.

The creation/annihilation operators under a PseudoQuantum transformation have the form:

$$w'_{k\beta b}(p, \lambda) = c_{k1\beta}\, w_{1\beta}(p, \lambda) + c_{k2\beta}\, w_{2\beta}(p, \lambda) \qquad A.26)$$
$$z'_{k\beta b}{}^{\dagger}(p, \lambda)) = c'_{k1\beta}\, z^{\dagger}{}_{1\beta}(p, \lambda) + c'_{k2\beta}\, z^{\dagger}{}_{2\beta}(p, \lambda)$$

where

$$\begin{aligned} c_{11\beta} &= \cos(\theta_{1\beta} - \theta_{2\beta})/\sin\theta_{1\beta} \\ c_{12\beta} &= \sin(\theta_{1\beta} - \theta_{2\beta})/\sin\theta_{1\beta} \\ c'_{11\beta} &= \cos(\theta_{1\beta} - \theta_{2\beta})/\cos\theta_{1\beta} \\ c'_{12\beta} &= -\sin(\theta_{1\beta} - \theta_{2\beta})/\cos\theta_{1\beta} \\ c_{21\beta} &= \cos\theta_{2\beta} \\ c_{22\beta} &= \sin\theta_{2\beta} \\ c'_{21\beta} &= \sin\theta_{2\beta} \\ c'_{22\beta} &= \cos\theta_{2\beta} \end{aligned} \qquad (A.27)$$

The commutation relations of the creation/annihilation operators are

$$\begin{aligned} [w_{j\beta b}(p, \lambda), w^{\dagger}{}_{j'\beta'b'}(p', \lambda')] &= (1 - \delta_{jj'})\delta_{\beta\beta'}\,\delta_{b\,b'}\,\delta_{\lambda\lambda'}\,\delta^{r-1}(\mathbf{p} - \mathbf{p'}) \qquad (A.28) \\ [w_{j\beta b}(p, \lambda), w_{j'\beta'b'}(p'\, \lambda')] &= 0 \\ [w^{\dagger}{}_{j\beta b}(p, \lambda), w^{\dagger}{}_{j'\beta'b'}(p'\, \lambda')] &= 0 \\ [z_{j\beta b}(p, \lambda), z^{\dagger}{}_{j'\beta'b'}(p'\, \lambda')] &= (1 - \delta_{jj'})\delta_{\beta\beta'}\,\delta_{b\,b'}\,\delta_{\lambda\lambda'}\delta^{r-1}(\mathbf{p} - \mathbf{p'}) \\ [z_{j\beta b}(p, \lambda), z_{j'\beta'b'}(p'\, \lambda')] &= 0 \\ [z^{\dagger}{}_{j\beta b}(p, \lambda), z^{\dagger}{}_{j'\beta'b'}(p'\, \lambda')] &= 0 \\ [z_{j\beta b}(p, \lambda), w^{\dagger}{}_{j'\beta'b'}(p'\, \lambda')] &= 0 \\ [z_{j\beta b}(p, \lambda), w_{j'\beta'b'}(p'\, \lambda')] &= 0 \\ [z^{\dagger}{}_{j\beta b}(p, \lambda), w^{\dagger}{}_{j'\beta'b'}(p'\, \lambda')] &= 0 \end{aligned}$$

for j, j' = 1, 2.

Appendix B. PseudoQuantum Paper

This papers are reprinted with the kind permission of Physical Review D and Il Nuovo Cimento..

S. Blaha, "New Framework for Gauge Field Theories", IL Nuovo Cimento **49A**, 113 (1979).

S. Blaha, Phys. Rev. D**11**, 2921 (1975).

IL NUOVO CIMENTO VOL. 49 A, N. 1 1 Gennaio 1979

New Framework for Gauge Field Theories (*).

S. BLAHA
Physics Department, Williams College - Williamstown, Ma. 01267

(ricevuto l'8 Agosto 1978)

Summary. — We formulate gauge theories within the framework of a generalization of quantum field theory. In particular, we discuss models of electrodynamics and of Yang-Mills theories, a model of the strong interaction with higher-order derivatives and quark confinement and a renormalizable model of pure quantum gravity with Einstein Lagrangian. In the case of electrodynamics we show that two models are possible: one with predictions which are identical to QED and one which is a quantum action-at-a-distance model of electrodynamics. In the case of Yang-Mills theories we can construct a model which is identical in predictions to any conventional model, or a quantum action-at-a-distance model. In the second case it is possible to eliminate all loops of Yang-Mills particles (in all gauges) in a manner consistent with unitarity. A variation of Yang-Mills models exists in our formulation which has higher-order derivative field equations. It is unitary and has positive probabilities. It can be used to construct a model of the strong interactions which has a linear potential and manifest quark confinement. Finally we show how to construct an action-at-a-distance model of pure quantum gravity (whose classical limit is the dynamics of the Einstein Lagrangian) coupled to an external classical source. The model is trivially renormalizable.

1. – Introduction.

Because of the absence of an acceptable physical interpretation of conventional quantum field theory in the case of curved nonstatic space-time, we

(*) Supported in part by Research Corporation and the National Science Foundation.

recently developed a modified formulation of quantum field theory (¹). We showed it had a unique physical particle interpretation in nonstatic space-time. We described the flat–space-time formulation for the cases of scalar particles and spin–one-half fermions. In both cases we found that it was possible to construct a model which was equivalent in predictions to any model of the usual formulation. However, it was also possible to construct models which had no analogue in the usual formulation. In these models the quantum exchange of a « particle » did not take place via a Feynman propagator. Rather, the propagator G could be chosen to be any of a continuum of possibilities ranging from the Feynman propagator to the principal-value (half advanced-half retarded) propagator. The choice is parametrized by an angle θ, whose specification is equivalent to a choice of boundary condition

$$G(x-y) = \sin^2\theta\, G_F(x-y) + \cos^2\theta\, CG_F^*C^{-1}, \tag{1}$$

where G_F is the usual Feynman propagator with G_F^* its complex conjugate and C is an appropriate charge conjugation matrix.

The new degree of freedom, represented by θ, leads to new possibilities for the formulation of flat–space-time quantum field theory models which will be especially evident in the case of gauge theories.

In sect. **2** we shall explore models of quantum electrodynamics which occur within the framework of our formulation. We shall show that one model is completely equivalent to QED in its predictions. In addition, we shall show that a quantum action-at-a-distance electrodynamics is also possible.

In sect. **3** we describe model Yang-Mills theories and show that a model equivalent to any conventional model can be formulated as well as a quantum action-at-a-distance model. In the second case all non-Abelian boson loops can be eliminated (in all gauges) in a manner which is consistent with unitarity.

In sect. **4** we describe a non-Abelian model of quark confinement based on higher-derivative field equations for which a unitary physical S-matrix is obtained by the choice of principal-value propagators. This demonstrates the utility of principal-value propagators for widening the class of unitary quantum field theories to include higher-derivative theories.

In sect. **5** we describe a model of pure quantum gravity based on the Einstein Lagrangian which is trivially renormalizable, if gravitons propagate via principal-value propagators. (No graviton loops.) This illustrates the potential of the principal-value propagator to ameliorate renormalization problems by allowing one to limit the self-interactions of quantum fields.

In view of the unfamiliar features of principal-value propagators, we shall

---

(¹) S. BLAHA: *Nuovo Cimento*, **49** A, 35 (1978).

devote the remainder of this section to a discussion of their properties in relation to causality and analyticity.

Among the first appearances of principal-value, or half advanced-half retarded, propagators was in Feynman's space-time approach to quantum electrodynamics [2], where he observed that one is naturally led to such a propagator for the photon, if one considers a quantum one-photon exchange between two electrons wherein the photons propagate, as in classical electrodynamics, via retarded propagators. Since one cannot distinguish between the process where a photon propagates from electron A to B and the process where the photon propagates from B to A for short time intervals one must sum the two amplitudes and a principal-value propagator results. This observation illustrates the absence of a direct relation between the propagator of the classical field and the propagator of the corresponding quantum field. In particular, it is perfectly possible for a quantum field to have a principal-value propagator, and yet have the classical field propagate via a retarded propagator. This can be understood within the framework of the absorber model of Feynman and Wheeler [3]. For a quantum process, where a finite number of quanta is exchanged, the effective propagator of the quanta cannot have this aspect of its character changed. But in a classical situation, where infinite numbers of quanta are involved, the effective propagator of the quanta can be changed through interaction with the absorber in the manner outlined in ref. [3]. We therefore conclude that the propagation of quanta via principal-value propagators does not imply that macroscopic causality (as represented by retarded propagators) is lost. This remark is relevant to our models of the strong interaction and quantum gravity discussed later.

So far as microscopic causality is concerned, it will be seen that principal-value propagators are completely consistent with the vanishing of commutators of field operators for spacelike distances. Thus neither macroscopic nor microscopic causality is necessarily inconsistent with the use of quantum principal-value propagators.

The use of principal-value propagators does lead to a different analytic structure of S-matrix amplitudes. Amplitudes are now piecewise analytic. However, several authors [4] have shown that such amplitudes are not necessarily inconsistent with experiment—even for the case of the pion-nucleon dispersion relations. In particular, considering the absence of any direct relation between the analytic properties of quark-quark scattering amplitudes,

---

(2) R. Feynman: *Phys. Rev.* **76**, 769 (1949), footnote (7).

(3) J. Wheeler and R. Feynman: *Rev. Mod. Phys.*, **17**, 157 (1945); **21**, 425 (1949); F. Hoyle and J. Narlikar: *Ann. of Phys.*, **54**, 207 (1969); **62**, 44 (1971).

(4) R. E. Cutkowsky, P. V. Landshoff, D. I. Olive and J. C. Polkinghorne: *Nucl. Phys.*, **12** B, 281 (1969); T. D. Lee and G. C. Wick: *Phys. Rev. D*, **2**, 1033 (1970); M. Gundzik and E. C. G. Sudarshan: *Phys. Rev. D*, **6**, 798 (1972); H. Rechenberg and E. C. G. Sudarshan: *Nuovo Cimento*, **14** A, 299 (1973).

and the analytic properties of the scattering amplitudes of their bound states one cannot rule out the possibility of principal-value propagators for color gluons. In the case of quantum gravity, another potential application of principal-value propagators, the analytic structure of scattering amplitudes due to graviton exchange is unknown and likely to remain so. In the absence of such information, the renormalizability of pure quantum gravity, and the compatibility with Mach's principle, resulting from the use of principal-value propagators for gravitons is encouraging.

Perhaps the most important question facing field theory models with principal-value propagators is unitarity. The appendix contains a detailed discussion of this issue. The physical S-matrix is shown to be unitary. In addition the value of using infinite-momentum frame variables to compute S-matrix elements is pointed out.

2. – Model electrodynamics.

In this section we shall describe a model electrodynamics which is identical to quantum electrodynamics in its predictions. We shall also discuss a model for quantum action-at-a-distance electrodynamics.

The Lagrangian density of our models (1) is

$$\mathcal{L} = -\tfrac{1}{2} F^1_{\mu\nu} F^{2\mu\nu} - \tfrac{1}{4} F^1_{\mu\nu} F^{1\mu\nu} + \tilde{\psi}_2 \gamma^0 (i(\gamma\cdot\nabla) - e_0(\gamma\cdot A_2) - m)\psi_2 + (\tilde{\psi}_1 - \tilde{\psi}_2)\gamma^0 (i(\gamma\cdot\nabla) - m)(\psi_1 - \psi_2)\,, \tag{2}$$

where we have introduced two electromagnetic fields, $A_{1\mu}$ and $A_{2\mu}$, so that

$$F^i_{\mu\nu} = \partial_\nu A_{i\mu} - \partial_\mu A_{i\nu}\,, \tag{3}$$

for $i = 1, 2$, and two fields, ψ_1 and ψ_2 for electrons. The field $\tilde{\psi}_i$ is related to the transpose of the charge conjugate of ψ_i by (1)

$$\tilde{\psi}_i = \psi_i^{cT} \gamma^0 C^T\,, \tag{4}$$

where C is a charge conjugation matrix. We follow the notation and conventions of ref. (5). The free field theory for fermions has been developed in detail in ref. (1).

(5) J. Bjorken and S. Drell: *Relativistic Quantum Fields* (New York, N. Y., 1965).

Before discussing the free field theory of the photons we note the invariance of $\mathscr{L}$ under a restricted gauge transformation of the second kind,

$$\psi_2 \to \psi_2 \exp[i\Lambda], \tag{5}$$

$$\bar{\psi}_2 \to \bar{\psi}_2 \exp[-i\Lambda], \tag{6}$$

$$e_0 A_{2\mu} \to e_0 A_{2\mu} - \partial_\mu \Lambda, \tag{7}$$

$$\psi_1 \to \psi_1 + [\exp[i\Lambda] - 1]\psi_2, \tag{8}$$

$$\bar{\psi}_1 \to \bar{\psi}_1 + [\exp[-i\Lambda] - 1]\bar{\psi}_2. \tag{9}$$

The associated conserved current is

$$J_\mu = \bar{\psi}_2 \gamma^0 \gamma_\mu \psi_2. \tag{10}$$

It should be noted that the Lagrangian is also invariant under an independent gauge transformation, $A_{1\mu} \to A_{1\mu} - \partial_\mu \Lambda_1$.

Upon varying the action associated with $\mathscr{L}$, we find the field equations

$$\partial^\mu F^1_{\mu\nu} = \partial^\mu F^2_{\mu\nu}, \tag{11}$$

$$\partial^\nu F^2_{\mu\nu} = e_0 \bar{\psi}_2 \gamma^0 \gamma_\nu \psi_2, \tag{12}$$

$$(i(\gamma\cdot\nabla) - m)\psi_1 = (i(\gamma\cdot\nabla) - m)\psi_2, \tag{13}$$

$$(i(\gamma\cdot\nabla) - e_0(\gamma\cdot A_2) - m)\psi_2 = 0. \tag{14}$$

Equations (12) and (14) demonstrate the equivalence of our model to the usual model of electrodynamics at the level of c-number fields.

We now turn to the case of free photons. From the Lagrangian we can identify the canonical momentum conjugate to $A_{1\mu}$ as

$$\pi_{1\mu} = F^2_{0\mu} - F^1_{0\mu}, \tag{15}$$

while the momentum conjugate to $A_{2\mu}$ is

$$\pi_{2\mu} = F^1_{0\mu}. \tag{16}$$

Since the free electromagnetic Lagrangian is invariant under independent gauge transformations of $A_{1\mu}$ and $A_{2\mu}$ the choice of gauges for $A_{1\mu}$ and $A_{2\mu}$ are also independent. We shall second quantize the fields in the joint Coulomb gauge

$$\vec{\nabla}\cdot\vec{A}_1 = \vec{\nabla}\cdot\vec{A}_2 = 0. \tag{17}$$

The resulting equal-time commutation relations are

$$[F^1_{0i}(\vec{x}, t), A_{1j}(\vec{y}, t)] = 0 , \tag{18}$$

$$[F^1_{0i}(\vec{x}, t), A_{2j}(\vec{y}, t)] = i\delta^{tr}_{ij}(\vec{x} - \vec{y}) , \tag{19}$$

$$[F^2_{0i}(\vec{x}, t), A_{2j}(\vec{y}, t)] = i\delta^{tr}_{ij}(\vec{x} - \vec{y}) , \tag{20}$$

$$[F^2_{0i}(\vec{x}, t), A_{1j}(\vec{y}, t)] = i\delta^{tr}_{ij}(\vec{x} - \vec{y}) , \tag{21}$$

with

$$\delta^{tr}_{ij}(\vec{x} - \vec{y}) = \int \frac{d^3k}{(2\pi)^3} \exp[-i\vec{k}\cdot(\vec{x} - \vec{y})](\delta_{ij} - k_i k_j/|\vec{k}|^2) . \tag{22}$$

The field operator expansions are completely analogous to the expansion of a free scalar field discussed in ref. (1). The major difference is the necessary presence of polarization vectors in the electromagnetic-field expansions

$$\vec{A}_1(x) = \int d^3k \sum_{\lambda=1}^{2} \vec{\varepsilon}(k, \lambda)[f_k(x)(\cos(\theta_1 - \theta_2)a_{1k\lambda} + \sin(\theta_1 - \theta_2)a_{2k\lambda})/\sin\theta_1 + \\ + f_k^*(x)(\cos(\theta_1 - \theta_2)a^\dagger_{1k\lambda} - \sin(\theta_1 - \theta_2)a^\dagger_{2k\lambda})/\cos\theta_1] \tag{23}$$

and

$$\vec{A}_2(x) = \int d^3k \sum_{\lambda=1}^{2} \vec{\varepsilon}(k, \lambda)[f_k(x)(\cos\theta_2 a_{2k\lambda} + \sin\theta_2 a_{1k\lambda}) + \\ + f_k^*(x)(\sin\theta_2 a^\dagger_{2k\lambda} - \cos\theta_2 a^\dagger_{1k\lambda})] , \tag{24}$$

where θ_1 and θ_2 are arbitrary angles. As discussed in ref. (1) the choice of these angles represents a choice of boundary conditions for the free-field Green's functions. If the Fourier coefficient operators satisfy

$$[a_{ik\lambda}, a_{jk'\lambda'}] = [a^\dagger_{ik\lambda}, a^\dagger_{jk'\lambda'}] = 0 , \tag{25}$$

$$[a_{ik\lambda}, a^\dagger_{jk'\lambda'}] = (1 - \delta_{ij})\delta_{\lambda\lambda'}\delta^3(\vec{k} - \vec{k}') , \tag{26}$$

for $i, j = 1, 2$, then the equal-time commutation relations will be satisfied for arbitrary θ_1 and θ_2. In addition the Hamiltonian which is the 0-0 component of the energy-momentum tensor derived from the Lagrangian, has the form

$$H = \int d^3k \sum_{\lambda=1}^{2} \omega_k(a^\dagger_{2k\lambda}a_{1k\lambda} + a_{2k\lambda}a^\dagger_{1k\lambda}) , \tag{27}$$

independent of θ_1 and θ_2. There is an analogy (6) between the quantization

(6) S. BLAHA: *Phys. Rev. D*, **17**, 994 (1978).

of the mode amplitudes, eqs. (25) and (26), and the co-ordinates of an assembly of one-dimensional harmonic oscillators just as in the second quantization of conventional field theory. We shall define photon vacua in the present case in a manner consistent with the procedure of ref. ([1]) and in analogy with the simple harmonic-oscillator model of ref. ([6]):

$$a_{1k\lambda}|0\rangle_2 = a^\dagger_{1k\lambda}|0\rangle_2 = 0\,, \tag{28}$$

$$a_{2k\lambda}|0\rangle_2 \neq 0\,, \qquad a^\dagger_{2k\lambda}|0\rangle_2 \neq 0\,, \tag{29}$$

for one vacuum, $|0\rangle_2$, and

$$a_{2k\lambda}|0\rangle_1 = a^\dagger_{2k\lambda}|0\rangle_1 = 0\,, \tag{30}$$

$$a_{1k\lambda}|0\rangle_1 \neq 0\,, \qquad a^\dagger_{1k\lambda}|0\rangle_1 \neq 0\,, \tag{31}$$

for the other vacuum, $|0\rangle_1$, for all k and λ. The vacua are related to each other by a Dirac-metric operator in a manner familiar from the scalar case. The metric operator γ is necessary in view of the commutation relations of the Fourier-coefficient operators, eqs. (25) and (26). A one-photon ket is defined by

$$|k\lambda\rangle = a^\dagger_{2k\lambda}|0\rangle_2\,, \tag{32}$$

while the dual one-photon bra is defined by

$$\langle k\lambda| = {}_1\langle 0|a_{1k\lambda}\,, \tag{33}$$

so that

$$\langle k'\lambda'|k\lambda\rangle = \delta_{\lambda\lambda'}\delta^3(\vec{k}-\vec{k}')\,, \tag{34}$$

by means of eq. (26) and ${}_1\langle 0|0\rangle_2 = 1$. The metric operator is introduced in order to relate the dual of a ket to its Hermitian conjugate:

$$\langle k\lambda| = (|k\lambda\rangle)^\dagger\gamma\,. \tag{35}$$

Consequently the metric operator is defined to have the properties

$$\gamma|0\rangle_1 = |0\rangle_2\,, \tag{36}$$

$$\gamma^{-1}a_{1k\lambda}\gamma = a_{2k\lambda}\,, \tag{37}$$

$$\gamma^{-1}a_{2k\lambda}\gamma = a_{1k\lambda} \tag{38}$$

with $\gamma = \gamma^\dagger = \gamma^{-1}$. It has a form similar to the metric operator of the scalar case which is given in eq. (36) of ref. ([1]). The definition of many-photon bras and kets is a direct generalization of eqs. (32), (33) and (35).

The time-ordered propagator which will later be of use in the development of perturbation theory is

$$(39)\qquad G^{22}_{\mu\nu}(x-y) = -i\,{}_1\langle 0|T(A_{2\mu}(x)A_{2\nu}(y))|0\rangle_2 = \\ = \sin^2\theta_2\, G_{F\mu\nu}(x-y) + \cos^2\theta_2\, G^*_{F\mu\nu}(x-y)\,,$$

where $G_{F\mu\nu}(x-y)$ is the usual Feynman propagator for photons, while G^*_F is its complex conjugate. Another nonzero free-field propagator, $G^{12}_{\mu\nu}$ is defined as the vacuum expectation value of the time-ordered product of $A_{1\mu}(x)$ and $A_{2\nu}(y)$. It depends on both θ_1 and θ_2. It does not appear in perturbation theory.

Equation (39) shows that $G^{22}_{\mu\nu}$ is a Feynman propagator if $\theta_2 = \pi/2$, and a principal-value propagator if $\theta_2 = \pi/4$. Thus

$$(40)\qquad G^{22}_{\mu\nu}(x-y)_{\theta_2=\pi/4} = g_{\mu\nu}\int \frac{d^4k}{(2\pi)^4}\exp[-ik\cdot(x-y)]\,\frac{P}{k^2}\,,$$

where

$$(41)\qquad \frac{P}{k^2} = \frac{1}{2}\left(\frac{1}{k^2+i\varepsilon} + \frac{1}{k^2-i\varepsilon}\right),$$

in the Feynman gauge. This Green's function has previously appeared in action-at-a-distance formulations (3) of classical electrodynamics. The generality of our formulation of quantum field theory allows it to appear as a special case. This does not seem to be possible within the framework of the conventional formulation (7). If we choose $\theta_2 = \pi/4$ and proceed to develop perturbation theory for the full interacting theory then we obtain a quantum-action-at-a-distance theory. We obtain quantum exchange of energy and momentum to which we associate lines in Feynman diagrams. The only difference is in the analytic structure associated with the « epsilontics » of the pole locations. Scattering amplitudes are piecewise analytic in this case (4). More importantly, quanta associated with the electromagnetic field do not appear in asymptotic states, if unitarity is to be maintained. Thus the electromagnetic field does not have its own degrees of freedom, but is constrained to have its source in the charged matter fields which are present. The physical expectations of a quantum action-at-a-distance electrodynamics are therefore satisfied. Except for the absence of « photons » from asymptotic states, and the principal-value nature of the electromagnetic-field propagator, the perturbation theory rules, diagrams and combinatorics are identical to those of QED, although they must be unitarized using the method described in the appendix.

(7) R. FEYNMAN: *Science*, **153**, 699 (1966).

We now turn to the Feynman case ($\theta_2 = \pi/2$). The field $\vec{A}_2$ takes the form

$$\vec{A}_2 = \int \mathrm{d}^3k \sum_{\lambda=1}^{2} \vec{\varepsilon}(k, \lambda)[f_k(x) a_{1k\lambda} + f_k^*(x) a_{2k\lambda}^\dagger] , \tag{42}$$

in the Coulomb gauge. Note that the form of eq. (42) and the commutators, eq. (25) and (26) imply the free-field identity

$$_1\langle 0|A_{2i_1}(x_1) A_{2i_2}(x_2) \dots A_{2i_n}(x_n)|0\rangle_2 = \langle 0|A_{i_1}(x_1) A_{i_2}(x_2) \dots A_{i_n}(x_n)|0\rangle \tag{43}$$

for all spatial indices $i_1, i_2, \dots, i_n$, and all n for which the right-hand side is the vacuum expectation value of n free fields in the usual formulation. This identity is the basis of our claim that the Feynman case of our formulation is *completely* equivalent to QED in its predictions. An important part of establishing the correspondence is the LSZ reduction formulae for photons: for an in photon $(k\lambda)$ we have

$$\begin{aligned} \langle \beta \text{ out}|\alpha(k\lambda) \text{ in}\rangle &= \langle \beta \text{ out}|a_{2k\lambda \text{in}}^\dagger|\alpha \text{ in}\rangle = \\ &= \langle \beta - (k\lambda) \text{ out}|\alpha \text{ in}\rangle - \frac{i}{\sqrt{Z_3}} \int \mathrm{d}^4x \, A_{k\lambda}^\mu(x) \overleftarrow{\Box} \langle \beta \text{ out}|A_{2\mu}(x)|\alpha \text{ in}\rangle \end{aligned} \tag{44}$$

and for an out photon $(k\lambda)$ we have

$$\begin{aligned} \langle \beta(k\lambda) \text{ out}|\alpha \text{ in}\rangle &= \langle \beta \text{ out}|a_{1k\lambda \text{out}}|\alpha \text{ in}\rangle = \\ &= \langle \beta \text{ out}|\alpha - (k\lambda) \text{ in}\rangle - \frac{i}{\sqrt{Z_3}} \int \mathrm{d}^4x \langle \beta \text{ out}|A_{2\mu}(x)|\alpha \text{ in}\rangle \overleftarrow{\Box} A_{k\lambda}^{\mu *}(x) \end{aligned} \tag{45}$$

in the notation of ref. ([5]).

The second quantization of the free-electron field is described in detail in ref. ([1]). We assume an in and out electron field formalism is developed along those lines. Our formulation of spin-one-half fermion theory also allows one to choose Feynman propagators or principal-value propagators for fermions. In order to establish a model equivalent to QED we choose Feynman propagators for the electrons.

The perturbation theory for our model is based on the interaction Lagrangian

$$\mathscr{L}_{\text{int}} = - e_0 \tilde{\psi}_2 \gamma^0 (\gamma \cdot A_2) \psi_2 . \tag{46}$$

We shall need the LSZ reduction formulae for electrons

$$\begin{aligned} \langle \beta \text{ out}|(ps)\alpha \text{ in}\rangle &= \langle \beta - (ps) \text{ out}|\alpha \text{ in}\rangle - \\ &- \frac{i}{\sqrt{Z_2}} \int \mathrm{d}^4x \langle \beta \text{ out}|\tilde{\psi}_2(x)\gamma^0|\alpha \text{ in}\rangle (-\overleftarrow{i\nabla} - m) U_{ps}(x) \end{aligned} \tag{47}$$

and

$$(48)\qquad \langle\beta(ps)\,\text{out}|\alpha\,\text{in}\rangle = \langle\beta\,\text{out}|\alpha-(ps)\,\text{in}\rangle - \frac{i}{\sqrt{Z_2}}\int \mathrm{d}^4x\,\overline{U}_{ps}(x)(i\overrightarrow{\nabla}-m)\langle\beta\,\text{out}|\psi_2(x)|\alpha\,\text{in}\rangle$$

in the notation of ref. (5). To remove a positron from an in state, we use

$$(49)\qquad \langle\beta\,\text{out}|(ps)\alpha\,\text{in}\rangle = \langle\beta-(ps)\,\text{out}|\alpha\,\text{in}\rangle + \frac{i}{\sqrt{Z_2}}\int \mathrm{d}^4x\,\overline{V}_{\overline{ps}}(x)(i\overrightarrow{\nabla}-m)\langle\beta\,\text{out}|\psi_2(x)|\alpha\,\text{in}\rangle$$

and, to remove a positron from an out state, we use

$$(50)\qquad \langle\beta(ps)\,\text{out}|\alpha\,\text{in}\rangle = \langle\beta\,\text{out}|\alpha-(ps)\,\text{in}\rangle + \frac{i}{\sqrt{Z_2}}\int \mathrm{d}^4x\langle\beta\,\text{out}|\tilde{\psi}_2\gamma^0|\alpha\,\text{in}\rangle(-i\overleftarrow{\nabla}-m)\,V_{\overline{ps}}(x)\,.$$

In view of the interaction Lagrangian and eqs. (47)-(50) it is clear that only time-ordered products of $\psi_{2\text{in}}$ and $\tilde{\psi}_{2\text{in}}$ appear in the perturbation theory expansion. The Wick theorem, which applies in the present case, reduces all vacuum expectation values of time-ordered products of $\psi_{2\text{in}}$ and $\tilde{\psi}_{2\text{in}}$ to products of the time-ordered two-point function

$$(51)\qquad iS(x-y) = {}_1\langle 0|T\big(\psi_2(x)\,\tilde{\psi}_2(y)\gamma^0\big)|0\rangle_2$$

$$(52)\qquad = i\int\frac{\mathrm{d}^4p}{(2\pi)^4}\exp[-ip\cdot(x-y)]\,\frac{\not{p}+m}{p^2-m^2+i\varepsilon}\,,$$

where eqs. (51) is in the notation of ref. (1).

The above considerations and the familiar development of the S-matrix as an expansion in vacuum expectation values of time-ordered products of in-field operators lead us to assert that our model electrodynamics is identical to QED in its predictions. In particular

$$(53)\qquad {}_1\langle 0|T\Big(A_{2\mu_1\text{in}}(x_1)A_{2\mu_2\text{in}}(x_2)\ldots\psi_{2\text{in}}(y_1)\psi_{2\text{in}}(y_2)\ldots\tilde{\psi}_{2\text{in}}(z_1)\tilde{\psi}_{2\text{in}}(z_2)\ldots$$
$$\ldots\exp\Big[i\int_{-\infty}^{\infty}\mathrm{d}t\,L_{\text{int}}\Big]\Big)|0\rangle_2 =$$
$$= \langle 0|T\Big(A_{\mu_1\text{in}}(x_1)A_{\mu_2\text{in}}(x_2)\ldots\psi_{\text{in}}(y_1)\psi_{\text{in}}(y_2)\ldots\bar{\psi}_{\text{in}}(z_1)\bar{\psi}_{\text{in}}(z_2)\ldots\exp\Big[i\int_{-\infty}^{\infty}\mathrm{d}t\,L_{\text{int}}^{\text{QED}}\Big]\Big)|0\rangle\,,$$

where the right-hand side is the corresponding time-ordered product of the conventional formulation of QED with

$$L_{\text{int}}^{\text{QED}} = -i:\int d^3x\, e_0 \bar{\psi}_{\text{in}}(\gamma\cdot A_{\text{in}})\psi_{\text{in}}: . \tag{54}$$

Thus, the perturbation theory of our model electrodynamics is term by term equivalent to QED—the equivalent QED term being obtained by ignoring the subscripts 1 and 2 on fields and vacua, and letting $\tilde{\psi}_2\gamma^0 \to \bar{\psi}$. Feynman diagrams can be appropriated without modification for use in our formulation. We conclude with a quote from Feynman (7): « It always seems odd to me that the fundamental laws of physics can appear in so many different forms that are not apparently identical at first.... Theories of the known which are described by different physical ideas may be equivalent in all their predictions and hence scientifically indistinguishable. However they are not psychologically identical when one is trying to move from that base into the unknown. For different views suggest different modifications ... » In that spirit we turn to the investigation of non-Abelian theories.

3. – Yang-Mills models.

The models we have hitherto investigated are members of a class whose Lagrangians have the form

$$\mathscr{L} = \mathscr{L}_0(q_2) + \mathscr{L}_1(q_1 - q_2) , \tag{55}$$

where $\mathscr{L}_0$ has a form which is essentially identical to a Lagrangian of the conventional formulation. For example, in the case of electrodynamics the Lagrangian of eq. (2) can be put in the form of eq. (55), if we let

$$\mathscr{L}_0(A_{2\mu}, \psi_2) = -\tfrac{1}{4} F^2_{\mu\nu}F^{2\mu\nu} + \tilde{\psi}_2\gamma^0\big(i(\gamma\cdot\nabla) - e_0(\gamma\cdot A_2) - m\big)\psi_2 \tag{56}$$

and

$$\begin{aligned}\mathscr{L}_1(A_{1\mu} - A_{2\mu}, \psi_1 - \psi_2) = \\ = \tfrac{1}{4}(F^1_{\mu\nu} - F^2_{\mu\nu})^2 + (\tilde{\psi}_2 - \tilde{\psi}_1)\gamma^0\big(i(\gamma\cdot\nabla) - m\big)(\psi_1 - \psi_2) .\end{aligned} \tag{57}$$

At the level of c-number fields it is clear from eqs. (56) and (57) that $A_{2\mu}$ and ψ_2 reproduce the usual electrodynamics. At the level of quantum fields one can opt to have a model reproducing the conventional quantum theory—we call it the Feynman case. It maintains the implied decoupling of $A_{1\mu} - A_{2\mu}$ and $\psi_1 - \psi_2$ from $A_{2\mu}$ and ψ_2. On the other hand, we can take advantage of the larger space of states in the larger-manifestly-indefinite-metric theory and couple the fields together by taking advantage of degrees of freedom in the

Fourier expansion of asymptotic fields and the definition of the vacuum. Since we « perturb around » the asymptotic fields, perturbation theory will reflect this coupling. Our principal-value or quantum action-at-a-distance models are based on this possibility.

We shall use the form of eq. (55) as an ansatz to generate Yang-Mills models in our formulation. This leads to the gauge field Lagrangian

$$\mathscr{L} = -\tfrac{1}{4}\boldsymbol{F}_{2\mu\nu}\cdot\boldsymbol{F}_2^{\mu\nu} + \tfrac{1}{4}(\boldsymbol{G}_{1\mu\nu} - \boldsymbol{G}_{2\mu\nu})\cdot(\boldsymbol{G}_1^{\mu\nu} - \boldsymbol{G}_2^{\mu\nu}), \tag{58}$$

where

$$\boldsymbol{F}_{2\mu\nu} = \partial_\mu \boldsymbol{A}_{2\nu} - \partial_\nu \boldsymbol{A}_{2\mu} + g\boldsymbol{A}_{2\mu}\times\boldsymbol{A}_{2\nu} \tag{59}$$

and

$$\boldsymbol{G}_{i\mu\nu} = \partial_\mu \boldsymbol{A}_{i\nu} - \partial_\nu \boldsymbol{A}_{i\mu} \tag{60}$$

for $i = 1, 2$. Under a local gauge transformation, $S = S(x)$,

$$A_{2\mu} \to A'_{2\mu} = S^{-1}A_{2\mu}S + \frac{i}{g}S^{-1}\partial_\mu S, \tag{61}$$

$$A_{1\mu} \to A'_{1\mu} = A_{1\mu} + A'_{2\mu} - A_{2\mu}, \tag{62}$$

the Lagrangian $\mathscr{L}$ is invariant, where $A_{i\mu} = \boldsymbol{A}_{i\mu}\cdot\boldsymbol{T}$ for $i = 1, 2$ with $\boldsymbol{T}$ a vector composed of matrices representing generators of the gauge group. There is an additional invariance under the local transformation, $\boldsymbol{A}_{1\mu} \to \boldsymbol{A}_{1\mu} + \partial_\mu\boldsymbol{\Lambda}$, which is a straightforward generalization of the Abelian gauge transformation.

The field equations derived from eq. (58), upon varying the action with respect to $\boldsymbol{A}_{1\mu}$ and $\boldsymbol{A}_{2\mu}$, are

$$\partial^\mu(\boldsymbol{G}_{1\mu\nu} - \boldsymbol{G}_{2\mu\nu}) = 0 \tag{63}$$

and

$$(\partial^\mu + g\boldsymbol{A}_2^\mu\times)\boldsymbol{F}_{2\mu\nu} = 0\,. \tag{64}$$

The canonical momentum conjugate to $\boldsymbol{A}_{2\mu}$ is

$$\boldsymbol{\Pi}_{2\mu} = \boldsymbol{F}_{2\mu 0} + \boldsymbol{G}_{1\mu 0} - \boldsymbol{G}_{2\mu 0} \tag{65}$$

and the momentum conjugate to $\boldsymbol{A}_{1\mu}$ is

$$\boldsymbol{\Pi}_{1\mu} = \boldsymbol{G}_{2\mu 0} - \boldsymbol{G}_{1\mu 0}\,. \tag{66}$$

We choose the Coulomb gauge to implement the field quantization. Due to the form of the Lagrangian, we can treat $\boldsymbol{A}_{2\mu}$ and $\boldsymbol{A}_{3\mu} = \boldsymbol{A}_{1\mu} - \boldsymbol{A}_{2\mu}$ as independent fields. Choosing the Coulomb gauge for $A_{2\mu}$, $\vec{\nabla}\cdot\vec{A}_2 = 0$, we can isolate the independent field quantities and establish their equal-time commutation relations in the well-known manner (*). We also can choose to work in the Coulomb gauge of $\boldsymbol{A}_{3\mu}$, $\vec{\nabla}\cdot\vec{A}_3 = 0$ which is equivalent to $\vec{\nabla}\cdot\vec{A}_1 = 0$. This possibility follows from the invariance of $\mathscr{L}$ under the local « Abelian » transformation of $A_{3\mu}$ or $A_{1\mu}$ mentioned above. The resulting equal-time commutation relations are equivalent to

$$[G^{\mathrm{T}}_{\alpha 0ia}(x), A_{\beta jb}(y)] = i(1 - \delta_{\alpha 1}\delta_{\beta 1})\delta_{ab}\Delta^{\mathrm{tr}}_{ij}(x - y)\,, \tag{67}$$

where $\alpha, \beta = 1, 2$; i and j are spatial- and a and b are internal-symmetry indices. G^{T} is the transverse part of G. One can now proceed to linearize the field equations. The Fourier expansions of the solutions of the linearized equations have the form of eqs. (23) and (24), if an internal-symmetry index is appended to the Fourier component operators. The discussion then reduces to the Abelian case considered in the last section. In particular the Green's function of the linearized model which is relevant to perturbation theory is

$$G^{22}_{ab\mu\nu}(x - y) = - i\langle 0|T\big(A_{2\mu a}(x)\,A_{2\nu b}(y)\big)|0\rangle_2\,. \tag{68}$$

It can be expressed as a linear combination of a Feynman propagator and its complex conjugate as in eq. (39). The arbitrary angle can be chosen to give Feynman propagators or principal-value propagators.

The perturbation theory is based on the linearized model. In the Feynman case the LSZ reduction of asymptotic non-Abelian quanta is essentially given in eqs. (44) and (45), if appropriate internal-symmetry indices are introduced. Only fields of « type 2 » are introduced by the LSZ reduction. Furthermore, the cubic and quartic terms of the Lagrangian « interaction » term also involve only fields of type 2. If we restrict all couplings to other fields—such as fermions—to involve only $\boldsymbol{A}_{2\mu}$, then only time-ordered products of $\boldsymbol{A}_{2\mu}$ appear in the perturbation theory expansion of the S-matrix. The consequence of this development is that one has constructed a model which makes predictions which are identical to a conventional formulation based on a Lagrangian of the usual type

$$\mathscr{L} = -\tfrac{1}{4}\boldsymbol{F}_{\mu\nu}\cdot\boldsymbol{F}^{\mu\nu}\,. \tag{69}$$

This can be verified in a path integral approach based on the Lagrangian of eq. (58).

The point of our formulation is that one can make other choices—such as the principal-value propagator choice. (Note that the path integral formulation

---

(*) E. Abers and B. W. Lee: *Phys. Rep.*, **9** C, 1 (1973).

only has a formal validity in these cases, since the Wick rotation to Euclidean co-ordinates is highly nontrivial.) The choice of principal-value propagators has several important consequences.

In the appendix we show how to calculate the unitary physical S-matrix in this case. We show that loops of principal-value propagators do not contribute to the absorptive parts of S-matrix elements. As a result Faddeev-Popov ghost loops are not needed to maintain unitarity (in any gauge). We also point out a radical possibility: a unitary S-matrix can be defined in which all non-Abelian boson loops are excluded. Thus the non-Abelian boson sector consists solely of tree diagrams. This is a quantum analogue of the absence of self-interactions in classical action-at-a-distance theories.

4. – Non-Abelain model of the strong interactions.

Several years ago a non-Abelian model of the strong interaction was constructed (9) which had manifest quark confinement and a linear potential. The model was based on higher-order field equations. Such equations lead to well-known unitarity problems, if the usual quantization program is implemented. The author chose to avoid the unitarity problems through an *ad hoc* quantization procedure which gave principal-value propagators. The result was a model having unconventional analyticity properties. Despite the successful confinement scheme in this model, there was some reason for uneasiness, due to the unusual form of the Lagrangian (where two sets of fields were associated with the gluons) and the *ad hoc* method of quantization.

The new framework we have developed in ref. (1) and this paper eliminates these sources of concern. Our formulation is based on fundamental ground—the need for an acceptable physical particle interpretation of quantum field theory in flat and curved space-time. We shall see that the two-field Lagrangian of the strong-interaction model lies within the general framework we have established. The principal-value propagators, required to have a unitary model, will also be obtained in a straightforward manner.

Since the strong-interaction model is described in some detail in ref. (9), we shall only outline the aspects necessary to connect that model with the present work.

The strong-interaction Lagrangian departs slightly from the ansatz of eq. (55) (mainly in the replacement $\boldsymbol{G}^1_{\mu\nu}\cdot\boldsymbol{G}^{1\mu\nu} \to \Lambda^2\boldsymbol{A}_{1\mu}\cdot\boldsymbol{A}_1^\mu$):

$$\mathscr{L} = -\tfrac{1}{2}\boldsymbol{F}^1_{\mu\nu}\cdot\boldsymbol{F}^{2\mu\nu} - \tfrac{1}{2}\Lambda^2\boldsymbol{A}_{1\mu}\cdot\boldsymbol{A}_1^\mu + \tilde{\psi}_2\gamma^0\big(i(\gamma\cdot\nabla) + g(\gamma\cdot A_2) - m\big)\psi_2 - (\tilde{\psi}_1 - \tilde{\psi}_2)\gamma^0\big(i(\gamma\cdot\nabla) - m\big)(\psi_1 - \psi_2)\,. \tag{70}$$

---

(9) S. BLAHA: *Phys. Rev. D*, **10**, 4268 (1974); **11**, 2921 (1975). It should be noted that the claim that loops of principal-value propagators are identically zero is not true. In the case of one-loop diagrams only the real part is zero.

where ψ_1 and ψ_2 are the quark fields and

$$F^2_{\mu\nu} = \partial_\mu \boldsymbol{A}_{2\nu} - \partial_\nu \boldsymbol{A}_{2\mu} + g\boldsymbol{A}_{2\mu} \times \boldsymbol{A}_{2\nu} \tag{71}$$

and

$$\boldsymbol{F}^1_{\mu\nu} = D_\mu \boldsymbol{A}_{1\nu} - D_\nu \boldsymbol{A}_{1\mu} \tag{72}$$

with the covariant derivative defined by

$$D_\mu = \partial_\mu + g\boldsymbol{A}_{2\mu} \times . \tag{73}$$

The Lagrangian is invariant under the local color SU_3 gauge transformation

$$\psi_2 \to S^{-1}\psi_2 , \tag{74}$$

$$\psi_1 \to \psi_1 + (S^{-1} - I)\psi_2 , \tag{75}$$

$$A_{1\mu} \to S^{-1} A_{1\mu} S , \tag{76}$$

$$A_{2\mu} \to S^{-1} A_{2\mu} S + \frac{i}{g} S^{-1} \partial_\mu S , \tag{77}$$

with $A_{i\mu} = \boldsymbol{A}_{i\mu} \cdot \boldsymbol{T}$, where $\boldsymbol{T}$ is a vector composed of matrices representing SU_3 generators. The equations of motion and the Coulomb gauge quantization are discussed in detail in ref. (9). (The subscripts, 1 and 2, on the gluon fields are reversed in ref. (9) relative to the present development.)

The essential feature of the model can be seen in the linearized field equations

$$\Box \boldsymbol{A}_{2\mu} = -\Lambda^2 \boldsymbol{A}_{1\mu} , \tag{78}$$

$$\Box \boldsymbol{A}_{1\mu} = -g\boldsymbol{J}_\mu , \tag{79}$$

where $\boldsymbol{J}_\nu$ is the color quark current. Equations (78) and (79) show an apparent mass term, $\Lambda^2(\boldsymbol{A}_{1\mu})^2$, actually leads to a higher-order derivative field equation. The implications of this feature for quark confinement are discussed in ref. (9). The discussion is based on the assumption that all gluon propagators are principal-value propagators. We shall now show that the propagators for color gluons in the linearized model can be principal-value propagators.

We work in the Coulomb gauge, $\vec{\nabla} \cdot \vec{\boldsymbol{A}}_2 = 0$. A mode expansion of the gluon fields must be consistent with the free linearized field equations and the equal-time commutation relations,

$$[F^{\alpha T}_{0ia}(x), A_{\beta jb}(y)] = i\delta_{ab}(1 - \delta_{\alpha\beta})\delta^{tr}_{ij}(\vec{x} - \vec{y}) \tag{80}$$

(cf. eqs. (65) and (66) of ref. (9)). These conditions and the requirement of a unitary theory imply the Fourier expansions

$$\vec{A}_{1a}(x) = \int \frac{d^3k}{\sqrt{2}} \sum_{\lambda=1}^{2} \vec{\varepsilon}(k, \lambda)[(a_{1k\lambda a} + a_{2k\lambda a}) f_k(x) + (a^\dagger_{1k\lambda a} + a^\dagger_{2k\lambda a}) f^*_k(x)] , \tag{81}$$

$$\vec{A}_{2a}(x) = \int \frac{d^3k}{\sqrt{2}} \sum_{\lambda=1}^{2} \vec{\varepsilon}(k, \lambda)[(a_{2k\lambda a} - a_{1k\lambda a}) f_k(x) + (a^\dagger_{2k\lambda a} - a^\dagger_{1k\lambda a}) f^*_k(x)] + \tag{82}$$
$$+ A^2 \int \frac{d^4k}{\sqrt{2}} 2\omega_k \theta(k_0) \delta'(k^2) \sum_{\lambda=1}^{2} \vec{\varepsilon}(k, \lambda)[(a_{1k\lambda a} + a_{2k\lambda a}) f_k(x) + (a^\dagger_{1k\lambda a} - a^\dagger_{2k\lambda a}) f^*_k(x)] ,$$

where $\delta'(k^2) = \partial\delta(k^2)/\partial k^2$ and $\omega_k = |\vec{k}|$. From eqs. (81) and (82) we can determine the Green's functions of the linearized model:

$$iG^{12}_{\mu\nu ab}(x - y) = {}_1\langle 0|T(A_{1\mu a}(x) A_{2\nu b}(y))|0\rangle_2 = \tag{83}$$
$$= -i\delta_{ab} \int \frac{d^4k}{(2\pi)^4} \exp[-ik\cdot(x - y)] r^{12}_{\mu\nu} P \frac{1}{k^2}$$

and

$$iG^{22}_{\mu\nu ab}(x - y) = {}_1\langle 0|T(A_{2\mu a}(x) A_{2\nu b}(y))|0\rangle_2 = \tag{84}$$
$$= -iA^2 \delta_{ab} \int \frac{d^4k}{(2\pi)^4} \exp[-ik\cdot(x - y)] r^{22}_{\mu\nu} P \frac{1}{k^4} ,$$

where

$$P \frac{1}{k^{2N}} = \frac{1}{2} \left[\frac{1}{(k^2 + i\varepsilon)^N} + \frac{1}{(k^2 - i\varepsilon)^N} \right] \tag{85}$$

and where $r^{12}_{\mu\nu}$ and $r^{22}_{\mu\nu}$ are gauge-dependent tensors. While the path integral formulation is only of formal value for the present model, it can be used to determine the form of the Green's functions for any choice of gauge. It can also be used to generate the perturbation theory rules for the model. The character of the perturbation theory is described in ref. (9). Confinement arises through the Schwinger mechanism in a manner reminiscent of the two-dimensional Schwinger model. The mass scale is set by A. The Lagrangian of eq. (70) leads to an interaction between quarks which has the linear potential r as its Coulomb potential in lowest order. The r potential has had notable success in the explication of the charmonium system. A r^{-1} term may also be needed to successfully describe charmonium. Such a term can be introduced in our model by adding another interaction term

$$\mathcal{L}_c = g' \bar{\psi}_2 \gamma^0 (\gamma \cdot A_1) \psi_2 \tag{86}$$

to eq. (70). Since $A_{1\mu}$ transforms homogeneously under a gauge transformation, the gauge invariance of the Lagrangian is not altered by the addition of this term.

In conclusion, we have established the basis for a manifestly quark-confining model of the strong interaction within the framework of our formulation of quantum field theory. This example illustrates the value of our formulation in the construction of unitary quantum field theories with higher-order derivative field equations. Another interesting application of this approach would be to quantum gravity which may have its renormalization problems ameliorated by introducing higher-order derivatives. Certain formal similarities, and this possibility, have led the author to propose a unified model of vierbein gravity and the strong interaction ([10]). The higher-order derivative gravity terms which lead to power counting renormalizability for the quantum gravity sector are linked to the higher-order derivative strong-interaction sector terms which lead to quark confinement. The recent discovery of important spin effects in high-energy p-p scattering ([11]) is suggestive in view of the presence of strong spin-spin interactions in unified models of the strong-interaction and vierbein gravity.

5. – Model of quantum gravity.

In this section we develop a trivially renormalizable model of quantum gravity (based on the Einstein Lagrangian) which is coupled to an external classical source. In addition to obtaining a physically interesting model, we shall see the utility of principal-value propagators in ameliorating divergence problems. The self-interactions of the « gravitons » will be reduced if principal-value propagators are used. This corresponds to the absence of self-interactions in classical action-at-a-distance models.

The model which we shall develop lies within the framework established above. It can be described as quantum action-at-a-distance gravity. We begin with the Einstein action

$$I = -2\varkappa^{-2}\int d^4x\, g^{\frac{1}{2}} R \tag{87}$$

with $\varkappa = 32\pi G$, where G is Newton's constant, $g = \det g_{\mu\nu}$, $R = g^{\mu\nu}R^{\lambda}_{\mu\nu\lambda}$ and where

$$R^{\alpha}_{\mu\nu\lambda} = \partial_\nu \Gamma^{\alpha}_{\mu\lambda} + \Gamma^{\alpha}_{\nu\beta}\Gamma^{\beta}_{\mu\lambda} - (\nu \Leftrightarrow \lambda) \tag{88}$$

and

$$\Gamma^{\alpha}_{\mu\nu} = \frac{1}{2} g^{\alpha\beta}[\partial_\mu g_{\nu\beta} + \partial_\nu g_{\mu\beta} - \partial_\beta g_{\mu\nu}]\,. \tag{89}$$

([10]) S. BLAHA: *Lett. Nuovo Cimento*, **18**, 60 (1977).

([11]) J. R. O'FALLON, L. G. RATNER, P. F. SCHULTZ, K. ABE, R. C. FERNOW, A. D. KRISCH, T. A. MULERA, A. J. SATTHOUSE, B. SANDLER, K. M. TERWILLIGER, D. G. CRABB and P. H. HANSEN: *Phys. Rev. Lett.*, **39**, 733 (1977).

Our approach is analogous to Feynman's covariant quantization around a flat background field (¹²). We therefore let

$$g_{\mu\nu} = \eta_{\mu\nu} + \varkappa h^2_{\mu\nu}, \tag{90}$$

where $\eta_{\mu\nu} = (-1, 1, 1, 1)$. We then introduce a second tensor field, $h^1_{\mu\nu}$; and choose the action for our model to be (cf. eq. (55))

$$I = \int d^4x [\tfrac{1}{2}(h^3_{\mu\nu,\sigma})^2 - (h^3_\mu)^2 - \tfrac{1}{2}(h^3_{,\mu})^2 - 2\varkappa^{-2} g^{\frac{1}{2}} R], \tag{91}$$

where $h^3_{\mu\nu} = h^1_{\mu\nu} - h^2_{\mu\nu}$, $h^3_\mu = h^3_{\mu\nu,\nu}$, $h^3_{,\mu} = h_{\nu\nu,\mu} = \partial_\mu h$ and where the last term on the right-hand side of eq. (91) is expanded in $h^2_{\mu\nu}$ by using eq. (90). The gauge invariance of the Lagrangian is maintained by requiring $h^1_{\mu\nu}$ and $h^2_{\mu\nu}$ to satisfy the same transformation law

$$h^i_{\mu\nu} \to h^{i\prime}_{\mu\nu} = h^i_{\mu\nu} - \partial_\nu \varepsilon_\mu - \partial_\mu \varepsilon_\nu, \tag{92}$$

where ε_μ are four small, but otherwise arbitrary, functions of the co-ordinates. We introduce a de Donder harmonic gauge fixing term

$$\mathscr{L}_D = -\frac{1}{2\gamma}\left(h^2_\mu - \frac{1}{2}h^2_{,\mu}\right)^2 + \frac{1}{2\gamma}\left(h^3_\mu - \frac{1}{2}h^3_{,\mu}\right)^2, \tag{93}$$

in order to obtain a regular kinetic matrix from the quadratic terms of the augmented Lagrangian. The quadratic terms are (if we let $\gamma = \frac{1}{2}$)

$$\mathscr{L}_{\text{quad}} = -h^1_{\alpha\beta,\mu} h^2_{\alpha\beta,\mu} + \tfrac{1}{2} h^1_{,\mu} h^2_{,\mu} + \tfrac{1}{2}(h^1_{\alpha\beta,\mu})^2 - \tfrac{1}{4}(h^1_{,\mu})^2 . \tag{94}$$

These terms plus the higher-order « interaction » terms can be introduced into a suitable path integral (¹³) expression and the perturbation theory rules can be developed. The path integral approach is only of formal significance, in general, in our models. It gives the correct algebraic expressions and combinational rules. The nontriviality of Wick rotation to Euclidean co-ordinates in any case, but the Feynman case, precludes attributing anything more than formal value to the path integral approach. Of course, the applicability of the path integral formalism to the Feynman case, and the absence of any difference at the algebraic level (before integrations) between the Feynman and principal-value cases, means that the path integral formalism also describes the combinatorics of the principal-value case.

(¹²) R. FEYNMAN: *Acta Phys. Polonica*, **24**, 697 (1963).
(¹³) V. N. POPOV and L. FADEEV: Kiev Report. No. ITP 67-36 (1967).

In order to have a nontrivial model in the principal-value case, we shall assume the gravitational field is coupled to an external classical source. Note that covariance requires that only $h^2_{\mu\nu}$ couples to that source. This fact, plus the absence of $h^1_{\mu\nu}$ in higher-order terms, implies that the only free-field propagator necessary to develop perturbation theory is the vacuum expectation value of the time-ordered product of free $h^2_{\mu\nu}$ fields:

$$(95)\qquad iG^{22}_{\mu\nu,\varrho\sigma} = {}_1\langle 0|T\big(h^2_{\mu\nu}(x)h^2_{\varrho\sigma}(y)\big)|0\rangle_2 = \\ = -\frac{i}{2}(\eta_{\mu\varrho}\eta_{\nu\sigma}+\eta_{\mu\sigma}\eta_{\nu\varrho}-\eta_{\mu\nu}\eta_{\varrho\sigma})\int\frac{\mathrm{d}^4k}{(2\pi)^4}\frac{\exp[-ik\cdot(x-y)]}{k^2}\,.$$

If one chooses the propagator in eq. (95) to be a Feynman propagator, then the usual model of quantum gravity results with its attendant renormalizability problems.

We shall consider the case, in which G^{22} is a principal-value propagator. One can establish this case by following a canonical quantization procedure which is completely analogous to those of models considered earlier. The propagation of « gravitons » by principal-value propagators has important consequences for perturbation theory. All graviton loops can be excluded from the physical S-matrix without leading to unitarity problems (cf. appendix). As a result, only tree diagrams occur and unitarity requires that all gravitons are absorbed—either within trees or on the external classical source. Thus we have a model quantum gravity with no renormalization problem. In the classical limit the model becomes Einstein's classical theory of gravity ([14]). There is, of course, the question of whether the classical limit is a retarded model of gravity or not. The answer lies in cosmology—is the absorber mechanism operative?

An action-at-a distance model of gravity with absorber seems to be very much in the spirit of Mach's principle. The usual approach—where the gravitational field is treated as having its own degrees of freedom—almost invariably leads to the surrepitious introduction of absolute space. The reason is simple. The solution of the field equations for a localized mass distribution is asymptotically flat. The only apparent way to avoid this problem is to say Mach restricts us to the class of closed-universe solutions.

In an action-at-a-distance gravity ([15]) the metric field exists only in the presence of matter, so that any asymptotic flatness of solutions is a mathematical

---

([14]) There is an opinion which is sometimes expressed that a tree diagram model is equivalent to a classical theory. This is not true. The correct view is that the classical limit of a tree diagram model is the corresponding classical theory. Cf. S. DESER's paper in *Quantum Gravity*, edited by C. J. ISHAM, R. PENROSE and D. W. SCIAMA (Oxford, 1975).

([15]) A. WHITEHEAD: *The Principle of Relativity* (Cambridge, 1922).

artifact. The metric properties of space are a consequence of the presence of mass-energy. (Consistent with this view, we note that mass-energy must be present to measure the metric properties of space.)

The ultimate realization of Mach's principle from the present viewpoint requires that inertia be the result of the gravitational effects of distant matter. Preliminary work ([16]) in this direction can be easily incorporated within the framework of an action-at-a-distance model of gravity. In fact, Sciama's model of the inertial effects of the distant stars displays a close analogy to the Feynman-Wheeler absorber model.

In closing we remark that the introduction of quantum matter fields in our action-at-a-distance quantum gravity appears to result in a nonrenormalizable model. It seems that a higher-order derivative Lagrangian of the type mentioned in the last section may be necessary for a fully renormalizable quantum gravity. Of course, that would also be a quantum action-at-a-distance model.

Appendix

In this appendix we define a unitary physical S-matrix for action-at-a-distance quantum field theories. The usual formal definition of the S-matrix is unacceptable because it introduces negative-metric states which lead to negative probabilities.

We begin by defining the set of physical asymptotic states to be those states of positive metric which do not contain quanta of action-at-a-distance fields. In an action-at-a-distance electrodynamics the set of physical states includes all electron and positron states not containing « action at a distance » photons. The problem is to define a unitary S-matrix taking « in » physical states to « out » physical states.

We choose to use a variation of Bogoliubov's procedure ([17]) for defining a unitary physical S-matrix as implemented by SUDARSHAN and co-workers ([18]). The starting point is the expansion of the S-matrix in « old fashioned » perturbation theory:

$$S_{\beta\alpha} = \delta_{\beta\alpha} - i(2\pi)^4\delta^4(P_\beta - P_\alpha)T_{\beta\alpha} \tag{A.1}$$

with

$$T_{\beta\alpha} = \langle\beta|H_I|\alpha\rangle + \sum_N \frac{\langle\beta|H_I|N\rangle\langle N|H_I|\alpha\rangle}{E_\alpha - E_N + i\varepsilon} + \dots, \tag{A.2}$$

([16]) D. W. SCIAMA: *Roy. Astron. Soc. Month. Not.*, **113**, 34 (1953).

([17]) N. N. BOGOLIUBOV: *Annales of Invitational Conference on High-Energy Physics*, CERN (1958).

([18]) E. C. G. SUDARSHAN: *Fields and Quanta*, **2**, 175 (1972); C. A. NELSON: Louisiana State University preprint (1972); J. L. RICHARD: *Phys. Rev. D*, **7**, 3617 (1973); C. C. CHIANG: University Göteborg preprint (1972), and references therein.

where H_I is the interaction Hamiltonian. We wish to modify eq. (A.2) so that states with «action-at-a-distance quanta» do not contribute to the absorptive part of amplitudes. This would allow unitarity to be maintained under the restriction of unitarity sums to the set of physical states. SUDARSHAN and co-workers ([18]) have shown that this can be done by taking the energy denominator factor of unphysical states in principal value:

$$(\text{A.3}) \qquad T_{\beta\alpha} = \langle\beta|H_I|\alpha\rangle + \sum_p \frac{\langle\beta|H_I|p\rangle\langle p|H_I|\alpha\rangle}{E_\alpha - E_p + i\varepsilon} + \sum_u \langle\beta|H_I|u\rangle\langle u|H_I|\alpha\rangle P\frac{1}{E_\alpha - E_u} + \dots,$$

where p labels physical states and u labels unphysical states (*i.e.* those containing action-at-a-distance quanta in our case). Equation (A.3) defines the physical T-matrix in our models. It could serve as the basis for calculating S-matrix elements.

However, we would like to re-express this S-matrix in terms of covariant perturbation theory. In order to do this we shall take advantage of Weinberg's study ([19]) of the infinite-momentum frame limit of old-fashioned perturbation theory, and Chang and Ma's realization of infinite-momentum frame results by a change of variable ([20]).

Since the difference between the physical T-matrix and the unmodified T-matrix lies only in the character of the singularity in the energy denominator, the algebraic development of Weinberg, and his power counting arguments, in particular, can be taken over without change to our case. (We also will ignore possible complications due to interchanging the order of integration and the $P\to\infty$ limit. Actually we define our T-matrix to be the result of this procedure.) The result is that Weinberg's rules apply in the present case also —except that Weinberg's rule (*d*) must be modified, so that an energy denominator is taken in principal value, if the corresponding intermediate state is unphysical—*i.e.* contains action-at-a-distance quanta.

CHANG and MA showed that it was not necessary to go to the infinite-momentum frame limit in order to realize Weinberg's rules. They showed that the expression of the usual Feynman rules in terms of infinite momentum frame variables, $\eta = p^0 + p^3$ and $s = p^0 - p^3$ for each momentum p^μ, led to the same results as Weinberg's rules. This fact can be used to advantage in the case of action-at-a-distance models. S-matrix elements can be calculated in our models from the usual Feynman rules in the following way: i) for a given diagram follow the usual Feynman rules to obtain a S-matrix contribution using infinite-momentum frame variables *à la* Chang and Ma—in particular, associate Feynman propagators with action-at-a-distance particle lines; ii) evaluate the «energy» integrals by complex integration; iii) the result will be a series of terms corresponding to different intermediate states in old-fashioned perturbation theory language. Those denominators corresponding to intermediate states with principal-value quanta should be taken in principal value. Those denominators corresponding to physical intermediate states should remain unmodified.

([19]) S. WEINBERG: *Phys. Rev.*, **150**, 1313 (1966).

([20]) S. J. CHANG and S. K. MA: *Phys. Rev.*, **180**, 1506 (1969).

Let us examine the results of this procedure for some simple cases. First, consider the pole diagram for two-particle–to–two-particle scattering. If the pole corresponds to a principal-value particle the amplitude will be proportional to

$$P\frac{1}{q^2-m^2}\,, \tag{A.4}$$

using the above rules ([21]). This result also follows from explicit evaluation of the diagram in canonical perturbation theory, using eq. (1) of the text for $\theta = \pi/4$.

Next consider a second-order self-energy correction due to the emission and absorption of a principal-value quantum by a particle (propagating via Feynman propagators). Our modified Feynman rules require us to evaluate an integral

$$I = i\int \mathrm{d}^4k\,\frac{1}{k^2-m^2+i\varepsilon}\,\frac{1}{(p-k)^2-\mu^2+i\varepsilon}\,, \tag{A.5}$$

using-infinite-momentum frame variables:

$$\eta' = k^0+k^3 \qquad s' = k^0-k^3 \qquad \vec{q} = (k^1, k^2) \tag{A.6}$$

and $p = (s, \eta, \vec{0})$. The result is

$$I = \frac{\pi}{2}\int \mathrm{d}^2q\int\limits_0^\eta \frac{\mathrm{d}\eta'}{\eta'(\eta-\eta')}\,\frac{1}{s-(\vec{q}^{\,2}+\mu^2)/(\eta-\eta')-(\vec{q}^{\,2}+m^2)/\eta'+i\varepsilon} \tag{A.7}$$

which must be modified to

$$I' = \frac{\pi}{2}\int \mathrm{d}^2q\int\limits_0^\eta \frac{\mathrm{d}\eta'}{\eta'(\eta-\eta')}\,P\,\frac{1}{s-(\boldsymbol{q}^2+\mu^2)/(\eta-\eta')-(\boldsymbol{q}^2+m^2)/\eta'}\,. \tag{A.8}$$

In this case we see that $2I' = I + I^*$. The propagation of the particle which would ordinarily propagate via Feynman propagators is manifestly different in states where virtual action-at-a-distance quanta are present. This is necessary in order to define a unitary physical S-matrix. Thus the physical S-matrix only agrees with the canonically defined S-matrix for one–principal-value particle intermediate states and states without principal value particles.

If we consider a non-Abelian action-at-a-distance gauge theory coupled to an external classical source, it is possible to define the physical S-matrix in

---

([21]) Compare to eqs. (5) and (6) of ref. ([9]).

a different manner from the above. Consider the usual definition of the S-matrix and in particular the set of tree diagrams with no external principal-value particle lines. It is easy to verify that the absorptive part of any tree diagram is zero *in the physical region* since

$$\text{Abs}\left[P\frac{1}{k^2-m^2}\right]=0\,. \tag{A.9}$$

As a result the set of tree diagrams defines a unitary (gauge invariant) S-matrix. (This should be contrasted with the usual theory where taking the absorptive part of a tree diagram introduces states containing non-Abelian bosons which in turn introduce loops and Fadeev-Popov ghosts.) In the present case no Fadeev-Popov ghost loops are needed to maintain unitarity ([22]). Thus we wind up with a loopless, tree diagram model. The possibility of limiting the self-interaction of fields in this way allows us to develop a renormalizable model of quantum gravity.

Finally we note that the lack of contributions to unitarity sums from intermediate states containing principal-value particles allows us to avoid the introduction of Fadeev-Popov ghosts in *all* non-Abelian action-at-a-distance models.

([22]) Cf. ref. ([12]) and B. W. Lee and J. Zinn-Justin: *Phys. Rev. D*, **5**, 3121 (1972).

● RIASSUNTO (*)

Si formulano teorie di gauge nel sistema di una generalizzazione della teoria quantistica dei campi. In particolare si discutono modelli di teorie di elettrodinamica e di Yang-Mills, un modello dell'interazione forte con derivate di ordine più alto e confinamento dei quark, e un modello rinormalizzabile di gravità quantistica pura con una Lagrangiana di Einstein. Nel caso dell'elettrodinamica si mostra che due modelli sono possibili: uno con predizioni che sono identiche a QED e uno che è un modello quantistico di azione a distanza dell'elettrodinamica. Nel caso delle teorie di Yang-Mills si può costruire un modello che è identico per quanto riguarda le predizioni a qualsiasi modello convenzionale o modello di azione a distanza. Nel secondo caso è possibile eliminare tutti i cappi di particelle di Yang-Mills (in tutti i gauge) in una maniera consistente con l'unitarietà. Esiste una variazione dei modelli di Yang-Mills nella nostra formulazione che ha equazioni di campo con derivate ad ordine più alto. È unitario ed ha probabilità positive. Può essere usato per costruire un modello d'interazioni forti che ha un potenziale lineare e manifesta confinamento dei quark. Infine si mostra come costruire un modello di azione a distanza della gravità quantistica pura (il cui limite classico è la dinamica della Lagrangiana di Einstein) accoppiato ad una sorgente esterna classica. Il modello è grossolanamente rinormalizzabile.

(*) *Traduzione a cura della Redazione.*

Новый подход к калибровочным теориям поля.

Резюме (*). — Мы формулируем калибовочные теории в рамках обобщения квантовой теории поля. В частности, мы обсуждаем модели электродинамики и теорий Янга-Миллса, модель сильных взаимодействий с производными высших порядков и удержанием кварков и перенормируемую модель для чистой квантовой гравитации с Лагранжианом Эйнштейна. Мы показываем, что в случае электродинамики возможны две модели: одна модель имеет предсказания, которые идентичны предсказаниям квантовой электродинамики, и другая модель, которая представляет модель элзктродинамики с квантовым действием на расстоянии. В случае теорий Янга-Миллса мы можем сконструировать модель, которая идентична по предсказаниям любой общепринятой модели или модели с квантовым действием на расстоянии. Во втором случае имеется возможность исключить все петли для частиц Янга-Миллса (во всех калибровках). Наша формулировка содержит изменение моделей Янга-Миллса, которое включает уравнения поля с производными высших порядков. Наша модель является унитарной и имеет положительные вероятности. Наша формулировка может быть использована для конструирования модели сильных взаимодействий, которая имеет линейный потенциал и обеспечивает удержание кварков. Мы показываем, как сконструировать модель действия на расстоянии для квантовой гравитации, связанной с внешним классическим источником. Предложенная модель является перенормируемой.

(*) *Переведено редакцией.*

Direttore responsabile: CARLO CASTAGNOLI

Stampato in Bologna dalla Tipografia Compositori coi tipi della Tipografia Monograf
Questo fascicolo è stato licenziato dai torchi il 17-I-1979

PHYSICAL REVIEW D VOLUME 11, NUMBER 10 15 MAY 1975

Second-quantized non-Abelian field theory for hadrons with quark confinement and scaling deep-inelastic structure functions*

Stephen Blaha
Laboratory of Nuclear Studies, Cornell University, Ithaca, New York 14853
(Received 30 December 1974)

A four-dimensional second-quantized field theory with quarks bound by "colored" non-Abelian gluons is described which has the following properties: (1) the only physical particles are color singlets composed solely of quarks, (2) the deep-inelastic structure functions have Bjorken scaling, (3) gluon loops and Faddeev-Popov ghost loops are identically zero in any gauge, (4) Regge trajectories are apparently linear on a Chew-Frautschi plot, and (5) constituent motion within hadrons can be nonrelativistic.

I. INTRODUCTION

After a period of some skepticism the possibility that hadronic interactions might be understood within the framework of quantum field theory is again being seriously considered.[1] This is partly the result of the psychological climate created by the apparently successful unification of weak and electromagnetic interactions in a renormalizable field theory and partly the result of a greater appreciation of the variety of phenomena which can occur in field theories.

In this article we shall describe a field-theoretic model of hadron binding which has two major features: (1) Hadrons only occur as quark-antiquark or three-quark bound states, and (2) quarks behave as quasifree particles within hadrons. We assume that the suggestions of an internal symmetry called color[2] are correct and that the strong interaction consists of the exchange of colored Yang-Mills gluons. The nature of the interaction allows only color singlet states to occur in the gauge-invariant physical particle spectrum and consequently the first feature will be realized by choosing the color group to be SU(3). Since the (Schwinger) mechanism which produces this result is an infrared phenomenon, the second feature is not precluded and the model is essentially free in the ultraviolet region of the quark sector.

Our model is a non-Abelian version of a recently investigated Abelian field theory which had quark confinement and scaling electroproduction structure functions.[3] In that theory the free propagator of the massless gluon field embodying the quark-quark interaction was proportional to

$$\lambda^2/k^4, \tag{1}$$

where λ is a constant with the dimensions of mass and k is the gluon four-momentum. As a result the Schwinger mechanism[4] manifestly occurred, and it was shown that any charged particle was totally screened by vacuum polarization effects. In addition, explicit calculations of the deep-inelastic electroproduction structure functions in perturbation theory were in agreement with Bjorken scaling with corrections of $O(q^{-4})$, where q is the virtual photon four-momentum. These features of the Abelian model will also be shown to be true in the non-Abelian version. In addition, we shall argue that the quarks can be nonrelativistic within hadrons and that the spectrum of states has linearly rising Regge trajectories.

In spite of these salutary properties an interaction of the form of Eq. (1) could be questioned because of well-known[5] indefinite-metric difficulties which result in the violation of unitarity. While an optimist may hope that the nonappearance of colored gluons in asymptotic (color singlet) states might eliminate unitarity problems it is almost certain that the approximation techniques which will necessarily be used to find the bound states will lead to the occurrence of negative-metric states. Whether these states are "real" or artifacts of the approximation will not be clear. In view of this we suggested[3] that the gluon propagator be taken in principal value rather than as a Feynman propagator:

$$P\frac{\lambda^2}{k^4} \equiv \frac{\lambda^2}{2}\left[\frac{1}{(k^2+i\epsilon)^2}+\frac{1}{(k^2-i\epsilon)^2}\right]. \tag{2}$$

As a result unitarity is maintained order by order in perturbation theory. Gluons do not appear in asymptotic states. All components of the vector-gluon propagator are "Coulombized" and the gluon field reduced to the embodiment of a direct quark interaction. There are a number of other decided advantages to principal-value propagators in the present context: (1) no color singlet states composed solely of gluons, (2) the elimination of substantial infrared divergences, (3) the suppression of corrections to Bjorken scaling in the electroproduction structure functions by a factor of q^2

vis-à-vis the corresponding Feynman-propagator result which sets the stage for precocious scaling, and (4) the elimination of closed loops of vector gluons and thus the elimination of Faddeev-Popov ghost loops.

In Sec. II we give a brief recapitulation of the Abelian model. In Sec. III we describe the canonical properties of the non-Abelian model. In Sec. IV we describe the qualitative features of the model and describe an approximation technique which appears to be naturally adopted to "solving" the theory. We shall restrict our discussion to the color binding interaction and defer the introduction of other interactions to a later work. The properties of the bound states in the non-Abelian model are currently under study and will be the subject of the next report.

II. ABELIAN MODEL

The possibility that the physical particle spectrum of a field theory consisted only of neutral states and did not include states of charged fields was first investigated in massless two-dimensional quantum electrodynamics.[6] In that case the absence of the "electron" from the gauge-invariant physical particle spectrum was directly related to the acquisition of a mass by the photon via the Schwinger mechanism. The Schwinger mechanism was manifest in the lowest-order contribution to the vacuum polarization (Fig. 1), and, taking account of the dimensionality of the coupling constant, $e \sim$ mass, could almost be considered a consequence of dimensional analysis. These vacuum polarization effects led to the total screening of the "electronic" charge, and, as a result, the "electron" was removed from the gauge-invariant physical particle spectrum. Our Abelian and non-Abelian models will display a similar pattern of events.

The Lagrangian of the Abelian model contains two gluon fields, $A_\mu^1(x)$ and $A_\mu^2(x)$, and the quark field $\psi(x)$:

$$\mathcal{L} = -\tfrac{1}{2}F^1_{\mu\nu}F^2_{\mu\nu} - \tfrac{1}{2}\lambda^2 A^2_\mu A^2_\mu + \bar{\psi}(i\not\nabla - g\not{A}^1 - m)\psi\ , \tag{3}$$

where for typographic convenience we denote the inner product of four vectors, $a\cdot b = a_\mu b_\mu = a_0 b_0 - \vec{a}\cdot\vec{b}$ throughout, λ is a constant with the dimensions of mass, g is dimensionless, and $F^i_{\mu\nu} = \partial_\nu A^i_\mu - \partial_\mu A^i_\nu$.

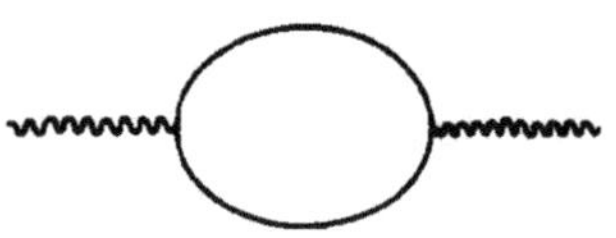

FIG. 1. A vacuum polarization diagram.

Following the canonical procedure we find the equations of motion;

$$\partial_\mu F^1_{\mu\nu} + \lambda^2 A^2_\nu = 0\ , \tag{4}$$

$$\partial_\mu F^2_{\mu\nu} + gJ_\nu = 0\ , \tag{5}$$

$$(i\not\nabla - g\not{A}^1 - m)\psi = 0\ , \tag{6}$$

and nonzero equal-time commutation relations [in the Coulomb gauge $\vec{\nabla}\cdot\vec{A}^1 = 0$; note $\partial_\mu A^2_\mu = 0$ by Eq. (4)]

$$[F^1_{0i}(x), A^2_j(y)] = i\Delta^{tr}_{ij}(x-y)\ , \tag{7}$$

$$[F^2_{0i}(x), A^1_j(y)] = i\Delta^{tr}_{ij}(x-y)\ , \tag{8}$$

with $i,j = 1,2,3$ and

$$\Delta^{tr}_{ij}(x-y) = \int \frac{d^3k}{(2\pi)^3}\, e^{i\vec{k}\cdot(\vec{x}-\vec{y})}\left(\delta_{ij} - \frac{k_i k_j}{|\vec{k}|^2}\right). \tag{9}$$

It is clear from the equations of motion, Eqs. (4) and (5), tnat A^2_μ may be eliminated to obtain

$$\Box\partial_\mu F^1_{\mu\nu} + g\lambda^2 J_\nu = 0\ . \tag{10}$$

The form of the quark-gluon interaction and Eq. (10) show that only the Green's function of A^1_μ is relevant to quark-quark scattering. The perturbation theory rules of QED may be used if the photon propagator is replaced with the gluon propagator for A^1_μ:

$$iG^{11}_{\mu\nu}(k) = \frac{i\lambda^2(g_{\mu\nu} - \chi k_\mu k_\nu/k^2)}{k^4}\ , \tag{11}$$

where χ is constant, determined by the gauge choice.

In Ref. 3 we showed that choosing $G^{11}_{\mu\nu}$ to be a principal-value propagator allowed us to develop a perturbation theory which was unitary order by order:

$$G^{11}_{\mu\nu}(k^2) \equiv \tfrac{1}{2}[G^{11}_{\mu\nu}(k^2 + i\epsilon) + G^{11}_{\mu\nu}(k^2 - i\epsilon)]\ . \tag{12}$$

In addition, the equivalent of the Nambu representation of a Feynman diagram was given and some features of the perturbation theory discussed. Of particular interest was a calculation of the deep-inelastic electroproduction structure functions which scaled in the Bjorken limit. Leading corrections to scaling were of $O(q^{-4})$ as $q^2 \to \infty$ with q being the virtual photon four-momentum, and were given by the diagrams of Fig. 2(b), 2(c), and 2(d). This is to be contrasted with the logarithmic deviations from scaling found in pseudoscalar or vector meson models previously studied.[7]

The Schwinger mechanism manifestly occurred in low orders of perturbation theory. As a result quarks (and all charged objects) are removed from the gauge-invariant spectrum of physical states. The total screening of charge can be seen from the following argument.[3] Consider a spatially bounded system of charge density ρ. The total charge is

$$Q = \int d^3x\, \rho(x) \tag{13}$$

$$= \frac{-1}{g\lambda^2} \int d^3x\, \Box \nabla^2 A_0^1 \tag{14}$$

using the equations of motion in the Coulomb gauge. By Gauss's law

$$Q = \frac{-1}{g\lambda^2} \int d\vec{S} \cdot \vec{\nabla} \Box A_0^1 . \tag{15}$$

From the definition of a Green's function, we have

$$A_0^1(x) \equiv \int d^4y\, G_{00}^{11}(x-y)\rho(y) \tag{16}$$

in the Coulomb gauge. If, for simplicity, we choose ρ to describe a static point quark charge and use the free gluon propagator [Eq. (11)], then $Q \neq 0$. However, if we take account of the effect of vacuum polarization processes (the Schwinger mechanism) we find $A_0^1(x)$ is a monotonically decreasing function of $|\vec{x}|$ for large $|\vec{x}|$ and consequently $Q = 0$ in the limit where the integration surface is taken to infinity in Eq. (15). Thus the spectrum of physical states does not include states of nonzero charge. In the next section we shall show that the proof of quark confinement is essentially the same in the non-Abelian model.

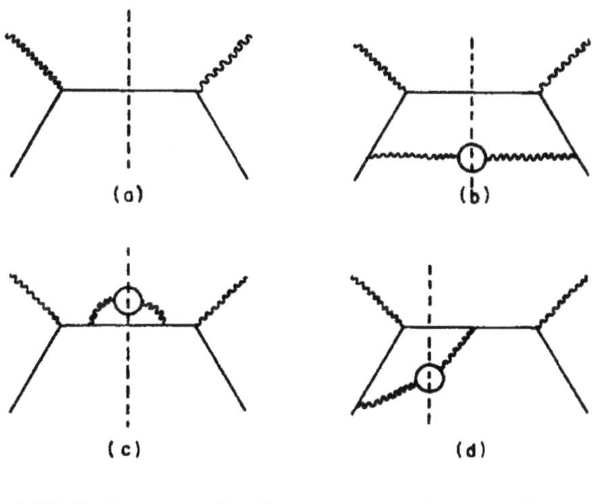

FIG. 2. Lowest-order diagrams contributing to the inelastic electroproduction structure functions. The dashed lines indicate the only contributions to the electroproduction structure functions of the absorptive part of the forward virtual Compton scattering diagram. External "wiggly" lines represent photons while internal "wiggly" lines represent gluons.

III. NON-ABELIAN MODEL

The non-Abelian model for the color sector of hadronic interactions is a direct generalization of the model of the last section.[8] There are two colored Yang-Mills fields, $A_{\mu a}^1(x)$ and $A_{\mu a}^2(x)$, which when regarded as vectors in the adjoint representation of the color group are denoted $\underline{A}_\mu^1$ and $\underline{A}_\mu^2$. The Lagrangian is

$$\begin{aligned} \mathcal{L} = {} & \tfrac{1}{2}\underline{F}_{\mu\nu}^1 \cdot \underline{F}_{\mu\nu}^2 - \tfrac{1}{2}\underline{F}_{\mu\nu}^2 \cdot (\partial_\mu \underline{A}_\nu^1 - \partial_\nu \underline{A}_\mu^1 + g\underline{A}_\mu^1 \times \underline{A}_\nu^1) \\ & - \tfrac{1}{2}\underline{F}_{\mu\nu}^1 \cdot (\partial_\mu \underline{A}_\nu^2 - \partial_\nu \underline{A}_\mu^2 + g\underline{A}_\mu^1 \times \underline{A}_\nu^2 - g\underline{A}_\nu^1 \times \underline{A}_\mu^2) \\ & - \tfrac{1}{2}\lambda^2 \underline{A}_\mu^2 \cdot \underline{A}_\mu^2 + \bar{\psi}(i\nabla\!\!\!/ + g A\!\!\!/^1 - m)\psi \end{aligned} \tag{17}$$

$$= \mathcal{L}_0 + \bar{\psi}(i\nabla\!\!\!/ + g A\!\!\!/^1 - m)\psi , \tag{18}$$

with ψ being the quark field.

It is invariant under the local gauge transformation

$$\psi' = S^{-1}\psi , \tag{19}$$

$$A_\mu^{1\prime} = S^{-1}A_\mu^1 S + \frac{i}{g} S^{-1}\partial_\mu S , \tag{20}$$

$$A_\mu^{2\prime} = S^{-1}A_\mu^2 S , \tag{21}$$

$$F_{\mu\nu}^{1\prime} = S^{-1}F_{\mu\nu}^1 S , \tag{22}$$

$$F_{\mu\nu}^{2\prime} = S^{-1}F_{\mu\nu}^2 S , \tag{23}$$

where S is an element in the gauge group G [which is color SU(3) in our case], and A_μ^1 is a matrix in the defining representation of G formed from

$$A_\mu^1 = \underline{A}_\mu^1 \cdot \underline{T} . \tag{24}$$

T_a is a matrix in the defining representation of G satisfying

$$[T_a, T_b] = i t_{abc} T_c , \tag{25}$$

and $\underline{T}$ is a vector formed from such matrices. We note that the homogeneity of the gauge transformation of A_μ^2 allows a mass term to occur in $\mathcal{L}$ without breaking the gauge symmetry. We shall see that the natural gauge-fixing term to add to the Lagrangian has the form

$$-\frac{1}{\beta}\partial_\mu \underline{A}_\mu^1 \cdot \partial_\nu \underline{A}_\nu^2 . \tag{26}$$

The Euler-Lagrange equations of motion are obtained in the canonical manner:

$$(\partial_\mu + g\underline{A}_\mu^1 \times)\underline{F}_{\mu\nu}^1 - \lambda^2 \underline{A}_\nu^2 = 0 , \tag{27}$$

$$(\partial_\mu + g\underline{A}_\mu^1 \times)\underline{F}_{\mu\nu}^2 + g\underline{A}_\mu^2 \times \underline{F}_{\mu\nu}^1 + g\underline{J}_\nu = 0 , \tag{28}$$

$$\underline{F}_{\mu\nu}^1 = \partial_\mu \underline{A}_\nu^1 - \partial_\nu \underline{A}_\mu^1 + g\underline{A}_\mu^1 \times \underline{A}_\nu^1 , \tag{29}$$

$$\underline{F}_{\mu\nu}^2 = \partial_\mu \underline{A}_\nu^2 - \partial_\nu \underline{A}_\mu^2 + g\underline{A}_\mu^1 \times \underline{A}_\nu^2 - g\underline{A}_\nu^1 \times \underline{A}_\mu^2 , \tag{30}$$

$$(i\nabla\!\!\!/ + g A\!\!\!/^1 - m)\psi = 0 . \tag{31}$$

The antisymmetry of $\underline{F}_{\mu\nu}^1$ and $\underline{F}_{\mu\nu}^2$ leads to two conservation laws,

$$\partial_\nu(g\underline{A}^1_\mu\times\underline{F}^1_{\mu\nu}-\lambda^2\underline{A}^2_\nu)=0\,, \tag{32}$$

$$\partial_\nu(\underline{A}^1_\mu\times\underline{F}^2_{\mu\nu}+\underline{A}^2_\mu\times\underline{F}^1_{\mu\nu}+\underline{J}_\nu)=0\,, \tag{33}$$

which can be reexpressed as

$$(\partial_\nu+g\underline{A}^1_\nu\times)\underline{A}^2_\nu=0 \tag{34}$$

and

$$(\partial_\nu+g\underline{A}^1_\nu\times)\underline{J}_\nu=0 \tag{35}$$

using the equations of motion. The first of these relations acts in effect as a gauge-fixing term for A^2_μ if a gauge is chosen for A^1_μ. The second relation has the familiar form of current-conservation equations in conventional Yang-Mills theories.

We turn now to the derivation of the perturbation-theory rules in the gluon sector. We consider the vacuum-vacuum transition amplitude in the presence of external sources[9]:

$$W(\underline{J}^1_\mu,\underline{J}^2_\mu)=\int\prod_x dA^1_\mu\,dA^2_\nu\exp\left[i\int d^4x\left(\mathcal{L}_0-\frac{1}{\beta}\partial_\mu\underline{A}^1_\mu\cdot\partial_\nu\underline{A}^2_\nu+\underline{A}^1_\mu\cdot\underline{J}^1_\mu+\underline{A}^2_\mu\cdot\underline{J}^2_\mu\right)\right]. \tag{36}$$

After some functional translations we find

$$W(\underline{J}^1_\mu,\underline{J}^2_\nu)\equiv\exp\left\{-i\int d^4x\,d^4y\left[\underline{J}^1_\mu(x)\cdot G^{12}_{\mu\nu}(x-y)\cdot\underline{J}^2_\nu(y)+\tfrac{1}{2}\underline{J}^1_\mu(x)\cdot G^{11}_{\mu\nu}(x-y)\cdot\underline{J}^1_\nu(y)\right]\right\}, \tag{37}$$

where we have dropped an irrelevant factor independent of $\underline{J}^1_\mu$ and $\underline{J}^2_\mu$ on the right-hand side, and

$$G^{12}_{\mu\nu ab}(x)=-\delta_{ab}\int\frac{d^4k\,e^{-ik\cdot x}}{(2\pi)^4k^2}\left[g_{\mu\nu}+(\beta-1)\frac{k_\mu k_\nu}{k^2}\right] \tag{38}$$

and

$$G^{11}_{\mu\nu ab}(x)=\frac{\lambda^2\delta_{ab}}{(2\pi)^4}\int\frac{d^4k\,e^{-ik\cdot x}}{k^4}\left[g_{\mu\nu}+(\beta^2-1)\frac{k_\mu k_\nu}{k^2}\right], \tag{39}$$

with a and b labeling color indices. The free propagators corresponding to the time-ordered products are

$$\langle TA^1_{\mu a}(x)A^1_{\nu b}(y)\rangle=iG^{11}_{\mu\nu ab}(x-y) \tag{40}$$

and

$$\langle TA^1_{\mu a}(x)A^2_{\nu b}(y)\rangle=iG^{12}_{\mu\nu ab}(x-y). \tag{41}$$

The somewhat unusual Green's functions of Eqs. (40) and (41) have their origin in the canonical equal-time commutation relations which we shall now find.

From Eqs. (27)–(30) we obtain the equations of motion

$$\partial_0\underline{A}^1_k=\underline{F}^1_{0k}+\partial_k\underline{A}^1_0+g\underline{A}^1_k\times\underline{A}^1_0\,, \tag{42}$$

$$\partial_0\underline{A}^2_k=\underline{F}^2_{0k}+\partial_k\underline{A}^2_0+g\underline{A}^2_k\times\underline{A}^1_0-g\underline{A}^2_0\times\underline{A}^1_k\,, \tag{43}$$

$$\partial_0\underline{F}^1_{0k}=(\partial_i+g\underline{A}_i\times)\underline{F}^1_{ik}-g\underline{A}^1_0\times\underline{F}^1_{0k}+\lambda^2\underline{A}^2_k\,, \tag{44}$$

$$\partial_0\underline{F}^2_{0k}=(\partial_i+g\underline{A}^1_i\times)\underline{F}^2_{ik}-g\underline{A}^1_0\times\underline{F}^2_{0k}-g\underline{A}^2_\mu\times\underline{F}^1_{\mu k}-g\underline{J}_k\,, \tag{45}$$

and equations of constraint

$$\underline{F}^1_{ik}=\partial_i\underline{A}^1_k-\partial_k\underline{A}^1_i+g\underline{A}^1_i\times\underline{A}^1_k\,, \tag{46}$$

$$\underline{F}^2_{ik}=\partial_i\underline{A}^2_k-\partial_k\underline{A}^2_i+g\underline{A}^1_i\times\underline{A}^2_k-g\underline{A}^1_k\times\underline{A}^2_i\,, \tag{47}$$

$$(\partial_i+g\underline{A}^1_i\times)\underline{F}^1_{i0}+\lambda^2\underline{A}^2_0=0\,, \tag{48}$$

$$(\partial_i+g\underline{A}^1_i\times)\underline{F}^2_{i0}+g\underline{A}^2_i\times\underline{F}^1_{i0}-g\underline{J}_0=0\,. \tag{49}$$

The Lagrangian indicates that the canonical momenta are

$$\underline{\Pi}^1_j=\underline{F}^2_{0j} \tag{50}$$

and

$$\underline{\Pi}^2_j=\underline{F}^1_{0j}\,, \tag{51}$$

for $j=1,2,3$ with $\underline{\Pi}^i_j$ conjugate to $\underline{A}^i_j$, and $\underline{A}^i_0$ having no conjugate momentum for $i=1,2$. However, the equations of constraint indicate that not all components are independent. We now find the independent components. Let us define

$$\underline{F}^a_{0i}=\underline{F}^{aT}_{0i}+\underline{F}^{aL}_{0i} \tag{52}$$

and

$$\underline{F}^{aL}_{0i}=\partial_i\underline{\phi}^a\,, \tag{53}$$

where

$$\partial_i\underline{F}^{aT}_{0i}=0\,. \tag{54}$$

Then Eq. (48) gives

$$(\partial_i+g\underline{A}^1_i\times)\partial_i\underline{\phi}^1-\lambda^2\underline{A}^2_0=-g\underline{A}^1_i\times\underline{F}^{1T}_{i0} \tag{55}$$

and Eq. (49) gives

$$(\partial_i+g\underline{A}^1_i\times)\partial_i\underline{\phi}^2+g\underline{A}^2_i\times\partial_i\underline{\phi}^1=g\underline{A}^1_i\times\underline{F}^{2T}_{i0}+g\underline{A}^2_i\times\underline{F}^{1T}_{0i}-g\underline{J}_0\,. \tag{56}$$

Rewriting Eqs. (42) and (43) after taking the divergence with respect to spatial components gives

$$(\partial_0+g\underline{A}^1_0\times)\partial_k\underline{A}^1_k=(\partial_k+g\underline{A}^1_k\times)\partial_k\underline{A}^1_0+\partial_k\partial_k\underline{\phi}^1 \tag{57}$$

and

$$(\partial_0+g\underline{A}^1_0\times)\partial_k\underline{A}^2_k+g\underline{A}^2_0\times\partial_k\underline{A}^1_k=\partial_k\partial_k\underline{A}^2_0+g\underline{A}^2_k\times\partial_k\underline{A}^1_0+g\underline{A}^1_k\times\partial_k\underline{A}^2_0+\partial_k\partial_k\underline{\phi}^2 \tag{58}$$

If we choose the Coulomb gauge, $\vec{\nabla}\cdot\underline{A}^1=0$, then

$$\partial_k\partial_k\underline{A}_0^1+g\underline{A}_k^1\times\partial_k\underline{A}_0^1+\partial_k\partial_k\underline{\phi}^1=0 \tag{59}$$

and

$$(\partial_0+g\underline{A}_0^1\times)\partial_k\underline{A}_k^2-\partial_k\partial_k\underline{A}_0^2-g\underline{A}_k^2\times\partial_k\underline{A}_0^1-g\underline{A}_k^1\times\partial_k\underline{A}_0^2 -\partial_k\partial_k\phi^2=0\,, \tag{60}$$

thus determining $\underline{A}_0^1$ and $\underline{A}_0^2$. Suppose we now define

$$\vec{\underline{A}}^2=\vec{\underline{A}}^{2T}+\vec{\underline{A}}^{2L}\,, \tag{61}$$

$$\vec{\underline{A}}^{2L}=\vec{\nabla}\underline{\phi}^3\,, \tag{62}$$

with

$$\vec{\nabla}\cdot\vec{\underline{A}}^{2T}=0 \tag{63}$$

Taking the divergence of Eq. (44) leads to our final equation for dependent variables

$$\lambda^2\partial_k\partial_k\underline{\phi}^3=\partial_0\partial_k\partial_k\underline{\phi}^1+g\partial_k(\underline{A}_\mu^1\times\underline{F}_{\mu k}^1)\,. \tag{64}$$

The independent dynamical variables are thus seen to be F_{0i}^{1T}, F_{0i}^{2T}, A_i^{1T}, and A_i^{2T}. Their equal-time commutation relations are

$$[F_{0ia}^{1T}(x),A_{jb}^2(y)]=i\delta_{ab}\Delta_{ij}^{tr}(x-y)\,, \tag{65}$$

$$[F_{0ia}^{2T}(x),A_{jb}^1(y)]=i\delta_{ab}\Delta_{ij}^{tr}(x-y)\,, \tag{66}$$

with $i,j=1,2,3$, Δ_{ij}^{tr} given by Eq. (9), and a and b are color indices. All other commutators of the forms $[A^1,A^1]$, $[A^2,A^2]$, $[F^1,F^1]$, $[F^2,F^2]$, $[F^1,F^2]$ are zero.

We return to our development of perturbation-theory rules. The cubic and quartic gluon vertices of our model are given by (see Fig. 3)

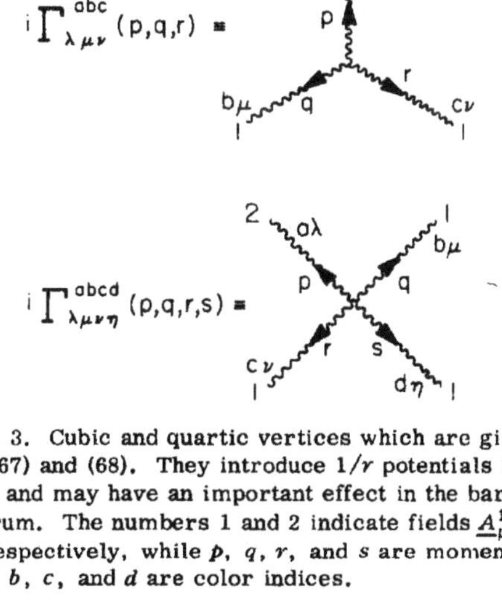

FIG. 3. Cubic and quartic vertices which are given in Eqs. (67) and (68). They introduce $1/r$ potentials in the model and may have an important effect in the baryon spectrum. The numbers 1 and 2 indicate fields $\underline{A}_\mu^1$ and $\underline{A}_\mu^2$, respectively, while p, q, r, and s are momenta, and a, b, c, and d are color indices.

$$i\Gamma^{abc}_{\lambda\mu\nu}(p,q,r)=t^{abc}[g_{\nu\lambda}(r_\mu-p_\mu)+g_{\mu\lambda}(p_\nu-q_\nu)+g_{\mu\nu}(q_\lambda-r_\lambda)]\,, \tag{67}$$

with $p+q+r=0$, and

$$i\Gamma^{abcd}_{\lambda\mu\nu\eta}(p,q,r,s)=-it^{abf}t^{cdf}(g_{\lambda\mu}g_{\nu\eta}-g_{\eta\mu}g_{\lambda\nu})-it^{acf}t^{bdf}(g_{\lambda\eta}g_{\nu\mu}-g_{\nu\lambda}g_{\mu\eta})-it^{adf}t^{bcf}(g_{\lambda\eta}g_{\nu\mu}-g_{\lambda\mu}g_{\nu\eta})\,, \tag{68}$$

with $p+q+r+s=0$.

The Faddeev-Popov ghost loops will not be relevant to our line of development so we omit their discussion. The necessity for their introduction[10] is closely related to the requirement of unitarity in Yang-Mills theories. In the present model unitarity will be necessarily violated irrespective of the ghost loops if the Green's functions [Eqs. (38) and (39)] pole ambiguities are resolved by using Feynman's $i\epsilon$ procedure. To avoid unitarity violation we have suggested an alternative procedure where the Green's function singularities are taken in principal value,

$$G^{kL}_{\mu\nu ab}(k^2)=\tfrac{1}{2}[G^{kL}_{\mu\nu ab}(k^2+i\epsilon)+G^{kL}_{\mu\nu ab}(k^2-i\epsilon)]\,, \tag{69}$$

in momentum space (cf. the Appendix). This choice has the advantage stated in the Introduction. The effects are the same as in the Abelian model[3] and may be summarized as: (1) Only states composed solely of quarks contribute to unitarity sums, (2) gluons do not appear in asymptotic states, (3) unitarity is achieved but at the price of possible advanced effects whose range is limited to hadronic dimensions and thus apparently unobservable, and (4) nonscaling corrections to Bjorken scaling in the deep-inelastic electroproduction structure functions are suppressed by a factor of q^2 vis-à-vis the corresponding result using Feynman propagators with q being the virtual photon four-momentum.

A novel feature of the use of principal-value propagators in non-Abelian models is the elimination of closed loops composed solely of gluons. If we consider a subdiagram consisting of a gluon loop with p lines, then Eq. (51) of Ref. 3 gives the Feynman parameter representation

$$I=\int_{-\infty}^{\infty}\prod_{j=1}^{p}\alpha_j\,d\alpha_j\,\frac{\epsilon(\alpha_1\alpha_2\cdots\alpha_p C)}{C^2}\,Ne^{iD/C}\,, \tag{70}$$

where C is a polynomial consisting of Feynman parameters only, while D contains scalar products of external momenta, N symbolizes appropriate numerator factors, and $\epsilon(\alpha)=\pm1$ if $\alpha\gtrless0$. Since N can be written as a sum of terms each of which is homogeneous in the Feynman parameters, we can take N to be homogeneous without loss of generality. Then scaling all parameters with u, assuming

$$N(u\alpha_1, u\alpha_2, \ldots, u\alpha_p) = u^r N(\alpha_1, \alpha_2, \ldots, \alpha_p), \quad (71)$$

with r an integer, and using

$$\int_0^\infty \frac{du}{u}\, \delta\left(1 - \frac{|\alpha_1 + \alpha_2 + \cdots + \alpha_p|}{u}\right) = 1 \quad (72)$$

we find

$$I = \Gamma(r + 2p - 2L) \times \int_{-\infty}^{\infty} \frac{\prod_{j=1}^{p} \alpha_j \, d\alpha_j \epsilon(\alpha_1\alpha_2 \cdots \alpha_p C) N \delta\left(1 - \left|\sum_k \alpha_k\right|\right)}{C^2(-iD/C)^{r+2p-2L}}, \quad (73)$$

with L = number of loops = 1. Suppose we let $\alpha_j \to -\alpha_j$ for all j in I. Then we find $I = -I$ or

$$I = 0. \quad (74)$$

Thus any closed loop containing only principal-value propagators is zero. Since Faddeev-Popov ghosts appear only in closed loops and consistency[11] requires we use principal-value propagators for them if we use such propagators for gluons, we see that ghosts do not appear in our model. Physically we can understand this result if we remember that ghost loops were introduced to cure problems arising from contributions to unitarity sums of "opened" gluon loops.[10] In our model "opened" loops do not contribute to unitarity sums in any case so the raison d'être for ghosts is lacking.

We now derive the Ward-Takahashi-Slavnov identities using functional methods. Since we take our gluon propagators in principal value it might appear that our use of functional techniques is unjustified. We shall take the view that the functional representation of the vacuum-vacuum transition amplitude embodies the combinatorics of perturbation theory and acts as a generating function for identities, such as the Ward-Takahashi-Slavnov identities. Thus, questions of convergence of functional integrals are irrelevant—the important question is whether identities are valid in perturbation theory.

We define $W(J)$, the vacuum-vacuum transition amplitude, by

$$W(J) = \int \prod_x dA_\mu^1 \, dA_\mu^2 \, d\psi \, d\bar{\psi} \exp\left(i \int \bar{\mathcal{L}}\, dx\right), \quad (75)$$

with

$$\bar{\mathcal{L}} = \mathcal{L} - \frac{1}{\beta} \partial_\mu \underline{A}_\mu^1 \cdot \partial_\nu \underline{A}_\nu^2 + \underline{A}_\mu^1 \cdot \underline{J}_\mu^1 + \underline{A}_\mu^2 \cdot \underline{J}_\mu^2 + \bar{\psi}\eta + \bar{\eta}\psi,$$

with $\mathcal{L}$ given by Eq. (17). Under the infinitesimal gauge variation

$$\underline{A}_\mu^1 \to \underline{A}_\mu^1 - (\partial_\mu + g\underline{A}_\mu^1 \times)\, \underline{\theta}, \quad (76)$$

$$\underline{A}_\mu^2 \to \underline{A}_\mu^2 - g\underline{A}_\mu^2 \times \underline{\theta}, \quad (77)$$

$$\psi \to \psi - ig\theta\psi, \quad (78)$$

$$\bar{\psi} \to \bar{\psi} + ig\bar{\psi}\theta, \quad (79)$$

with $\theta = \underline{T} \cdot \underline{\theta}$, $\mathcal{L}$ is invariant but the remaining terms in $\bar{\mathcal{L}}$ lead to

$$\delta\bar{\mathcal{L}} \equiv \frac{1}{\beta}[(\partial_\mu + g\underline{A}_\mu^1 \times)\partial_\nu\partial_\mu \underline{A}_\nu^2 + g\underline{A}_\nu^2 \times \partial_\mu\partial_\nu \underline{A}_\mu^1] \cdot \underline{\theta} - (\partial_\mu + g\underline{A}_\mu^1 \times)\underline{J}_\mu^1 \cdot \underline{\theta} - g\underline{J}_\mu^2 \times \underline{A}_\mu^2 \cdot \underline{\theta} + ig\bar{\psi}\theta\eta - ig\bar{\eta}\theta\psi. \quad (80)$$

Since a transformation of the integration variables does not change the value of the functional integral, the variation of W with respect to θ can be taken to be zero and our equivalent of the Ward-Takahashi-Slavnov identity is

$$\left\{\frac{1}{\beta}\left[D_\nu\left(\frac{\delta}{i\delta \underline{J}_\alpha^1}\right)\partial_\nu\partial_\mu \frac{\delta}{i\delta \underline{J}_\mu^2} + g\frac{\delta}{i\delta \underline{J}_\nu^2} \times \partial_\nu\partial_\mu \frac{\delta}{i\delta \underline{J}_\mu^1}\right] + D_\mu\left(\frac{\delta}{i\delta \underline{J}_\alpha^1}\right)J_\mu^1 - g\underline{J}_\mu^2 \times \frac{\delta}{i\delta \underline{J}_\mu^2} + g\,\underline{T}\eta\frac{\delta}{\delta\eta} - g\bar{\eta}\underline{T}\frac{\delta}{\delta\bar{\eta}}\right\} W = 0, \quad (81)$$

with

$$D_\mu\left(\frac{\delta}{i\delta \underline{J}_\alpha^1}\right) = \partial_\mu + g\frac{\delta}{i\delta \underline{J}_\mu^1} \times. \quad (82)$$

In order to investigate the structure of the gluon propagators we shall obtain the proper vertex identity equivalent to Eq. (81). We focus on the novelties of the gluon sector and neglect the quark field terms in $\bar{\mathcal{L}}$ and Eq. (81). Let us define

$$W(J) = e^{iZ(J)}, \quad (83)$$

$$\underline{B}_\mu^i = -\frac{\delta Z(J)}{\delta \underline{J}_\mu^i}, \quad i = 1, 2 \quad (84)$$

$$\Gamma(B) = Z(J) + \int d^4x(\underline{J}_\mu^1 \cdot \underline{B}_\mu^1 + \underline{J}_\mu^2 \cdot \underline{B}_\mu^2), \quad (85)$$

where $\Gamma(B)$ is the generating functional of proper vertices. An immediate consequence is

$$\underline{J}^i_\mu = \frac{\delta\Gamma}{\delta \underline{B}^i_\mu}, \quad i=1,2 \tag{86}$$

and as a result Eq. (81) can be rewritten in the form

$$\frac{1}{\beta}\left[\Box\partial_\mu \underline{B}^2_\mu - g\underline{B}^1_\nu\times\partial_\nu\partial_\mu\underline{B}^2_\mu - g\underline{B}^2_\nu\times\partial_\mu\partial_\nu\underline{B}^1_\mu + g\frac{\delta}{i\delta\underline{J}^1_\nu}\times\partial_\nu\partial_\mu\underline{B}^2_\mu + g\frac{\delta}{i\delta\underline{J}^2_\nu}\times\partial_\nu\partial_\mu\underline{B}^1_\mu\right] - \partial_\mu\frac{\delta\Gamma}{\delta\underline{B}^1_\mu} + \underline{B}^1_\mu\times\frac{\delta\Gamma}{\delta\underline{B}^1_\mu} + \underline{B}^2_\mu\times\frac{\delta\Gamma}{\delta\underline{B}^2_\mu} = 0. \tag{87}$$

If we apply $\delta/\delta\underline{B}^1_\alpha$ to Eq. (87) and set $\underline{B}^i_\mu = 0$ afterwards, we find

$$-\partial_\mu \frac{\delta^2\Gamma}{\delta\underline{B}^1_\alpha\delta\underline{B}^1_\mu}\bigg|_{B^1=B^2=0} = 0. \tag{88}$$

The second-order functional derivative of Γ is the inverse of the full propagator $G^{11\prime}_{\mu\nu ab}$ and Eq. (88) implies that the proper part of $(G^{11\prime}_{\mu\nu ab})^{-1}$ is purely transverse. We note that the "free" propagator (Eq. 39) contribution to $(G^{11\prime}_{\mu\nu ab})^{-1}$ is not one-particle irreducible and thus not constrained by Eq. (88). Therefore we find the general form

$$G^{11\prime}_{\mu\nu ab}(k) = \delta_{ab}\left(g_{\mu\nu} - \frac{k_\mu k_\nu}{k^2}\right)G^{11}(k^2) + \delta_{ab}\beta^2\lambda^2\frac{k_\mu k_\nu}{k^6}, \tag{89}$$

so that the longitudinal part of the full propagator is not renormalized.

The longitudinal part of the full propagator $G^{12\prime}_{\mu\nu ab}(k)$ is also not renormalized. This may be seen by applying $\delta/\delta\underline{B}^2_\mu$ to Eq. (87) and setting $\underline{B}^i_\mu = 0$ afterwards:

$$\frac{1}{\beta}\Box\partial_\mu\delta^4(x-y) - \partial_\nu\frac{\delta^2\Gamma}{\delta\underline{B}^2_\mu\delta\underline{B}^1_\nu}\bigg|_{B^1=B^2=0} = 0. \tag{90}$$

This implies

$$(G^{12\prime}_{\mu\nu ab})^{-1} = \frac{\delta_{ab}(g_{\mu\nu} - k_\mu k_\nu/k^2)}{G^{12}} - \frac{k_\mu k_\nu\delta_{ab}}{\beta} \tag{91}$$

or

$$G^{12\prime}_{\mu\nu ab}(k) = \delta_{ab}\left(g_{\mu\nu} - \frac{k_\mu k_\nu}{k^2}\right)G^{12}(k) - \beta\delta_{ab}\frac{k_\mu k_\nu}{k^4}. \tag{92}$$

Having now developed the general form of the propagators we now will define the gluon vacuum polarization tensors,

$$\Pi^{11}_{\mu\nu ab}(k) = [G^{11\prime}_{\mu\nu ab}(k)]^{-1} - [G^{11}_{\mu\nu ab}(k)]^{-1}, \tag{93}$$

$$\Pi^{12}_{\mu\nu ab}(k) = [G^{12\prime}_{\mu\nu ab}(k)]^{-1} - [G^{12}_{\mu\nu ab}(k)]^{-1}, \tag{94}$$

which are transverse by our previous discussion:

$$k_\mu\Pi^{11}_{\mu\nu ab} = k_\mu\Pi^{12}_{\mu\nu ab} = 0. \tag{95}$$

Rather than write the Schwinger-Dyson equations for our polarization tensors we have given a diagrammatic representation in Fig. 4.

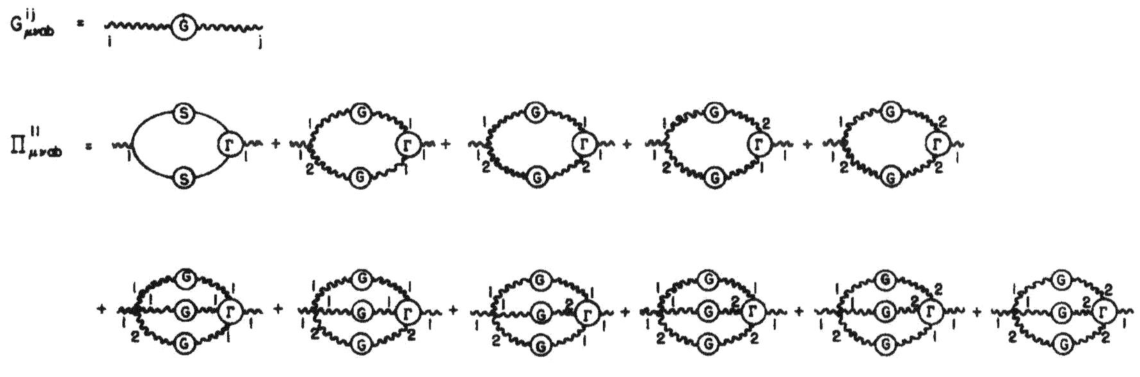

FIG. 4. Diagrammatic representation of the Schwinger-Dyson equation for the proper gluon self-energy, $\Pi^{11}_{\mu\nu ab}$. The numbers at the end of a gluon line specify whether $\underline{A}^1_\mu$ or $\underline{A}^2_\mu$ correspond to that end. The quark propagator is denoted S while Γ denotes the appropriate proper (one-particle irreducible) vertex function. A similar diagrammatic expression can be written for $\Pi^{12}_{\mu\nu ab}$.

IV. OBSERVATIONS

The Schwinger mechanism forces quark confinement to bound color singlet states in a manner which is identical to the Abelian case as described in Sec. II. In order to demonstrate that only color singlets exist in the gauge-invariant physical particle spectrum it is sufficient to show

$$\underline{Q}\psi_{\text{phys}}=0\,, \tag{96}$$

where

$$\underline{Q}=\int d^3x\,\underline{J}_0(x) \tag{97}$$

for any physical state ψ_{phys} corresponding to a spatially localized distribution of quarks. We consider a single static quark located at the origin and choose to work in the Coulomb gauge ($\vec{\nabla}\cdot\vec{A}^1=0$). Then the time components of the equations of motion [Eqs. (27) and (28)] lead to (at large distance)

$$\Box\nabla^2\underline{A}_0^1=g\lambda^2\underline{J}_0 \tag{98}$$

if we take into account the elimination of gluons' degrees of freedom through the choice of principal-value propagators and their consequent inability to act as sources. We may now repeat the arguments of Eqs. (13)–(16) for the Abelian case after noticing the occurrence of the Schwinger mechanism in the non-Abelian case which can be verified in low orders of perturbation theory for $\Pi^{11}_{\mu\nu ab}$. Thus the expectation value of the charge in the one-quark state is zero. Since the one-quark state is a charge eigenstate, we find Eq. (96) to be true in this case and more generally through the additivity of the charge operator. Thus only color singlet bound states of quarks are physical.[12]

While the infrared behavior of the theory leads to quark confinement, the ultraviolet behavior allows the quarks to appear quasifree. This is particularly noticeable when we take $\lambda^2=0$ in our Lagrangian and examine the corresponding perturbation theory. Taking $\lambda^2=0$ is equivalent to examining the short-distance behavior of the theory. The only diagrams which exist in this limit are given in Fig. 5. The quark sector of the theory is free. The only nontree structures are one-quark-loop diagrams for the scattering of gluons associated with A_μ^2 (which of course can only be generated by a hypothetical external source). (As a point of comparison we have shown in Fig. 6 the additional diagrams which would occur in the even that Feynman propagators were used—these diagrams necessarily involve gluon loops which principal-value propagators force to be zero.) The vital role of the $\lambda^2 A_\mu^2 A_\mu^2$ term in the Lagrangian in generating the interacting theory and the fact that λ^2 has the dimensions of $(\text{mass})^2$ allow a natural approximation procedure in this model. This is perhaps best seen within the context of deep-inelastic electroproduction. Just as in the Abelian case we find that the structure functions scale with leading corrections of $O(q^{-4})$, where q equals the virtual photon four-momentum. We can establish a parton picture of scattering wherein the photon is absorbed on one of the quasifree nucleon constituents [as in Fig. 2(a)] if $|q^2|\gg g^2\lambda^2$. Then leading corrections to such a picture [e.g., the diagrams of Fig. 2(b)–2(d)] will be suppressed by $(g^2\lambda^2/q^2)^2$. Thus the dimensional nature of the effective coupling constant allows a particularly simple picture to exist of the region of large spacelike virtual photon mass and the parton picture emerges as a natural approximation.

The k^{-4} form of the quark interaction also appears to have decidedly good features as far as the bound-state structure is concerned. Ignoring the numerator tensor (which does not affect our conclusions), we find the Fourier transform of the gluon propagator,

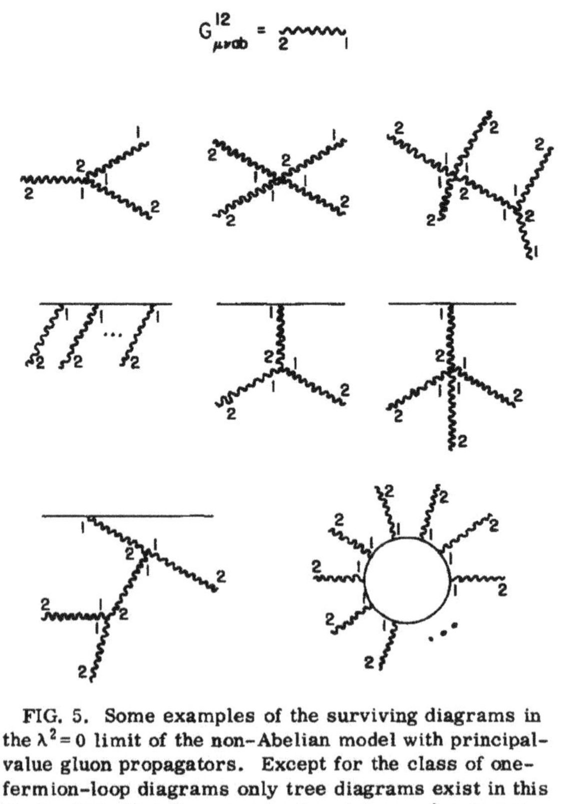

FIG. 5. Some examples of the surviving diagrams in the $\lambda^2=0$ limit of the non-Abelian model with principal-value gluon propagators. Except for the class of one-fermion-loop diagrams only tree diagrams exist in this limit. Note that there are no four (or more) external quark line diagrams and no two (or more) external $\underline{A}_\mu^1$ gluon "external" lines.

$$G(k) = \mathrm{P}\frac{1}{k^4}, \tag{99}$$

to be proportional to

$$\tilde{G}(x) = \theta(x^2). \tag{100}$$

Since $\tilde{G}$ has a smooth finite limit as $x^2 \to 0$, the short-distance limit, arguments can be made[13] that low-mass bound states can occur in this model. In addition, Dalitz[14] has pointed out that the linearity of trajectories on the Chew-Frautschi plot would follow from a flat-bottomed, smooth interaction—a criterion which is met by Eq. (100). [It is interesting to note that had we used a Feynman propagator rather than principal value, then $\tilde{G}$ would have been $\ln(x^2)$ and thus the general criterion just stated would not have been met. This would appear to be another point in favor of our choice of principal-value propagators.]

Another property which is desirable in the bound-state solutions is nonrelativistic motion of the bound-state constituents.[15] Again an interaction of the form of Eq. (99) appears to realize this feature—even in the strong-binding limit. To see this we shall first take account of the Schwinger mechanism and in the spirit of Hartree-Fock theory modify the quark interaction to

$$G'(k) = \mathrm{P}\frac{1}{(k^2-\mu^2)^2}. \tag{101}$$

If we now take Eq. (101) to be the Green's function for the effective gluon field and calculate the "Coulomb potential" of a static, point quark source located at the origin we find

$$\phi(r) = \frac{\phi_0}{\mu} e^{-\mu r}, \tag{102}$$

where ϕ_0 is a constant independent of μ. In the limit $\mu \to 0$ we find

$$\phi(r) \cong \phi_0\left(\frac{1}{\mu} - r + \cdots\right). \tag{103}$$

The first two terms of Eq. (103) correspond to choosing Eq. (99) rather than Eq. (101) as the gluon Green's function (in the limit $\mu \to 0$). Equation (102) includes vacuum polarization effects which damp the interaction at large distances. Thus Eq. (102) imperfectly reflects the possibility that a quark-antiquark pair can separate and induce another quark-antiquark pair to be created from the vacuum so that two color singlet mesons will result (presuming it is energetically favored). At shorter distances Eq. (102) appears to be a reasonable approximation. This exponential potential was studied within the framework of the Schrödinger equation in the strong-binding limit (ϕ_0/μ large) by Greenberg.[16] He showed that the average momentum of the bound constituent in the s state satisfied

$$\frac{p}{m} \sim \left(\frac{\mu}{m}\right)^{1/3} \tag{104}$$

with m being the quark mass. Thus for μ/m small the quark motion is self-consistently nonrelativistic.

In conclusion, we have shown that a four-dimensional, Lorentz-invariant second-quantized field theory of hadron binding is possible with scaling electroproduction structure functions, only zero-triality physical particle states, and, apparently, linearly rising Regge trajectories and nonrelativistic constituents. A detailed study of the bound states is now in progress.

ACKNOWLEDGMENT

I am grateful to the members of the Newman Laboratory for interesting conversations.

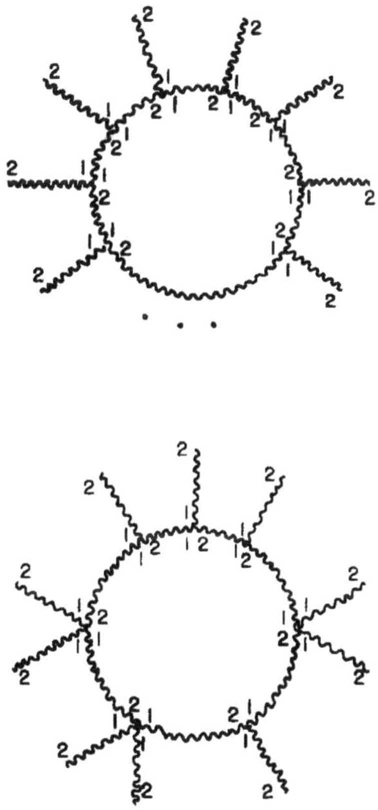

FIG. 6. Some additional diagrams which occur in the $\lambda^2 = 0$ limit of the non-Abelian model if Feynman gluon propagators are used. In addition, there will be Faddeev-Popov ghost-loop diagrams depending on the choice of gauge.

APPENDIX

In Ref. 3 semiclassical arguments based on Dirac's theory of constraints were given to introduce the use of principal-value propagators. We will now describe a second-quantized realization of those arguments for the case of a scalar Klein-Gordon field $\phi(x)$ with the Lagrangian

$$\mathcal{L} = \tfrac{1}{2}(\partial_\mu \varphi)^2 - \tfrac{1}{2}m^2\varphi^2. \tag{A1}$$

The generalization to vector gluons is immediate. The canonical equal-time commutation relations are

$$[\phi, \varphi] = [\dot{\phi}, \dot{\phi}] = 0\,, \tag{A2}$$

$$[\dot{\phi}(\vec{x}, t), \varphi(\vec{y}, t)] = -i\delta^3(\vec{x} - \vec{y}). \tag{A3}$$

If we expand $\phi(x)$ in plane waves,

$$\varphi(\vec{x}, t) = \sum_{\vec{k}} (A_{\vec{k}}^{-} e^{-ik\cdot x} + A_{\vec{k}}^{+} e^{ik\cdot x})\,, \tag{A4}$$

then the q-number Fourier components $A_{\vec{k}}^{\mp}$ must satisfy

$$[A_{\vec{k}}^{-}, A_{\vec{k}'}^{-}] = [A_{\vec{k}}^{+}, A_{\vec{k}}^{+}] = 0\,, \tag{A5}$$

$$[A_{\vec{k}}^{-}, A_{\vec{k}'}^{+}] = \delta^3(\vec{k}' - \vec{k}) \tag{A6}$$

for consistency with Eqs. (A2) and (A3). Now the time-ordered product satisfies

$$T(\varphi(x)\phi(y)) = \epsilon(x_0 - y_0)[\varphi(x), \phi(y)] + \{\varphi(x), \varphi(y)\}\,, \tag{A7}$$

with $\epsilon(x_0) = \pm 1$ for $x_0 \gtrless 0$ and $\{A, B\} = AB + BA$. The first term on the right-hand side is a c number completely determined by Eqs. (A5) and (A6). If the second q-number expression were zero, then we would obtain a principal-value propagator from Eq. (A7):

$$T(\varphi(x)\phi(y)) \equiv i \int \frac{d^4k}{(2\pi)^4} e^{-ik\cdot(x-y)}\, \mathrm{P}\, \frac{1}{(k^2 - m^2)}\,. \tag{A8}$$

We therefore require

$$\{\phi(x), \varphi(y)\} = 0\,, \tag{A9}$$

with the consequence

$$\{A_{\vec{k}}^{-}, A_{\vec{k}'}^{-}\} = \{A_{\vec{k}}^{+}, A_{\vec{k}'}^{+}\} = \{A_{\vec{k}}^{-}, A_{\vec{k}'}^{+}\} = 0\,. \tag{A10}$$

Equations (A5), (A6), and (A10) imply

$$A_{\vec{k}}^{-} A_{\vec{k}'}^{-} = A_{\vec{k}}^{+} A_{\vec{k}'}^{+} = 0\,, \tag{A11}$$

$$A_{\vec{k}}^{-} A_{\vec{k}'}^{+} = \tfrac{1}{2}\delta^3(\vec{k} - \vec{k}')\,, \tag{A12}$$

$$A_{\vec{k}}^{+} A_{\vec{k}'}^{-} = -\tfrac{1}{2}\delta^3(\vec{k} - \vec{k}') \tag{A13}$$

for all $\vec{k}$ and $\vec{k}'$. Thus quadratic terms in A and $A^\dagger$ are reduced to c numbers. It should further be noted that the multiplication rule is not associative.[17] In fact, the multiplication rules of the A and $A^\dagger$ operators in Eqs. (A11)–(A13) are realized by taking multiplication to be

$$UV = \tfrac{1}{2}[U, V] \tag{A14}$$

for U, V being any $A_{\vec{k}}^{-}$ or $A_{\vec{k}'}^{+}$. If we take an analogy to Lie-algebra theory seriously, where the adjoint representation of an algebra has a multiplication rule defined by commutators

$$\bar{U} * \bar{V} = [\bar{U}, \bar{V}] \tag{A15}$$

then we could call Eqs. (A11)–(A13) the adjoint representation of the Fourier components of φ.

The c-number nature of AA, $A^\dagger A^\dagger$, or $AA^\dagger$ can be understood physically in the following manner. Since the ϕ field has principal-value propagators it is not associated with a particle but is merely the embodiment of an interaction between other objects (which we have suppressed in our Lagrangian). Consequently an emission of a φ field quantum must be directly correlated with a subsequent absorption—it cannot propagate into empty space. The c-number nature of $AA^\dagger$ reflects this correlation between emission and absorption.

Finally, it should be noted that the existence of a vacuum is inconsistent with Eqs. (A11)—(A13).

---

*Work supported in part by the National Science Foundation.

[1] K. Johnson, Phys. Rev. D 6, 1101 (1972); C. M. Bender, J. E. Mandula, and G. S. Guralnik, Phys. Rev. Lett. 32, 1467 (1974); A. Chodos *et al.*, Phys. Rev. D 9, 3471 (1974); W. A. Bardeen *et al.*, *ibid.* 11, 1094 (1975); M. Creutz, *ibid.* 10, 1749 (1974); P. Vinciarelli, Nuovo Cimento Lett. 4, 905 (1972); R. Dashen, B. Hasslacher, and A. Neveu, Phys. Rev. D 10, 4114 (1974); 10, 4130 (1974); 10, 4138 (1974).

[2] Y. Nambu, in *Preludes in Theoretical Physics*, edited by A. de-Shalit, H. Feshbach, and L. Van Hove (North-Holland, Amsterdam, 1966), p. 133; H. J. Lipkin, Phys. Lett. 45B, 267 (1973).

[3] S. Blaha, Phys. Rev. D 10, 4268 (1974).

[4] J. Schwinger, Phys. Rev. 128, 2425 (1962).

[5] A. Pais and G. Uhlenbeck, Phys. Rev. 79, 145 (1950); J. Kiskis, Phys. Rev. D 11, 2178 (1975).

[6] A. Casher, J. Kogut, and L. Susskind, Phys. Rev. D 10, 732 (1974); J. Lowenstein and J. Swieca, Ann. Phys. (N.Y.) 68, 172 (1971).

[7] R. Jackiw and G. Preparata, Phys. Rev. Lett. 22, 975 (1969); S. Adler and W. Tung, *ibid.* 22, 978 (1969); S. Blaha, Phys. Rev. D 3, 510 (1971).

[8]The Lagrangian of Eq. (17) was first written by D. Sinclair as a generalization of the Abelian model of Ref. 3. An alternative non-Abelian model for quark confinement has been suggested by S. K. Kauffmann [Nucl. Phys. B87, 133 (1975)]. I am grateful to Dr. Kauffmann for sending me a copy of his paper prior to publication.

[9]E. Abers and B. W. Lee [Phys. Rep. 9C, 1 (1973)] provide a useful review of conventional Yang-Mills theories.

[10] R. P. Feynman, Acta Phys. Pol. 24, 697 (1963).

[11]B. W. Lee and J. Zinn-Justin [Phys. Rev. D 5, 3121 (1972)] point out that the $i\epsilon$ prescription in their Eq. (2.8) for the ghost loop is dictated by unitarity considerations.

[12]This does not preclude color-singlet states of the gluons from playing a role in the theory. They are not particles but can be exchanged between color-singlet quark states in scattering events. On naive dimensional grounds they should be most important in forward scattering. This leads to the possibility that the Pomeron might possibly be interpreted as a "two-gluon bound state". In the case of wide-angle scattering the predominant mechanism for large momentum transfer would appear to be constituent interchange due to the strong damping effects of k^{-4} propagators on momentum transfer.

[13]M. Böhm, H. Joos, and M. Krammer, in *Recent Developments in Mathematical Physics*, proceedings of the XII Schladming Conference (Acta Phys. Austriaca Suppl. XI), edited by P. Urban (Springer, New York, 1970), p. 3.

[14]R. H. Dalitz, a paper presented at the Topical Conference on Meson Spectroscopy, Philadelphia, 1968 (unpublished).

[15]H. J. Lipkin, Phys. Rep. 8C, 175 (1973).

[16]O. W. Greenberg, Phys. Rev. 147, 1077 (1966).

[17]Nonassociative field operators have been previously used by M. Günaydin and F. Gürsey, Phys. Rev. D 9, 3387 (1974).

Appendix C. Two-Tier and PseudoQuantum Formalisms

We are familiar with Quantum Theory, as it is usually developed, in our universe. In universes with higher space-time dimensions there is a need for a more expansive Quantum Field Theory. We presented this theory in earlier books on Two-Tier Quantum Theory[28] and PseudoQuantum Field Theory.[29] It resolved all divergences in ElectroWeak and Strong interaction theories. It also eliminated divergences in other types of Quantum Field Theories including theories with higher order derivatives and four fermion interactions.

In this book we focus on the higher space-time dimension spaces (universes) of Octonion Cosmology. In these universes we find Two-Tier Quantum Theory is *absolutely* necessary to avoid divergences in perturbation theory calculations. The universes of Octonion Cosmology have Feynman propagators and Perturbation Theory terms with integrations of the form:

$$\int d^n k \; f(k) \qquad \text{(C.1)}$$

where n = 4, 8, 10, 12, 14, 16, and 18 (and also a less interesting case n = 2.)

In eight space-time dimensions the second order single fermion loop vacuum polarization is sextic divergent: $\int d^8k/k^2 \sim k^6$. Two-Tier Quantum Theory eliminates divergences in Perturbation Theory with exponentiated Gaussian quadratic terms in momenta of the form:

$$\exp(-ak^2) \qquad \text{(C.2)}$$

where a is a constant. Two-Tier Quantum Theory is required for Perturbation Theory calculations in higher dimensions.

PseudoQuantum Field Theory is also needed in Octonion Cosmology. It is needed for proper quantization in arbitrary coordinate systems that might be relevant in higher dimension Octonion Cosmology spaces. For example, consider coordinate systems for non-static space-times where no time-like Killing vector exists.[30] It also enables canonical higher order derivative theories for quark confinement and other purposes. And it dovetails with Two-Tier Coordinates to "dress" bare fermion and boson particles.

---

[28] See Blaha (2005a) *Quantum Theory of the Third Kind.*

[29] Blaha (2018e) and earlier books.

[30] B. DeWitt, Phys. Rep. **19**, 295 (1975) and references therein. S. Blaha, "New Framework for Gauge Field Theories", IL Nuovo Cimento **49A**, 113 (1979).

C.1 Reasons For Two-Tier Quantum Theory

Originally Two-Tier coordinates were developed by this author to remove infinities that appear in perturbation theory calculations. We showed that the quantum smeared coordinates of Two-Tier Quantum Field Theory succeeded in removing all ultra-violet infinities in perturbation theory including the fermion triangle infinities.

Remarkably the high precision, "low" energy[31] *predictions of QED remained true in Two-Tier QED and thus remained consistent with experiment to a hitherto unsurpassed level of accuracy.* "Low" energy predictions in other quantum field theories also remained unchanged. At high energies, Two-Tier perturbation theory results are finite and consequently all ultra-violet infinities, to any order in perturbation theory, in *any number of space-time dimensions* were eliminated.

In addition to removing perturbation theory infinities, Two-Tier coordinates enable us to define finite theories of Quantum Gravity and 'non-renormalizable' quantum field theories based on polynomial Lagrangians, to tame vacuum fluctuations, to eliminate infinities associated with the Big Bang, and possibly to generate the explosive growth of the universe in its role as a type of Dark Energy.[32]

C.2 Two-Tier Features in 4-Dimensional Space-Time

Two-Tier Quantum Field Theory[33] is based on a new method[34] in the Calculus of Variations that uses two 'layers' of fields to introduce quantum coordinates. We shall consider this technique for the specific case of a massless vector field $Y^\mu(y)$ where the index μ ranges from 0 through 3. It is.analogous to the electromagnetic field.

The X^μ coordinate system, where it appears, has a c-number real part and a q-number imaginary part. Thus particle fields which are normally defined on real four-dimensional real space-time will now be defined on a "slightly" complex four-dimensional space-time:

$$X^\mu(y) = y^\mu + i\, Y^\mu(y)/M_c^2 \qquad \text{(C.3)}$$

where M_c is an extremely large mass of the order of the Planck mass or larger.

The $Y^\mu(y)$ field is a function of the space y coordinates. The real part of the space-time dimensions will be taken to be the space of real-valued y coordinates.[35]

The imaginary part of space-time coordinates is the massless $Y^\mu(y)$ vector quantum field that is suppressed by the very large mass scale. The effects of Quantum Dimensions only become appreciable in quantum field theory at energies of the order of M_c. At these energies exponential Gaussian factors in each particle (and ghost) propagator are generated by the Quantum Dimensions and serve to make *all*

[31] Relative to a mass scale that was perhaps of the order of the Planck mass.

[32] See Blaha (2017b) and earlier books for details. This section is basically a summary of some features.

[33] See Blaha (2005a), and Blaha (2002), for discussions of this new method to eliminate infinities in quantum field theory calculations.

[34] See Blaha (2005a)..

[35] In a deeper theory the real part might also be a quantum field that undergoes a condensation to generate c-number coordinates. We will not consider this possibility in this book.

perturbation theory calculations ultra-violet finite – including calculations in Quantum Gravity. Later we will see that the Two-Tier formalism may be extended directly to the universes of Octonion Cosmology with similar results – finiteness in Perturbation Theory.

The Two-Tier formalism introduces a new form of interaction that does not have the form of the simple polynomial interactions that have hitherto dominated quantum field theories. This form of interaction takes place via the composition of quantum fields and can be called a *Dimensional Interaction* or an *Interdimensional Interaction* since it affects particle behavior through Quantum Dimensions.

The basic *ansatz* of the Two-Tier formalism is to replace every appearance of a coordinate x in a quantum field with the variable

$$x^\mu \rightarrow X^\mu = (y^0, \mathbf{y} + \mathbf{Y}(y^0, \mathbf{y})/M_c^2) \tag{C.4}$$

where $\mathbf{Y}(y^0, \mathbf{y})$ is the spatial part of a free massless vector field with features that are identical to the free QED field in the Radiation gauge.

Then one finds that the momentum space free field Feynman propagators G(k) of all particles acquires a Gaussian factor exp(h(k)):

$$G(k) \rightarrow G(k)\exp(h(k)) \tag{C.5}$$

so that all perturbation theory diagrams are finite. The result is finite perturbative results for all calculations to any order in perturbation theory. Blaha (2005a) shows that Two-Tier theories are finite, Poincare covariant, and unitary. (See Blaha (2005a) for a complete discussion.)

C.3 Two-Tier Quantum Coordinates Formalism

In this section we will introduce the basic Two-Tier formalism. Taking the Lagrangian described in Blaha (2005a):

$$\mathscr{L}(y) = \mathscr{L}_F(X^\mu(y))J + \mathscr{L}_C(X^\mu(y), \partial X^\mu(y)/\partial y^\nu, y) \tag{C.6}$$

where

$$X^\mu(y) = y^\mu + i\, Y^\mu(y)/M_c^2 \tag{C.7}$$

with M_c being a large mass scale, $Y_\mu(y)$ a vector quantum field, and where J is the absolute value of the Jacobian of the transformation from X to y coordinates:

$$J = |\partial(X)/\partial(y)|$$

The Lagrangian term $\mathscr{L}_C$ is

$$\mathscr{L}_C = +\tfrac{1}{4} M_c^4 F^{\mu\nu}F_{\mu\nu}$$

with

$$\begin{aligned} F_{\mu\nu} &= \partial X_\mu/\partial y^\nu - \partial X_\nu/\partial y^\mu \\ &\equiv i\,(\partial Y_\mu/\partial y^\nu - \partial Y_\nu/\partial y^\mu)/M_c^2 \end{aligned} \tag{C.8}$$

The Lagrangian term $\mathscr{L}_F$ (X^μ(y)) contains the terms for scalar, fermion and other gauge terms in general. The sign in $\mathscr{L}_C$ is not negative – contrary to the conventional electromagnetic Lagrangian. The reason for this difference is that the quantum field part of X^μ is imaginary. Thus $\mathscr{L}_C$ ends up having the correct sign after taking account of the factor of i in the field strength $F_{\mu\nu}$.

Defining

$$F_{Y_{\mu\nu}} = (\partial Y_\mu/\partial y^\nu - \partial Y_\nu/\partial y^\mu)$$

we see the Lagrangian assumes the form of the conventional electromagnetic Lagrangian:

$$\mathscr{L}_C = -\tfrac{1}{4}\, F_Y{}^{\mu\nu} F_{Y_{\mu\nu}}$$

The action of this theory has the form

$$I = \int d^4y\, \mathscr{L}(y)$$

Since X^μ(y) has an imaginary part there would appear to be an issue with the conservation of probability (unitarity). *We show unitarity is not a problem later in this appendix in section C.8.*

C.4 Y^μ Gauge

The gauge invariance of the Lagrangian allows us to choose a convenient gauge. The gauge invariance of the full Lagrangian:

$$\mathscr{L}_S = L_F(\phi(X), \partial\phi/\partial X^\mu)\, J + \mathscr{L}_C(X^\mu(y), \partial X^\mu(y)/\partial y^\nu)$$

is based on the standard gauge invariance of $\mathscr{L}_C$, and the gauge invariance of $J\mathscr{L}_F$ in the form of translational invariance

$$X^\mu(y) \rightarrow X^\mu(y) + \delta X^\mu(y)$$

for the special case of a translation of X with the form of a gauge transformation:

$$\delta X^\mu(y) = \partial\Lambda(y)/\partial y_\mu$$

In this case we find

$$\int d^4y\, \Lambda(y)\, \partial\, [\, J\, \partial/\partial X^\mu\, \mathscr{T}_{F_{\mu\nu}}\,]/\partial y_\nu \;= 0 \qquad \text{(C.9)}$$

after a partial integration. Thus we have the differential conservation law:

$$\partial\, [\, J\, \partial \mathscr{T}_{F_{\mu\nu}}/\partial X^\mu]/\partial y_\nu \;= 0$$

since $\Lambda(y)$ is arbitrary. This conservation law is trivially obeyed:

$$\partial \mathscr{T}_{F\mu\nu}/\partial X^\mu = 0 \tag{C.10}$$

Thus translational invariance in the $\mathscr{L}_F$ sector together with standard gauge invariance in the $\mathscr{L}_C$ sector automatically guarantees Y field gauge invariance of the total Lagrangian. We use the separate invariance of each term of

$$L = \int d^4y\, [\mathscr{L}_F J + \mathscr{L}_C] = \int d^4X\, \mathscr{L}_F + \int d^4y\, \mathscr{L}_C = L_F + L_C$$

under a constant translation $X^\mu \to X^\mu + \delta X^\mu$ where δX^μ is constant. Then we consider a position dependent translation/gauge transformation, which taken together with the above equation, establishes the invariance under the position dependent translation/gauge transformation.

An alternate approach that leads to the same result is to start with the particle part of the Lagrangian $\mathscr{L}_F$ rewritten to be invariant under general coordinate transformations, as it must, when we generalize to include General Relativity. Since position dependent translations are a form of general coordinate transformation the full theory must be invariant under position dependent translations due to invariance under general coordinate transformations.

Having established invariance under gauge transformations we now choose to use the most convenient gauge – the radiation gauge[36]:

$$\partial Y^i/\partial y^i = 0 \tag{C.11}$$

where i = 1, 2, 3. In the absence of external sources, we set

$$Y^0 = 0$$

since Y^0 does not have a canonically conjugate momentum. A conventional treatment leads to the equal time commutation relations:

$$\begin{gathered}[Y^\mu(\mathbf{y}, y^0), Y^\nu(\mathbf{y}', y^0)] = [\pi^\mu(\mathbf{y}, y^0), \pi^\nu(\mathbf{y}', y^0)] = 0 \\ [\pi_j(\mathbf{y}, y^0), Y_k(\mathbf{y}', y^0)] = -i\,\delta^{tr}{}_{jk}(\mathbf{y} - \mathbf{y}')\end{gathered} \tag{C.12}$$

where (Note the locations of the j indexes above introduce a minus sign.)

$$\pi^k = \partial \mathscr{L}_C/\partial Y_k'$$

---

[36] It is also possible to quantize using an indefinite metric that preserves manifest Lorentz covariance as was done by Gupta and Bleuler for the electromagnetic field. We will use the Gupta-Bleuler approach later to establish covariance under special relativity later. Now we opt for manifest positivity and use the radiation gauge.

$$\pi^0 = 0$$

$$\delta^{tr}{}_{jk}(\mathbf{y} - \mathbf{y}') = \int d^3k\, e^{i\,\mathbf{k}\cdot(\mathbf{y}-\mathbf{y}')}(\delta_{jk} - k_j k_k/\mathbf{k}^2)/(2\pi)^3 \qquad \text{(C.13)}$$

$$Y_k' = \partial Y_k/\partial y^0$$

The Radiation gauge reveals the two degrees of freedom that are present in the vector potential. The Fourier expansion of the vector potential is:

$$Y^i(y) = \int d^3k\, N_0(k) \sum_{\lambda=1}^{2} \varepsilon^i(k, \lambda)[a(k,\lambda)\, e^{-ik\cdot y} + a^\dagger(k,\lambda)\, e^{ik\cdot y}] \qquad \text{(C.14)}$$

where

$$N_0(k) = [(2\pi)^3 2\omega_k]^{-½}$$

and (since m = 0)

$$\omega_k = (\mathbf{k}^2)^{½} = k^0$$

with $\vec{\epsilon}(k, \lambda)$ being the polarization unit vectors for $\lambda = 1,2$ and $k^\mu k_\mu = 0$.

The further development of Two-Tier theory is described in Part 3 of Blaha (2005a).

C.5 Two-Tier Uncertainty Principle

The Uncertainty Relation for Quantum Mechanics is based on the coordinate-momentum commutator. Similarly, in defining Quantum Coordinates we have established a commutation relation based on Y^μ. Its conjugate momentum is

$$P^\mu(y) = i\pi_Y{}^\mu(y)/M_c{}^2 \qquad \text{(C.15)}$$

where

$$\pi_Y{}^\mu(y) = -\, dY^\mu/(y)dy^0$$

In the Radiation gauge (eq. C.11) we see

$$\pi_Y{}^0(y) = 0 \qquad \text{(C.16)}$$

and

$$[X^0, P^0] = [y^0, p^0] = 0 \qquad \text{(C.17)}$$

The non-zero equal time commutation relation expressing a quantum Uncertainty Relation is

$$[P^i(\mathbf{y}, y^0)\, , Y^k(\mathbf{y}', y^0)] = -[\pi_Y{}^i(y),\, Y^k(y),]/M_c{}^4 \qquad \text{(C.18)}$$

$$= i\delta^{trik}(\mathbf{y} - \mathbf{y}')/M_c^4$$

using eq. C.12.

At low energy the impact of the Two-Tier Uncertainty Relation is diminished (more or less eliminated) by the factor M_c^4. Then Two-Tier Quantum Field Theory becomes ordinary Quantum Field Theory with the same results in Perturbation Theory. This limit is described in detail in Blaha (2005a), which appears in in Appendix B.

In the "low" energy limit the conventional Heisenberg Uncertainty Condition becomes evident as shown by Heitler (1954) and others. *Conventional Quantum Mechanics is a result of Second Quantization.*

C.6 Quantum Mechanics vs. Quantum Field Theory vs. Two-Tier Quantum Theory

Historically, Quantum Mechanics predated Quantum Field Theory, which, in turn, predates the Two-Tier Quantum Theory developed by the author in the early 2000's.

Logically, Two-Tier Quantum Theory, which embodies an Uncertainty Relation at ultrahigh energies, is the fundamental Quantum Theory. It is the predecessor of Quantum Field Theory, which only embodies the Heisenberg Uncertainty Relation. Quantum Field Theory is flawed by high energy divergences just as atomic physics, before the Bohr atom and Quantum Mechanics, was flawed by infinities in hydrogen atom models.

Quantum Field Theory (a "low" energy theory) implies Quantum Mechanics[37] as shown in Heitler[38] as well as elsewhere.

Quantum Mechanics is often treated as a complete, self-contained theory. On occasion paradoxes and ambiguities are found in Quantum Mechanics studies. They may be resolved within Quantum Mechanics. Any, that are not so resolved, should be considered within the framework of Quantum Field Theory, which is the ultimate forum for all quantum phenomena at "low" energy.

Note the trend of quantum theories, from the earliest at "low" energy theory, Quantum Mechanics, to the highest energy quantum theory: Two-Tier Quantum Theory,

Two-Tier Quantum Field Theory is a welcome extension of Quantum Theory to the deepest levels of Quantum Theory.

C.7 Two-Tier Perturbation Theory

The form of Two-Tier Perturbation Theory is similar to the Perturbation Theory of conventional Quantum Field Theory. It is described in *Quantum Theory of the Third Kind.* The reduction development for the U-matrix and the S-matrix are presented there.

---

[37] Quantum Mechanics is often treated as an independent theory. This practice may be valid mathematically but it is not valid physically. For example, many phenomena in atomic theory require explanation in terms of Quantum Field Theory.

[38] Heitler (1954).

C.8 Two-Tier Unitarity

The unitarity of Two-Tier Quantum Field Theory can be viewed in the cases of "low" energy phenomena and "high" energy phenomena with the scale mass of M_c separating the cases.

In the case of "low" energy phenomena the Two-Tier S-matrix gives results identical to conventional S-matrix results. Thus there is no conflict with unitarity for Two-Tier S-matrix results at "low" energy.

At "high" energy there are two potential issues: asymptotic states containing Y quanta; and conservation of probability. The first issue is resolved in the chapter 6 discussion in Appendix B of Blaha (2021i).

The second issue leads to a requirement to renormalize S-marix probability amplitudes such that their absolute values squared sum to unity – the unitarity condition. If we let S-matrix elements have the form S_{fi} where i represents an initial state and f represents one of its final states, then

$$\Sigma_n S_{nf}^* S_{ni} = \delta_{fi} \tag{C.19}$$

expresses the conventional unitarity condition. In Two-Tier Quantum Field Theory the unitarity condition for energies much less than M_c is the same as eq. C.19 since perturbation theory results are the same as conventional Quantum Field Theory. The sum over intermediate states n is restricted to states whose total energy is less than M_c by energy conservation.

For initial states i with energy E_i of the order of or greater than M_c, the intermediate state total energy and the final state total energy are also E_i by energy conservation. In this case the sum in eq. C.19 *may* be

$$\Sigma_n S_{nf}^* S_{ni} = g_i \delta_{fi} \tag{C.20}$$

where g_i is a constant term dependent on the state i. In this circumstance unitarity may be recovered by redefining S_{ni} and S_{nf} as

$$S_{ni}' = S_{ni}/\surd g_i \tag{C.21}$$

$$S_{nf}' = S_{nf}/\surd g_f \tag{C.22}$$

Then the *normalized* Two-Tier unitarity condition can be expressed as

$$\Sigma_n S_{nf}'^* S_{ni}' = \delta_{fi} \tag{C.23}$$

with the understanding that $g_i = 1$ for initial states with energies much less than M_c. Thus Two-Tier Quantum Field Theory satisfies an enhanced unitarity condition.

C.9 Two-Tier Quantum Space Theory

Quantum Space Theory is the theory of particles containing spaces internally. It was recently developed by the author.[39] This theory can directly put in the form of Two-Tier Quantum Space Theory by following the procedure described in section C.2. The structure of the Octonion Spaces Spectrum is not changed by this formulation.

C.10 CQ Mechanics – A Union of Classical and Quantum Mechanics

The author developed a larger theory, CQ Mechanics, containing both Quantum Mechanics and Classical Mechanics that is described in Blaha (2016f). In this theory a phenomenon can be described as classical in one limit and as quantum in another limit. The description is determined by an angle that specifies one, or the other, limiting case. Intermediate cases combining both quantum and classical are also present. The theory appears here in Appendix C.

This theory can bridge the classical and quantum regimes. We discuss some applications in Blaha (20016f): a generalized Feynman path integral formalism, a generalized Schrödinger equation, a generalized Boltzmann equation, the Fokker-Planck equation, a generalized approach to quantum and classical chaos, and to quantum entanglement as well as semi-quantum entanglement. Our formalism applies to Quantum Mechanics as well as the path integrals, the Fokker-Planck equation and the Boltzmann equation.

This "Mechanics" theory has an analogue at the field theoretic level, PseudoQuantum Field Theory that is briefly described next.

C.11 PseudoQuantum Field Theory

PseudoQuantum Field Theory[40] was developed by the author in the 1970's and presented in a series of articles including the papers below.

S. Blaha, "The Local Definition of Asymptotic Particle States", IL Nuovo Cimento **49A**, 35 (1979). It describes the PseudoQuantization of boson and fermion field theories for use in the quantization of fields in universes and the Megaverse.

S. Blaha, "New Framework for Gauge Field Theories", IL Nuovo Cimento **49A**, 113 (1979). It describes the PseudoQuantization of gauge field theories for the purposes of defining higher derivative field theories and for use in the quantization of fields in universes and the Megaverse.

PseudoQuantum Field Theory (and its Quantum Mechanics analogue CQ Mechanics[41]) originated in the need to second quantize in unusual coordinate systems, and in curved space-time coordinate systems. It is also very relevant for the canonical formulation of higher derivative field theories for quark confinement and other applications.

---

[39] Blaha (2021f) and (2021g).

[40] Appendix C describes PseudoQuantum Theory features in some detail.

[41] See Blaha (2016f) for details, which contains Blaha (2016f). CQ Mechanics encompasses both classical mechanics and quantum mechanics, and provides a method of rotating between them. It has applications to transitions between Quantum/Semi-Classical Entanglement, and Quantum/Classical Path Integrals, and Quantum/Classical Chaos.

PseudoQuantum Field Theory is formulated by duplicating all fields, both fermion and boson fields, in a "normal" Lagrangian theory. Scalar field theory provides a simple example that illustrates the PseudoQuantum Field Theory procedure.

We duplicate a scalar field generating two scalar fields: $\varphi_1(x)$ and $\varphi_2(x)$. We choose $\varphi_1(x)$ to have a zero equal time commutator with $d\varphi_1(x)/dx^0$ and $\varphi_2(x)$ to have a conventional equal time commutator with $d\varphi_2(x)/dx^0$. Conceptually $\varphi_1(x)$ is a "classical" field and $\varphi_2(x)$ is a quantum field. A Lagrangian that implements these choices of commutation relations is:

$$\mathcal{L} = \partial^\mu \varphi_1(x)\partial_\mu\varphi_2(x) - \tfrac{1}{2}\,\partial^\mu \varphi_1(x)\partial_\mu\varphi_1(x) - m_2^2\, \varphi_1(x)\varphi_2(x) + \tfrac{1}{2}\, m_1^2\, \varphi_1(x)^2 \qquad \text{(C.24)}$$

$$(\Box + m_2^2)\varphi_1(x) = 0 \qquad \text{(C.25)}$$

$$(\Box + m_2^2)\varphi_2(x) - (\Box + m_1^2)\varphi_1(x) = 0 \qquad \text{(C.26)}$$

The canonical momenta are

$$\pi_1 = d\varphi_2(x)/dt - d\varphi_1(x)/dt \qquad \text{(C.27)}$$
$$\pi_2 = d\varphi_1(x)/dt \qquad \text{(C.28)}$$

and the equal time commutation relations are

$$[\varphi_i(x), \pi_j(y)] = i\delta_{ij}\delta(\mathbf{x} - \mathbf{y}) \qquad \text{(C.29)}$$
$$[\varphi_i(x), \varphi_j(y)] = [\pi_i(x), \pi_j(y)] = 0 \qquad \text{(C.30)}$$

implying

$$[\varphi_1(x), d\varphi_1(y)/dt] = 0 \qquad \text{(C.31)}$$
$$[\varphi_2(x)\,, d\varphi_2(y)/dt] = i\delta(\mathbf{x} - \mathbf{y}) \qquad \text{(C.32)}$$
$$[\varphi_1(x)\,, d\varphi_2(y)/dt] = i\delta(\mathbf{x} - \mathbf{y}) \qquad \text{(C.33)}$$

C.11.1 Color Confinement

Appendix F contains a paper on a non-Abelian gauge field theory that gives explicit color confinement using a PseudoQuantum formulation that leads to a confining r-potential.

If we set $m_2 = 0$ in the above example, we see a fourth order field equation results

$$\Box^2\varphi_2(x) = 0 \qquad \text{(C.34)}$$

Color confinement results in the non-Abelian gauge field case in a similar manner.

C.11.2 PseudoQuantum Quantization for Non-Static Coordinate Systems

The PseudoQuantum formalism that was developed by the author for "normal" and non-static coordinate systems in the 1970s.

C.11.3 Advantages of PseudoQuantum Quantization

In this section we point out some of its advantages in a variety of field theory contexts that are relevant for Octonion Cosmology and Quantum Field Theory in general.

Some advantages of PseudoQuantum Field Theory are:

1. Quantization in any coordinate system in flat or curved space-times with an invariant definition of asymptotic particle states. *This is especially important for the higher dimension spaces of Octonion Cosmology.* An n particle asymptotic state in one coordinate system is a unitarily equivalent n particle asymptotic state in any other coordinate system. Therefore particle number is invariant under change of coordinate system. This is important for the Unified SuperStandard Theory in curved space-times. It is also important for quantization in higher dimensional spaces such as those of Octonion Cosmology. The method was developed in the late 1970's by the author to provide a quantization procedure which supports a unique particle interpretation of states in arbitrary non-static space-times where no global time-like coordinate (Killing vector) exists. PseudoQuantum Field Theory which we developed in a series of books[42] also can be formulated in the Octonion Spectrum of spaces. For example, we can use it to implement the Higgs Mechanism to generate particle masses and symmetry breaking.

2. PseudoQuantum Field Theory enables one to define Higgs particle dynamics in such a way that a non-zero vacuum expectation value cleanly separates from the quantum field part of the Higgs fields. This technique can be used in symmetry breaking mechanisms, mass generation, and possible generation of coupling constants as vacuum expectation values.

3. It supports the canonical definition of higher derivative field theories through the use of the Ostrogradski bootstrap. We have used it to construct a fourth order theory of the Strong interaction that has color confinement and a linear r potential. The potential part of this theory was used by the Cornell group to calculate the Charmonium spectrum. (See Blaha (2017b) for details.)

C.12 Combination of Two-Tier and PseudoQuantum Formalisms

These formalisms can be directly combined with no issues.

C.13 A Model Illustrating Scalar Field Quantization Using X^μ

We begin by considering the case of a scalar quantum field theory. We assume a real underlying y subspace. Since X^μ is a set of coordinates, we choose to define a scalar field ϕ as a function of X^μ, which, in turn, is a function of the y^ν coordinates. We will

[42] See Blaha (2017b) for the discussion of the PseudoQuantum field theory formalism for Higgs particles in our Extended Standard Model. See chapter 20 of Blaha (2017b), and earlier books, for a more detailed view than that presented here.

provisionally second quantize ϕ treating X^μ as c-number coordinates using a conventional approach.[43]

We assume a Lagrangian, with the momentum conjugate to ϕ:

$$\pi_\phi = \partial L_F / \partial \phi' \equiv \partial L_F / \partial(\partial\phi/\partial X^0) \tag{C.35}$$

Following the canonical quantization procedure, π and ϕ become Hermitian operators with equal time ($X^0 = X^{0\prime}$) commutation rules:

$$[\phi(X), \phi(X')] = [\pi_\phi(X), \pi_\phi(X')] = 0 \tag{C.36}$$
$$[\pi_\phi(X), \phi(X')] = -i\,\delta^3(\mathbf{X} - \mathbf{X}')$$

The standard Fourier expansion of the solution to the Klein-Gordon equation is:

$$\phi(X) = \int d^3p\, N_m(p)\, [a(p)\, e^{-ip\cdot X} + a^\dagger(p)\, e^{ip\cdot X}] \tag{C.36a}$$
$$= \int d^{3k}p\, N_m(p)\, [a(p) \exp(-ip\cdot(y + Y/M_c^2)) +$$
$$+ a^\dagger(p) \exp(ip\cdot(y + Y/M_c^2))] \qquad \text{C.36b)}$$

where

$$N_m(p) = [(2\pi)^3 2\omega_p]^{-1/2}$$

and

$$\omega_p = (\mathbf{p}^2 + m^2)^{1/2}$$

The commutation relations of the Fourier coefficient operators are:

$$[a(p), a^\dagger(p')] = \delta^3(\mathbf{p} - \mathbf{p}')$$
$$[a^\dagger(p), a^\dagger(p')] = [a(p), a(p')] = 0$$

The reader will recognize the quantization procedure is formally identical to the standard canonical quantization procedure of a free scalar quantum field.

In the case of spin ½, spin 1 and spin 2 fields the standard quantization procedure *in terms of the X coordinate system* can also be followed in a way similar to the procedure in standard texts.

C.14 Scalar Feynman Propagators

The momentum space free field Feynman propagators G…(k) of all particles and ghosts in all Two-Tier Quantum Field Theories acquires a Gaussian factor exp(h(k)):

---

[43] Some texts are: Bogoliubov, N. N., Shirkov, D. V., *Introduction to the Theory of Quantized Fields* (Wiley-Interscience Publishers Inc., New York, 1959); Bjorken, J. D., Drell, S. D., *Relativistic Quantum Fields* (McGraw-Hill, New York, 1965); Huang, K., *Quarks, Leptons & Gauge Fields Second Edition* (World Scientific, River Edge, NJ, 1992); Kaku, M., *Quantum Field Theory* (Oxford University Press, New York, 1993); Weinberg, S., *The Quantum Theory of Fields* (Cambridge University Press, New York, 1995).

$$G_{\ldots}(k) \rightarrow G_{\ldots}(k)\ \exp(h(k))$$

so that all perturbation theory diagrams are finite. The result is a finite perturbative result in all calculations to any order in perturbation theory. Blaha (2005a) shows that Two-Tier theories are finite, Poincare covariant, and unitary.

Appendix D. Book: The Origin of the Standard Model

3RD Edition Year 2007 EXCERPT

This is the third edition of *A Derivation of ElectroWeak Theory based on an Extension of Special Relativity; Black Hole Tachyons; & Tachyons of Any Spin.*

The Origin of the Standard Model:

The Genesis of Four Quark and Lepton Species, Parity Violation, the ElectroWeak Sector, Color SU(3)

Stephen Blaha, Ph. D.

Pingree-Hill Publishing

Pingree-Hill Publishing
P. O. Box 368
Auburn, NH 03032

ISBN: 0-9746958-8-2

This book is printed on acid free paper.

rev. 00/00/03

Prologue: Derivation of the Standard Model

Euclid's geometry is the example par excellence of the derivation of a theory from a small set of fundamental postulates phrased in terms of a set of semi-defined primitive terms such as straight edge and line. It might be possible to define a Lagrangian of some simplicity together with a procedure that would lead to a derivation of the Standard Model or some extension of the Standard Model. Given the current state of mathematical physics it is not clear that a "rigorous" derivation would be possible within the foreseeable future. In view of problems in the foundations of mathematics it is even not clear what constitutes a "rigorous" derivation.

However the attempt to find a deeper level of understanding of the Standard Model is undoubtedly a worthwhile goal for particle theorists and deserving of encouragement. One attempt at this deeper level is centered on Superstring theories. Another attempt is based on quantum field theories with broken symmetries.

While these approaches are to be encouraged it is also worthwhile to consider a totally different approach to an understanding of the Standard Model (and its extensions). This approach is based on taking the Standard Model as close to physical reality—none just an ad hoc theory that works. Therefore we derive the Standard Model from a small set of postulates. Rather than simply state the postulates and then derive the Standard Model we develop the features of the Standard Model starting from an extension of the Lorentz group to faster-than-light reference frames; then we generalize to a complex Lorentz group denoted L_c. L_c allows us to define four species of fermions. Then we proceed to consider a larger group U(8) of which L_c is a subgroup and show how the three known generations of fermions and their mass matrices are generated. The result is the Standard Model.

0. Tachyons Exist!

I accept the universe.
Margaret Fuller

Retort by Carlyle: "By God! She'd better."

In this prefatory chapter we establish the reality of tachyons with certitude inside Black Holes and show that many currently accepted theories contain tachyonic particles. Knowing that tachyons exist we can be confident that an acceptable formulation of tachyon quantum field theory must also exist.

0.1 Black Hole Tachyons

The conventional Schwarzschild solution of General Relativity has the metric

$$d\tau^2 = U(r)dt^2 - U(r)^{-1}dr^2 - r^2(d\theta^2 + \sin^2\theta d\varphi^2) \qquad (0.1)$$

It describes a non-rotating Black Hole if the mass M is within the Schwarzschild radius 2MG. The signs of the time and radial parts of the metric are reversed in the interior of a Black Hole where

$$U(r) = (1 - 2MG/r) < 0 \qquad (0.2)$$

.

We can establish a local inertial frame[44] at some point within a Black Hole and consider particle propagation within that reference frame. Particles moving only radially within that frame are tachyons. We see this when we consider the momentum of a particle of mass m. Its energy and radial momentum are[45]

$$p_0 = mE \qquad\qquad p^0 = mU^{-1}E \qquad (0.3)$$

and

$$p_r = -mU^{-1}dr/d\sigma \qquad\qquad p^r = m\, dr/d\sigma \qquad (0.4)$$

where σ is the proper time parameter. The mass condition for strictly radial motion is

[44] It is possible to establish a local inertial frame inside a Black Hole with a congruence since the number of positive signs in the Schwarzschild metric in the Black hole is the same as that of the Minkowski metric and the number of negative signs in the Schwarzschild metric is also the same as that of the Minkowski metric by eq. 0.1. See Turnbull (1961).

[45] See Blaha (2004) p. 71.

$$p^2 = p_0p^0 + p_rp^r = m^2E^2U^{-1} - m^2U^{-1}(dr/d\sigma)^2 = m^2 \qquad (0.5)$$

At any radial distance $r = r_a$ within the Schwarzschild radius, $U(r_a) < 0$ and thus a radially moving particle is a tachyon:

$$p^2 = m^2U^{-1}(E^2 - (dr/dp)^2) < 0 \qquad (0.6)$$

Since there are many Black Holes in the universe it is clear that tachyons exist in the universe in great quantity. Thus our development of the theory of faster-than-light motion and tachyons is justified, and not merely hypothetical.

0.2 Higgs Particles are Tachyonic

Higgs particles play an important role in Standard Model formulations. They implement spontaneous symmetry breaking and give masses to fermions and vector bosons. Higgs particles have "tachyonic" mass terms – m^2 terms with the 'wrong" sign. They do have a quartic interaction term that enables a stable vacuum state to be defined. Yet they are tachyonic and show that m^2 terms with the 'wrong" sign appear in Nature if the Higgs mechanism of the Standard Model is correct.

0.3 Tachyons Appear in Many Superstring Theories

There are many SuperString theories that contain tachyons. Typically the definition of the physical particle states of a SuperString theory exclude states containing tachyons. Nevertheless the presence of tachyon fields within a SuperString theory requires that a proper formulation of quantum tachyon fields exist.

0.4 Faster-Than-Light Phenomena on Earth

Faster-than-light phenomena exist not only in the cosmos but also on earth. Several authors[46] have pointed out the existence of macroscopic faster-than-light photonic tunneling with evanescent modes (purely imaginary wave numbers) in experiments on frustrated internal reflection with double prisms, undersized wave guides, and photonic band gap materials. Nimtz[47] points out the existence of experimentally measured group (signal) velocities of 2c and 8c where c is the speed of light.

Thus the Theory of Special Relativity would seem to need extension to speeds faster than the speed of light as we will proceed to do in this volume.

---

[46] G. Nimtz and A. A. Stahlhofen, "Macroscopic Violation of Special Relativity", arXiv:physics/07080681 and references therein.

[47] G. Nimtz, arXiv:physics/0204043.

1. Why go Beyond the Light Barrier?

1.1 Einstein's Argument against Faster-than-Light Particles

Einstein considered the possibility of particles traveling faster than the speed of light. He dismissed this possibility based on the simple argument that the energy of a particle would approach infinity if the speed of the particle approached the speed of light:

$$E = m/(1 - v^2)^{1/2}$$

where we let the speed of light c = 1. Thus

$$E \rightarrow \infty \quad \text{as} \quad v \rightarrow 1$$

However, if particles were such that $v > 1$ when they were created or experienced a quantum jump (perhaps in a particle interaction) from $v < 1$ to $v > 1$ then faster than light particles might exist.

Based on this reasoning a number of authors[48] explored the possibility of formulating a quantum field theory of a faster than light particle for which Feinberg coined the word "tachyon." These authors assumed conventional Lorentz invariance and as a result could only consider spin 0 particles.[49] They were unable to define a local canonical free tachyon quantum field theory. So the question remained open until the first edition of this book[50] where it was resolved.

1.2 Why Transformations Between Reference Frames with a Relative Speed Greater Than Light Speed?

Earlier attempts at a tachyon theory were based solely on the Lorentz group, which consists of transformations between reference frames whose relative speed is less than c.

However, if we imagine measuring the coordinates of an event in a reference frame on a sub-light particle, and also imagine measuring the coordinates of an event in a reference frame on a tachyon, then it is reasonable to expect that the coordinates of the event in each frame are mathematically related by a faster-than-light transformation.

---

[48] S. Tanaka, Prog. Theoret. Phys. (Kyoto) **24**, 171 (1960); O. M. P. Bilaniuk, V. K. Deshpanda and E. C. G.; Sudarshan, Am J. Phys. **30**, 718 (1962); G. Feinberg, Phys. Rev. **159**, 1089 (1967); M. E. Arons & E. C. G. Sudarshan, Phys. Rev. **173**, 1622 (1968).

[49] E. P. Wigner, Ann. Math. **40**, 149 (1939); Yu. M. Shirokov, Soviet Phys-JETP **6**, 664, 919, 929 (1958); ___ 7, 493 (1958); ___ Nuc. Phys. **15**, 1, 13 (1960).

[50] S. Blaha, *A Derivation of ElectroWeak Theory based on an Extension of Special Relativity; Black Hole Tachyons; & Tachyons of Any Spin.* (Pingree-Hill Publishing, Auburn, NH, 2006) which will be called the first edition.

Another situation which calls for transformations between reference frames with relative velocity greater than the speed of light is the case of two tachyons: for example one tachyon moving at 3c (three times the speed of light) and another tachyon moving at 7c (seven times the speed of light). Again it seems physically required that a faster than-light-transformation exists relating the coordinates of superluminal[51] reference frames residing on each tachyon.

The Lorentz group (actually the inhomogeneous Lorentz group) relates event coordinates for "normal" sub-light reference frames (i.e. the set of reference frames whose relative velocities have a magnitude less than c).

We will show that a similar set[52] of transformations relate sub-light and superluminal reference frames, and also relate two "superluminal" reference frames[53] whose relative speed is greater than the speed of light. In the next chapter we will see that there are two possible groups of these types of transformations. We will choose one of these groups, which we call the Left-handed extended Lorentz group, based on a derivation of features of the Standard Model that follow from it. The other group leads to a **non**-Standard Model that favors right-handed features such as right-handed doublets. Both groups are subgroups of the complex Lorentz group, denoted L_c, whose generators are the six Lorentz group generators and whose parameters are six complex numbers.

1.3 The Set of Reference Frames

Under Special Relativity the set of Lorentz reference frames consists of all reference frames that are related by a Lorentz transformation where the magnitude of the relative velocity is less than the speed of light.[54]

If we extend Special Relativity to include transformations between reference frames where the magnitude of the relative velocity is not limited by the speed of light, then we obtain the set of reference frames for the extended Lorentz groups that will be described in this book. We will consider specific examples of this larger set of reference frames later.

---

[51] A superluminal reference frame is a reference frame moving faster than the speed of light relative to another reference frame. Since the distinction is one of *relative* speeds a superluminal reference frame from the viewpoint of one observer may be a "normal" reference frame the viewpoint of another observer. For example suppose we have three reference frames. In one reference frame an observer, observer1, sees the other two frames moving at a speed of 3.5c and 4c respectively in the positive x direction. Both of these frames appear to be superluminal to observer1. However an observer, observer2, on the 3.5c reference frame will see observer1's frame as superluminal and the 4c reference frame as a "normal" reference frame. A third observer, observer3, on the 4c reference frame will see observer1's frame as superluminal and the 3.5c reference frame as a "normal" reference frame. *Thus the concept of superluminal reference frame is relative.*

[52] This possibility was considered in the first edition. In this edition we have changed the coordinate transformation group to a different group: one of two possible groups.

[53] Reference frames that appear to be moving faster than the speed of light in a third reference frame.

[54] Some efforts extend the set of Lorentz reference frames to include "infinite momentum frames" where the magnitude of the momentum $\rightarrow \infty$ as well. See for example S. Weinberg, Phys. Rev. **150**, 1313 (1966).

1.4 Impact on General Relativity

The larger set of reference frames allowed by our extended Lorentz group will not change most of the basic tenets of General Relativity with several exceptions: 1) the concept of a locally inertial coordinate system[55] that appears in the equivalence principle is broadened to include the larger set of coordinate systems of the extended Lorentz group;[56] and 2) coordinate transformations can be complex and not congruences with important consequences for the field equations. Conventional General Relativity remains a subsector of this wider formulation. We will consider extended General Relativity in detail in a future work.

---

[55] A Lorentz, or extended Lorentz, reference frame is a locally inertial coordinate system.

[56] Since this larger set of locally inertial coordinate systems it is natural to consider the possibility of complex extensions of General Relativity such as in Blaha (2004).

Appendix 1-A. Phenomena Beyond the Light Barrier

1-A.1 Superluminal (Faster-than-Light) Transformations

In this Appendix we will briefly survey some of the very different features of faster-than light physical phenomena. We will frame our discussion in terms of the two simple reference frames depicted in Fig. 1-A.1. The prime frame is moving at a speed v > c (the speed of light) in the positive x direction with respect to the unprimed reference frame.

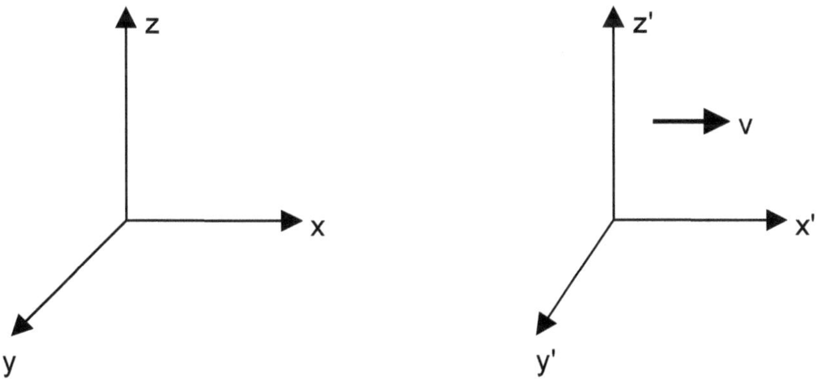

Figure 1-A.1. Two coordinate systems having a relative speed v in the x direction.

As shown later in the text (eq. 2.16 later) we define a superluminal (faster-than-light) transformation between coordinates in these reference frames with

$$\begin{aligned} t' &= \gamma_s(t - \beta x/c) \\ x' &= \gamma_s(x - \beta ct) \\ y' &= iy \\ z' &= iz \end{aligned} \tag{2.16}$$

where

$$\gamma_s = (\beta^2 - 1)^{-\frac{1}{2}} \tag{2.13}$$

and $\beta = v/c > 1$. The appearance of imaginary values for y' and z' is not a cause for alarm. An observer resident in the prime coordinate system will measure real y and z distances with a ruler. The only purpose of the factors of i is to relate the y and z coordinates to y' and z'. An observer in either coordinate system will view his/her coordinates as real.

The energy and momentum of a tachyon (faster-than-light) particle of mass m traveling at a speed v > c is

$$E = \gamma_s mc^2 \tag{1-A.1}$$

and

$$\mathbf{p} = m\gamma_s \mathbf{v} \tag{1-A.2}$$

Note that the tachyon defining condition is satisfied:

$$E^2 - c^2\mathbf{p}^2 = -m^2c^4 \tag{1-A.3}$$

Also note that in the limit $\beta \to \infty$ that

$$E = 0 \tag{1-A.4}$$

and

$$p = mc \tag{1-A.5}$$

where $p = |\mathbf{p}|$. Tachyons are always in motion. The minimal momentum of a tachyon is given by eq. 1-A.5. It corresponds to zero energy. It is the tachyon equivalent of Einstein's famous $E = mc^2$.

1-A.2 Length Dilations and Time Contractions

In ordinary Lorentz transformations a moving ruler will appear to be shorter in the direction of its motion when measured in another reference frame. This phenomenon is called *Lorentz contraction.*

Superluminal Length Dilation/Contraction

In the case of a superluminal transformation we find precisely the opposite effect, *superluminal length dilation*, is a possibility. Consider the case of the transformation of eq. 2.16 above (corresponding to Fig. 1-A.1), which relates the prime reference frame traveling at speed v in the positive x direction to the unprimed reference frame. A ruler perpendicular to the x-axis will have the same length in both reference frames if its endpoints are simultaneously measured – perhaps by photographing it. The y and z equations in eqs. 2.16 specify this fact up to an extraneous factor of i.

If the ruler is at rest in the prime reference frame and parallel to the x' axis, then a simultaneous measurement of its endpoints at the same time t_0 by an observer in the unprimed reference frame (perhaps by photographing it) will reveal both *length contraction and dilation* depending on the value of β. If the length is $L' = x'_2 - x'_1$ in the prime frame and $L = x_2 - x_1$ in the unprimed frame, then the equations:

$$x'_1 = \gamma_s(x_1 - \beta ct_0) \tag{1-A.7}$$
$$x'_2 = \gamma_s(x_2 - \beta ct_0) \tag{1-A.8}$$

imply

$$L' = \gamma_s L = (\beta^2 - 1)^{-1/2} L \tag{1-A.9}$$

Thus we have three cases:

Case 1: $\beta \in <1, \sqrt{2}>$: $L < L'$ Contraction (1-A.10)

Case 2: $\beta = \sqrt{2}$: $L = L'$ Equality (1-A.11)

Case 3: $\beta \in <\sqrt{2}, \infty>$: $L > L'$ Dilation (1-A.12)

Superluminal Time Contraction/Dilation

In the case of a superluminal transformation we find *superluminal time contraction* is a possibility. Consider again the case of the transformation of eq. 2.16 above corresponding to Fig. 1-A.1 relating the prime reference frame traveling at speed v in the positive x direction to the unprimed reference frame. Consider the time interval between two events occurring at the same point x'_0 in the prime reference frame. From the viewpoint of an observer in the unprimed frame the events take place at different points x_1 and x_2. If the time interval is $T' = t'_2 - t'_1$ in the prime frame and $T = t_2 - t_1$ in the unprimed frame, then the inverse of eqs. 2.16 give:

$$t_1 = \gamma_s(t'_1 + \beta x'_0/c) \tag{1-A.13}$$
$$t_2 = \gamma_s(t'_2 + \beta x'_0/c) \tag{1-A.14}$$

and imply

$$T = \gamma_s T' = (\beta^2 - 1)^{-\frac{1}{2}} T' \tag{1-A.15}$$

Again we have three cases:

Case 1: $\beta \in <1, \sqrt{2}>$: $T > T'$ Dilation (1-A.16)

Case 2: $\beta = \sqrt{2}$: $T = T'$ Equality (1-A.17)

Case 3: $\beta \in <\sqrt{2}, \infty>$: $T < T'$ Contraction (1-A.18)

The time interval in the unprimed frame can be less than, equal to, or greater than the time interval in the frame where the events take place at the same spatial point.

Thus superluminal transformations are more complex than Lorentz transformations with respect to space and time, dilation and contraction.

1-A.3 Tachyon Fission to More Massive Particles – Reverse Fission

Another way in which faster-than-light phenomena differ from sublight phenomena is particle fission. Normally when a particle or nucleus decays or fissions the masses of the particles produced by the decay are smaller than the mass of the original particle or nucleus. And energy is released. We are familiar with fission as the source of nuclear energy.

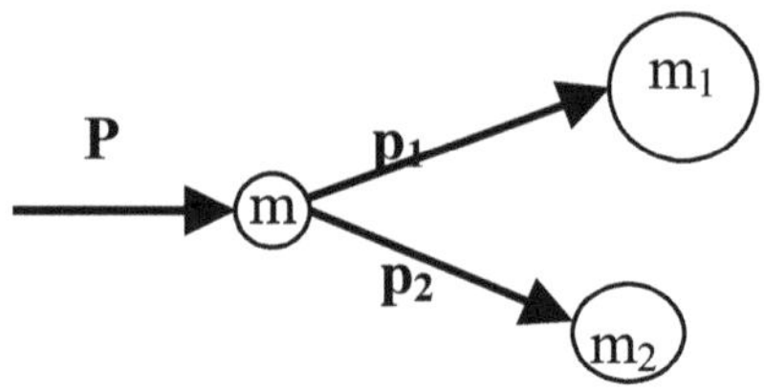

Figure 1-A.2. Two particle decay of a tachyon.

In the case of faster-than-light particles, tachyons, a much different possibility is present: a tachyon can decay into heavier tachyons: *a particle's spatial 3-momentum can be transformed into mass.* We will consider the specific case of a tachyon decaying into two particles to illustrate this possibility. (See Fig. 1-A.2.)

We will assume the initial tachyon has zero energy[57] and thus the tachyons emerging from the decay also have zero energy. The analysis is based on conservation of total energy and momentum.

Momentum conservation implies

$$\mathbf{P} = \mathbf{p}_1 + \mathbf{p}_2 \tag{1-A.19}$$

Since all energies are zero

$$\begin{aligned}(cP)^2 &= (c\mathbf{P})^2 = m^2\\ (cp_1)^2 &= (c\mathbf{p}_1)^2 = m_1^2\\ (cp_2)^2 &= (c\mathbf{p}_2)^2 = m_2^2\end{aligned} \tag{1-A.20}$$

where $P = |\mathbf{P}|$, $p_1 = |\mathbf{p}_1|$, and $p_2 = |\mathbf{p}_2|$. If we now square eq. 1-A.19 and use eqs. 1-A.20 we obtain

$$m^2 = m_1^2 + m_2^2 + 2m_1m_2 \cos\theta \tag{1-A.21}$$

where θ is the angle between the emerging particles momenta $\mathbf{p}_1$ and $\mathbf{p}_2$.

Eq. 1-A.21 has a number of interesting cases:

---

[57] If a particle has zero energy its velocity is infinite so the case considered is somewhat artificial. However the results would still be approximately true for very large velocities. The simplicity of the kinematics led us to consider this case.

Case $\theta = 0$:

$$m = m_1 + m_2 \tag{1-A.22}$$

The masses of the outgoing tachyons sum to the mass of the original tachyon.

Case $\theta = \pi/2$:

$$m^2 = m_1^2 + m_2^2 \tag{1-A.23}$$

The masses of each outgoing tachyon is less than the mass of the original tachyon.

Case $\theta = \pi$:

$$m^2 = (m_1 - m_2)^2 \tag{1-A.24}$$

In this case either $m_1 > m$ or $m_2 > m$. Thus one of the outgoing tachyons has a greater mass than the original tachyon. Mass is effectively created from the spatial momentum of the particle. This process is the inverse of normal particle decay or fission where the sum of the outgoing masses is always less than the original particle's mass and the difference is mass converted into energy in the form of additional photons via $E = mc^2$.

This last case, where one of the outgoing particles is more massive than the original particle, is not just for $\theta = \pi$. Since

$$\cos\theta = (m^2 - m_1^2 - m_2^2)/(2m_1m_2) \tag{1-A.25}$$

we see that *the sum of the outgoing tachyon masses is always greater than the original tachyon mass (except when $\theta = 0$)* since

$$\cos\theta = 1 + [m^2 - (m_1 + m_2)^2]/(2m_1m_2) \leq 1 \tag{1-A.26}$$

and thus

$$[m^2 - (m_1 + m_2)^2]/(2m_1m_2) \leq 0 \tag{1-A.27}$$

Note $m = m_1 + m_2$ only if $\theta = 0$.

Since we can transform the above discussion to the case of tachyons with a non-zero energy using an ordinary Lorentz transformation the above discussion in this subsection is general.

We therefore conclude that when a tachyon decays into two tachyons the sum of the masses of the produced tachyons is greater than the mass of the original tachyon except if the angle between the momenta of the produced tachyons is zero. In that case the sum of the masses of the produced tachyon equals the mass of the original tachyon.

Thus tachyons can engage in ***reverse fission*** *in which* ***momentum is converted into mass so the outgoing particles have a total mass greater than the incoming particle***. In the case of "normal" fission part of the mass of a particle can be converted to energy and the sum of the masses of the decay product particles is less than the mass of the original particle.

1-A.4 Light Chasing Faster-than-Light Particles?

Einstein told a story that he imagined positioning himself in a (Galilean) reference frame moving at the speed of light and seeing electromagnetic waves "frozen" in time so that they were no longer vibrating. This vision inspired him to reconsider the transformation laws between coordinate systems and to derive the theory of Special Relativity. In Special Relativity the speed of light is the same in all reference frames.

In this subsection we will consider a light pulse from the points of view of two reference frames whose relative speed v is greater than the speed of light. We will use the example considered earlier and add a pulse of light traveling in the positive x direction. (See Fig. 1-A.3.)

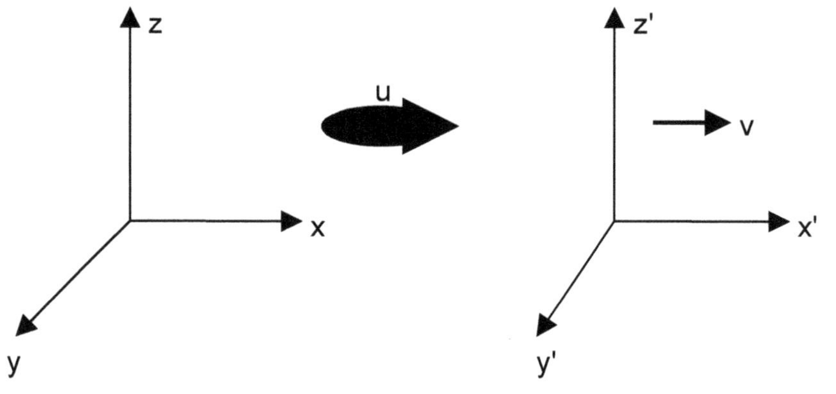

Figure 1-A.3. Two coordinate systems having a relative speed v in the x direction. A pulse of light is displayed as a thick arrow.

Eq. 2.17 (derived later) contains the general law for the addition of velocities in a situation such as depicted in Fig. 1-A.3. If we adapt it to the present example and let u be the speed of the pulse in the unprimed frame (temporarily forgetting it is a light pulse) we find eq. 2.17 implies

$$u' = (u - \beta c)/(1 - \beta u/c) \qquad (2.17a)$$

where $\beta = v/c > 1$, and u' is the speed of the pulse in the prime frame. Then if we set u = c we see that u' = c as well. *Thus our superluminal transformations preserve the constancy of the speed of light just like Lorentz transformations.*
As a result the pulse of light will intersect the z' axis eventually. However if superluminal transformations did not preserve the speed of light in all frames the pulse might never reach the z' axis. For example under a Galilean transformation the speed of the pulse would be u' = u – v = c – v and the pulse would actually be falling further and further behind the z' axis.

1-A.5 Electromagnetic Field of a Charged Tachyon – A Pancake Effect?

The electric field of a charge q at rest in a reference frame is:

$$\mathbf{E} = (q/(4\pi\varepsilon_0)\check{\mathbf{r}}/r^2 \tag{1-A.28}$$

in spherical coordinates where $\check{\mathbf{r}}$ is a unit vector in the radial direction.

Sublight Charged Particle

The electric and magnetic fields of a charge q moving in the positive x direction with speed v < c are

$$\mathbf{E} = (q/(4\pi\varepsilon_0)\check{\mathbf{r}}(1 - \beta^2)/[r^2(1 - \beta^2\sin^2\theta)^{\frac{3}{2}}] \tag{1-A.29}$$

$$\mathbf{B} = (q/(4\pi\varepsilon_0)\check{\mathbf{r}}\beta(1 - \beta^2)\sin\theta/[r^2(1 - \beta^2\sin^2\theta)^{\frac{3}{2}}] \tag{1-A.30}$$

where $\check{\mathbf{r}}$ is the radial unit vector, $\beta = v/c$, and θ is measured with respect to the polar axis which is taken to be the x axis. As $\beta \rightarrow 1$ the electric and magnetic fields develop a "pancake" form with large field strengths in the directions perpendicular to the direction of motion similar to the transverse fields of electromagnetic quanta. This feature is the basis of the Weizsäcker-Williams method of virtual quanta.

Charged Tachyon

The electric and magnetic fields of a tachyon of charge q moving in the positive x direction with speed v > c are

$$\mathbf{E} = (q/(4\pi\varepsilon_0)\check{\mathbf{r}}(\beta^2 - 1)/[r^2(\beta^2\sin^2\theta - 1)^{\frac{3}{2}}] \tag{1-A.31}$$

$$\mathbf{B} = (q/(4\pi\varepsilon_0)\check{\mathbf{r}}\beta(\beta^2 - 1)\sin\theta/[r^2(\beta^2\sin^2\theta - 1)^{\frac{3}{2}}] \tag{1-A.32}$$

where $\beta = v/c > 1$, and θ is again measured with respect to the polar axis which is taken to be the x axis. In the case of tachyons there are three cases of interest.

Case $\beta^2\sin^2\theta - 1 < 0$:

The electric and magnetic fields are pure imaginary and are excluded from the forward and backward cones surrounding the x axis defined by $|\sin\theta| < \beta^{-1}$.

Case $\beta^2\sin^2\theta - 1 = 0$:
The electric and magnetic fields are infinite. Thus the field strengths are infinite on a cone at the angle θ with respect to the x-axis. By comparison, a magnetic monopole only has a one-dimensional, singularity line extending from the monopole to infinity.

Case $\beta^2\sin^2\theta - 1 > 0$:
The electric and magnetic fields decrease in strength as $\sin^2\theta$ increases. Thus the regions of maximum field strength are the forward and backward cones where $|\sin\theta|$ is greater than but near β^{-1} in value. The pancake picture of the sublight charged particle does not hold for charged tachyons.

Are Tachyonic Cones in the Au-Au Scattering Quark-Gluon Plasma?

Cones have been observed in high energy Au-Au scattering in which a quark-gluon plasma is created. These cones have been attributed to a variety of causes such as hydrodynamically generated Mach cones, and Cherenkov radiation. The possibility exists that tachyonic excitations may transiently exist in the quark-gluon plasma and may, in part, explain the observed cones and dips. The above described cones in the case of a moving charged tachyon are remarkably similar in character. See the CERES collaboration paper arXiv:nucl-ex/0701023, and references therein, for experimental findings.

1-A.6 Superluminal (Tachyon) Physics is Different

The simple classical examples presented in this appendix demonstrate that superluminal physics has many interesting new features that are worthy of interest. Since tachyons exist in Black Holes, and, perhaps, in other contexts, their study is a worthwhile endeavor.

2. Lorentz Groups

2.0 Overview

This chapter discusses extensions of the Lorentz group of Special Relativity that includes transformations to reference frames moving at relative speeds greater than the speed of light in sections 2.1 through 2.4. We describe two groups that implement this extension: the Left-handed Extended Real Lorentz Group and the Right-handed Extended Real Lorentz Group. The difference in handedness is explored in detail in chapter 3 and ultimately leads to parity violation[58] in the Standard Model.

In section 2.5 we consider properties of the complex Lorentz Group, L_c. In chapter 3 we show that L_c boosts provide three spin ½ alternatives to the Dirac equation. The four spin ½ free dynamical equations correspond to two lepton species and two quark species. The quark fields have an inherent global SU(3) symmetry that leads us to postulate quark color SU(3) (a local Yang-Mills interaction) with quarks in the fundamental representation $\underline{3}$ of SU(3).

2.1 The Lorentz Group

The Lorentz group, with which we are familiar, relates the coordinates of an event in two coordinate systems that differ by a relative velocity whose magnitude is less than the speed of light. The inhomogeneous Lorentz group includes coordinate displacements. We will discuss the relationship between coordinates modulo displacements, and not consider the displacement generators, since they will be the same for both ordinary Lorentz group transformations, and L_c transformations. Thus we will consider the homogeneous Lorentz group, and its generalizations that include faster-than-light transformations.

The elements of the Lorentz group, $\Lambda(\mathbf{v})$, when treated as 4×4 matrices satisfy

$$\Lambda(\mathbf{v})^T G \Lambda(\mathbf{v}) = G \tag{2.1}$$

where G is the 4 x 4 diagonal matrix form of the metric diag(1, –1, –1, –1), the superscript "T" indicates the transpose, and $\mathbf{v}$ is the relative velocity of the reference frames. Note g^{00} is +1 in our metric. When $\Lambda(\mathbf{v})$ is a boost the components[59] of its matrix representation are real:

[58] Thus explaining the 50 year old question of the origin of parity violation discussed by T. D. Lee and C. N. Yang, Phys. Rev. **104**, 254 (1956). Parity violation originates in physics beyond the speed of light.

[59] We shall consider only the proper, orthochronous Lorentz group until section 2.3 where we extend the discussion to the other subgroups of the Lorentz group and extended Lorentz groups.

$$\Lambda(\mathbf{v}) = \begin{bmatrix} \gamma & -\gamma v_x & -\gamma v_y & -\gamma v_z \\ -\gamma v_x & 1 + (\gamma - 1)v_x{}^2/v^2 & (\gamma - 1)v_x v_y/v^2 & (\gamma - 1)v_x v_z/v^2 \\ -\gamma v_y & (\gamma - 1)v_x v_y/v^2 & 1 + (\gamma - 1)v_y{}^2/v^2 & (\gamma - 1)v_y v_z/v^2 \\ -\gamma v_z & (\gamma - 1)v_x v_z/v^2 & (\gamma - 1)v_y v_z/v^2 & 1 + (\gamma - 1)v_z{}^2/v^2 \end{bmatrix} \quad (2.2)$$

where $\gamma = (1 - v^2)^{-½}$, $c = 1$, $\mathbf{v} = (v_x, v_y, v_z)$, and $v = |\mathbf{v}|$. The boost transformation $\Lambda(\mathbf{v})$ can be expressed in the form

$$\Lambda(\mathbf{v}) = \exp[i\omega\hat{\mathbf{u}}\cdot\mathbf{K}] \quad (2.3)$$

where $\mathbf{v} = \hat{\mathbf{u}} \tanh \omega$, $\hat{\mathbf{u}}\bullet\hat{\mathbf{u}} = 1$, and $\mathbf{K}$ is the boost vector. Using the unit normalized velocity vector $\mathbf{u} = (u_x, u_y, u_z)$ the Lorentz boost matrix can be written:

$$\Lambda(\omega, \mathbf{u}) = \Lambda(\mathbf{v}) \quad (2.4)$$

$$= \begin{bmatrix} \cosh(\omega) & -\sinh(\omega)u_x & -\sinh(\omega)u_y & -\sinh(\omega)u_z \\ -\sinh(\omega)u_x & 1 + (\cosh(\omega) - 1)u_x{}^2 & (\cosh(\omega) - 1)u_x u_y & (\cosh(\omega) - 1)u_x u_z \\ -\sinh(\omega)u_y & (\cosh(\omega) - 1)u_x u_y & 1 + (\cosh(\omega) - 1)u_y{}^2 & (\cosh(\omega) - 1)u_y u_z \\ -\sinh(\omega)u_z & (\cosh(\omega) - 1)u_x u_z & (\cosh(\omega) - 1)u_y u_z & 1 + (\cosh(\omega) - 1)u_z{}^2 \end{bmatrix}$$

This definition of the general form of the proper, orthochronous, Lorentz boost matrix $\Lambda(\omega, \mathbf{u})$ will be used in subsequent sections to define extended faster-than-light boost transformations.

The vector form of a Lorentz boost transformation is

$$\mathbf{x}' = \mathbf{x} + (\gamma - 1)\mathbf{x}\cdot\mathbf{v}\ \mathbf{v}/v^2 - \gamma\mathbf{v}t$$
$$t' = \gamma(t - \mathbf{v}\cdot\mathbf{x}/c^2) \quad (2.5)$$

where $\gamma = (1 - \beta^2)^{-½}$ with $\beta = v/c = v$ (with $c = 1$).

2.2 Sets of Real Matrices Λ(ω, u) for Complex ω

Faster-than-light transformations will be implemented with specific complex forms of ω as we will see in the next section. Since

$$\cosh^2(z) - \sinh^2(z) = 1 \tag{2.6}$$

for any complex values z it is easy to show that

$$\Lambda(\omega, \mathbf{u})^T G \Lambda(\omega, \mathbf{u}) = G \tag{2.7}$$

for any complex ω. Eq. 2.7 implies Λ(ω, **u**) is a member of the Lorentz group or of the complex Lorentz group for complex ω.

For certain values of the imaginary part of ω the matrix Λ(ω, **u**) has a simple form, similar to that of Λ(ω, **u**) for real ω, but which generates boosts to relative speeds greater than the speed of light. Among these values are:

$$\omega \rightarrow \omega_{\pm} = \omega \pm i\pi/2 \tag{2.8}$$

In chapter 3 (cases 1 and 2 in section 3.7) we will see that the alternate choices $\omega_{\pm}$ correspond to specific physical situations when considered within the framework of the L_c group.

2.3 Extensions of the Lorentz Group to Faster-than-Light Transformations

In this section we will substitute $\omega_{\pm}$ for ω in Λ(ω, **u**) and then show that we obtain two sets of possible transformations to faster-than-light reference frames. One set of transformations where $\omega_L = \omega + i\pi/2$ will be called *left-handed superluminal boosts*. When they are combined with the Lorentz group we are led to the "left-handed" Standard Model. We denote members of this set, $\Lambda_L(\omega, \mathbf{u})$, with the subscript "L".

The other set of boosts where $\omega_R = \omega - i\pi/2$ will be called *right-handed superluminal boosts*. When they are combined with the Lorentz group they lead to a right-handed, unphysical, version of the Standard Model. We denote members of this set, $\Lambda_R(\omega, \mathbf{u})$, with the subscript "R".

Before considering faster-than-light boosts we note the relation between ω in a Lorentz boost Λ(ω, **u**), and the magnitude of the relative velocity v < 1, is

$$\mathbf{v} = \hat{\mathbf{u}} \tanh\omega \qquad\qquad \hat{\mathbf{u}}\bullet\hat{\mathbf{u}} = 1$$

$$\begin{aligned} \cosh(\omega) &= \gamma = (1 - v^2)^{-1/2} \\ \sinh(\omega) &= v\gamma = \beta\gamma \end{aligned} \tag{2.9}$$

where β = v = |**v**|.

Left-Handed Superluminal Transformations

Left-handed (proper orthochronous) superluminal boost transformations $\Lambda_L(\mathbf{v})$ have the same form as eq. 2.2 for ordinary (proper orthochronous) Lorentz boost transformations. However the magnitude of the relative velocity **v** is greater than the speed of light. Thus $\gamma = (1 - v^2)^{-1/2}$ is pure imaginary and $\Lambda_L(\mathbf{v})$ is complex.

$$\Lambda_L(\mathbf{v}) = \begin{bmatrix} \gamma & -\gamma v_x & -\gamma v_y & -\gamma v_z \\ -\gamma v_x & 1 + (\gamma - 1)v_x^2/v^2 & (\gamma - 1)v_x v_y/v^2 & (\gamma - 1)v_x v_z/v^2 \\ -\gamma v_y & (\gamma - 1)v_x v_y/v^2 & 1 + (\gamma - 1)v_y^2/v^2 & (\gamma - 1)v_y v_z/v^2 \\ -\gamma v_z & (\gamma - 1)v_x v_z/v^2 & (\gamma - 1)v_y v_z/v^2 & 1 + (\gamma - 1)v_z^2/v^2 \end{bmatrix} \tag{2.10}$$

This transformation raises several issues – the most prominent of which is the interpretation of the imaginary coordinates generated by the transformation. Imaginary coordinates would appear at first glance to be unphysical. However if we view the measurement of these quantities operationally: an observer measures distances with "rulers", and time with clocks, which both give real numeric values. Thus an observer in a coordinate system with imaginary coordinates will be "unaware" of the imaginary nature of these quantities. It is only when the observer's coordinate system is related to another coordinate system through a superluminal transformation as done above that the imaginary nature of the coordinates becomes evident. From this point of view imaginary coordinates pose no new conceptual issues.

It will become apparent that a related transformation defined by

$$E(\mathbf{v}) = i\Lambda_L(\mathbf{v}) \tag{2.10a}$$

should be used to define the coordinates generated by a superluminal boost transformation.[60] This definition of coordinates leads to the tachyonic Dirac equation for spin ½ tachyons. Therefore we define

$$X' = E(\mathbf{v})X = i\Lambda_L(\mathbf{v})X \tag{2.10b}$$

where X' and X are coordinates in the prime and unprimed reference frames respectively. X' and X are represented as column vectors.[61] If we consider the simple case of a relative velocity v in the x direction, then the E boost gives

---

[60] E is an upper case epsilon chosen to stand for Extended since the boosts in question are extended boosts (v > 1) that are not part of the Lorentz group.

$$t' = |\gamma|(t - \beta x)$$
$$x' = |\gamma|(x - \beta t)$$
$$y' = iy$$
$$z' = iz$$

where $|\gamma|$ is the absolute value of γ. Thus y' and z' are imaginary from the viewpoint of the unprimed coordinate system.

Eqs. 2.7 and 2.10a imply

$$E(\omega, \mathbf{u})^T G E(\omega, \mathbf{u}) = -G \tag{2.10c}$$

where $E(\omega, \mathbf{u}) \equiv E(\mathbf{v})$. Thus E inverts the sign of the metric tensor. As a result the invariant scalar product relation (in matrix form) is

$$XGY = X'(-G)Y' \tag{2.10d}$$

The metric tensor is represented by G. Most of the succeeding discussion will be phrased in terms of $\Lambda_L(\mathbf{v})$. Eq. 2.10a enables one to easily rephrase the discussion in terms of $E(\omega, \mathbf{u})$.

The Cosh-Sinh Representation of Left-Handed Superluminal Boosts

We will now develop the representation of left-handed superluminal boost transformations in terms of cosh(ω) and sinh(ω) for later use in our discussion of tachyons. We find that we must use a complex $\omega = \omega_L = \omega + i\pi/2$ to properly describe left-handed superluminal boosts. The relation between ω_L and v is different from eq. 2.9 for the case of left-handed superluminal boosts:

$$\cosh(\omega_L) = i \sinh(\omega) = -\gamma = i\gamma_s \tag{2.11}$$
$$\sinh(\omega_L) = i \cosh(\omega) = -\beta\gamma = i\beta\gamma_s$$

where $\beta = v > 1$ and $\boldsymbol{\omega \geq 0}$. Note

$$\sinh(\omega) = \gamma_s \tag{2.12}$$
$$\cosh(\omega) = \beta\gamma_s$$

with

$$\gamma_s = (\beta^2 - 1)^{-1/2} \tag{2.13}$$

Upon substituting ω_L for ω in eq. 2.4 we obtain another form for a left-handed superluminal transformation (equivalent to that of eq. 2.10):

[61] This definition of superluminal coordinate transformations is physically acceptable as long as allowance is made for the change in the definition of the invariant interval based on eq. 2.10c.

$\Lambda_L(\omega, \mathbf{u}) = \Lambda(\omega + i\pi/2, \mathbf{u}) = -iE(\omega, \mathbf{u})$

$$= \begin{bmatrix} \cosh(\omega_L) & -\sinh(\omega_L)u_x & -\sinh(\omega_L)u_y & -\sinh(\omega_L)u_z \\ -\sinh(\omega_L)u_x & 1+(\cosh(\omega_L)-1)u_x^2 & (\cosh(\omega_L)-1)u_xu_y & (\cosh(\omega_L)-1)u_xu_z \\ -\sinh(\omega_L)u_y & (\cosh(\omega_L)-1)u_xu_y & 1+(\cosh(\omega_L)-1)u_y^2 & (\cosh(\omega_L)-1)u_yu_z \\ -\sinh(\omega_L)u_z & (\cosh(\omega_L)-1)u_xu_z & (\cosh(\omega_L)-1)u_yu_z & 1+(\cosh(\omega_L)-1)u_z^2 \end{bmatrix}$$

$$= \begin{bmatrix} i\gamma_s & -i\beta\gamma_s u_x & -i\beta\gamma_s u_y & -i\beta\gamma_s u_z \\ -i\beta\gamma_s u_x & 1+(i\gamma_s-1)u_x^2 & (i\gamma_s-1)u_xu_y & (i\gamma_s-1)u_xu_z \\ -i\beta\gamma_s u_y & (i\gamma_s-1)u_xu_y & 1+(i\gamma_s-1)u_y^2 & (i\gamma_s-1)u_yu_z \\ -i\beta\gamma_s u_z & (i\gamma_s-1)u_xu_z & (i\gamma_s-1)u_yu_z & 1+(i\gamma_s-1)u_z^2 \end{bmatrix} = \Lambda_L(\mathbf{v}) \qquad (2.14)$$

by eq. 2.10.

A simple case that illustrates the nature of the left-handed superluminal boost is to assume the relative velocity is in the x direction. Then eq. 2.14 becomes

$$\Lambda_L(\omega, \mathbf{u} = (1,0,0)) = \begin{bmatrix} i\gamma_s & -i\beta\gamma_s & 0 & 0 \\ -i\beta\gamma_s & i\gamma_s & 0 & 0 \\ 0 & 0 & 1 & 0 \\ 0 & 0 & 0 & 1 \end{bmatrix} \qquad (2.15)$$

implementing the coordinate transformation:

$$X' = -i\Lambda_L(\omega, \mathbf{u} = (1,0,0))X = E(\omega, \mathbf{u} = (1,0,0))X$$

or

$$\begin{aligned} t' &= \gamma_s(t - \beta x) \\ x' &= \gamma_s(x - \beta t) \\ y' &= iy \\ z' &= iz \end{aligned} \qquad (2.16)$$

The addition rule for the x-component of velocity can be computed for infinitesimal displacements in space and time:

$$v_x' = \Delta x'/\Delta t' = (\Delta x\,\gamma_s - \Delta t\,\beta\gamma_s)/(\Delta t\,\gamma_s - \Delta x\,\beta\gamma_s)$$

$$= (v_x - \beta)/(1 - \beta v_x) \qquad (2.17)$$

(where $\beta = u$ is the relative speed) in the limit $\Delta t \rightarrow 0$ where the x component of a particle's velocity in the unprimed frame is $v_x = \Delta x/\Delta t$. $\Delta t'$ is determined by

$$\Delta t' = \Delta t\ \gamma_s(1 - \beta v_x) \tag{2.18}$$

Note the velocity of light is the same in the primed and unprimed reference frames. (If $v_x = 1$ then $v_x' = 1$.) *Thus left-handed superluminal transformations preserve the constancy of the speed of light in all reference frames.*

Further note that increasing the value of ω in $\Lambda_L(\omega, \mathbf{u})$ corresponds to decreasing the magnitude of the relative velocity v since

$$v = \text{cotanh}(\omega) \tag{2.19}$$

Thus when $v = 1$ then $\omega = \infty$, and when $\omega = 0$ then $v = \infty$.

General Velocity Transformation Law – Left-Handed Superluminal Boosts

The general velocity transformation law for a particle moving with velocity **v** in the unprimed reference frame and velocity **v'** in the primed reference frame is

$$\mathbf{v}' = \mathbf{w} + (\gamma - 1)\mathbf{w}\cdot\mathbf{v}\ \mathbf{w}/w^2 - \gamma\mathbf{w} \tag{2.20}$$

where **w** is the relative velocity of the primed reference frame with respect to the unprimed reference frame, and $\gamma = (1 - w^2)^{-1/2}$. Eq. 2.20 is obtained by calculating the derivative $d\mathbf{x}'/dt'$ using eqs. 2.5. The relative velocity **w** can be greater or less than the speed of light. Eq. 2.20 implies

$$v'^2 = 1 + (v^2 - 1)(1 - w^2)/(1 - \mathbf{w}\cdot\mathbf{v})^2 \tag{2.21}$$

The velocity transformation law can be used to determine the multiplication rules for Lorentz and extended Lorentz transformations (next subsection).

Left-Handed Transformations Multiplication Rules

In this subsection we will determine the multiplication rules of left-handed extended Lorentz boosts and show the Lorentz group is an invariant subgroup of the extended left-handed Lorentz group. To do this we will consider three reference frames: an unprimed frame, a “primed” frame moving with velocity **w** with respect to the unprimed frame, and a “double-primed” frame moving with velocity **v** with respect to the unprimed frame and velocity **v'** with respect to the primed frame. See Fig. 2.1.

A moment’s consideration reveals that **v'** is related to **v** by eqs. 2.20 and 2.21. (Think of the double-primed coordinate system as a particle or attached to a particle. In addition note that the transformation law from the unprimed to the double-primed reference frame must be the product of consecutive transformations (boosts) from the unprimed to the primed reference frames and then from the primed to the double-primed reference frames:

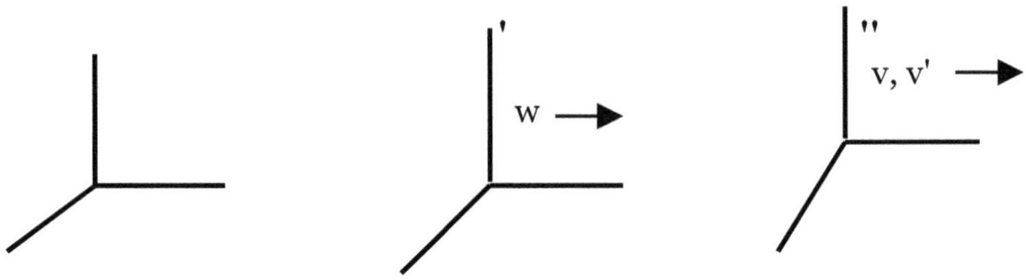

Figure 2.1. Three reference frames used to establish transformation multiplication rules.

$$\Lambda_?(\mathbf{v}) = \Lambda_?(\mathbf{v'})\Lambda_?(\mathbf{w}) \tag{2.22}$$

where the "?" subscripts indicate Lorentz or superluminal transformations (boosts) depending on the magnitude of the relative velocity in the transformation's parentheses.

We now consider the cases using eq. 2.21:

1) If $w > 1$ and $v' > 1$
   then eq. 2.21 implies $v < 1$ and thus $\Lambda_?(\mathbf{v})$ is a Lorentz transformation

$$\Lambda(\mathbf{v}) = \Lambda_L(\mathbf{v'})\Lambda_L(\mathbf{w}) \tag{2.23}$$

2) If $w > 1$, $v' < 1$
   then eq. 2.21 implies $v > 1$ and thus $\Lambda_?(\mathbf{v})$ is a superluminal transformation

$$\Lambda_L(\mathbf{v}) = \Lambda(\mathbf{v'})\Lambda_L(\mathbf{w}) \tag{2.24}$$

3) If $w < 1$, $v' > 1$
   then eq. 2.21 implies $v > 1$ and thus $\Lambda_?(\mathbf{v})$ is a superluminal transformation

$$\Lambda_L(\mathbf{v}) = \Lambda_L(\mathbf{v'})\Lambda(\mathbf{w}) \tag{2.25}$$

4) If $w < 1$, $v' < 1$
   then eq. 2.21 implies $v < 1$ and thus $\Lambda_?(\mathbf{v})$ is a Lorentz transformation

$$\Lambda(\mathbf{v}) = \Lambda(\mathbf{v'})\Lambda(\mathbf{w}) \tag{2.26}$$

where, in each above case, the transformation on the left side of the equation may be a boost or a combination of a boost and a spatial rotation. Thus we have obtained the multiplication rules for left-handed extended Lorentz transformations.

The inverse of a Lorentz boost (eq. 2.3) is

$$\Lambda^{-1}(\omega, \hat{\mathbf{u}}) = \exp[-i\omega\hat{\mathbf{u}}\cdot\mathbf{K}] \tag{2.27}$$

where $\boldsymbol{\omega} \geq \mathbf{0}$. It shows the inverse is generated by letting $\omega \rightarrow -\omega$. Note that since $v = \tanh \omega$, the effect of $\omega \rightarrow -\omega$ is to let $v \rightarrow -v$. In the case of superluminal left-handed boosts, since

$$\Lambda_L(\omega, \mathbf{u}) = \Lambda(\omega + i\pi/2, \mathbf{u}) = \exp[i(\omega + i\pi/2)\hat{\mathbf{u}}\cdot\mathbf{K}] \tag{2.28}$$

we find the inverse is

$$\Lambda_L^{-1}(\omega, \mathbf{u}) = \Lambda(-(\omega + i\pi/2), \mathbf{u}) = \exp[-i(\omega + i\pi/2)\hat{\mathbf{u}}\cdot\mathbf{K}] \tag{2.29}$$

where $\boldsymbol{\omega} \geq \mathbf{0}$. Since $\Lambda_L^{-1}(\omega, \mathbf{u})$ is not the Hermitian conjugate of $\Lambda_L(\omega, \mathbf{u})$, superluminal boosts are not unitary. However unitarity is not required since even complex Lorentz group elements satisfy the defining relation of the Lorentz group (eqs. 2.1 and 2.7). The effect of letting $\omega_L = \omega + i\pi/2 \rightarrow -(\omega + i\pi/2)$ is to let $v \rightarrow -v$ since

$$\tanh(\omega_L) = \beta = v \tag{2.30}$$

by eqs. 2.11.

We now turn to proving the Lorentz group is an invariant subgroup of the left-handed extended Lorentz group.

Theorem: The Lorentz group is an invariant subgroup of the left-handed extended Lorentz group.

Proof:

Consider the quantity

$$J = \Lambda_L(\omega_u, \mathbf{u})^{-1}\Lambda(\omega_v, \mathbf{v})\Lambda_L(\omega_u, \mathbf{u})$$

then

$$J = \Lambda_L(\omega_u, \mathbf{u})^{-1}\Lambda_L(\omega_z, \mathbf{z})$$
$$= \Lambda(\omega_w, \mathbf{w})$$

by eqs. 2.24 and 2.23 respectively for some ω_z, $\mathbf{z}$, ω_w and $\mathbf{w}$. Thus the Lorenz group is an invariant subgroup of the left-handed extended Lorentz group.

Right-Handed Superluminal Transformations

When we transform between reference frames using a right-handed[62] superluminal boost (where the magnitude of the relative velocity v is greater than c) the relation between ω and v also changes. The variable ω becomes $\omega_R = \omega - i\pi/2$ and

$$\cosh(\omega_R) = -i \sinh(\omega) = \gamma = -i\gamma_s \tag{2.31}$$
$$\sinh(\omega_R) = -i \cosh(\omega) = \beta\gamma = -i\beta\gamma_s \tag{2.32}$$

where $\beta = v > 1$ and $\boldsymbol{\omega \geq 0}$. Note (as before)

$$\sinh(\omega) = \gamma_s \tag{2.12}$$
$$\cosh(\omega) = \beta\gamma_s$$

with

$$\gamma_s = (\beta^2 - 1)^{-1/2} \tag{2.13}$$

Upon substituting ω_R for ω in eq. 2.4 we obtain the form of the right-handed superluminal boost and a corresponding transformation E_R:[63]

$$\Lambda_R(\omega, \mathbf{u}) = \Lambda(\omega - i\pi/2, \mathbf{u}) = -iE_R(\omega, \mathbf{u})$$

$$= \begin{bmatrix} -i\gamma_s & i\beta\gamma_s u_x & i\beta\gamma_s u_y & i\beta\gamma_s u_z \\ i\beta\gamma_s u_x & 1 + (-i\gamma_s - 1)u_x^2 & (-i\gamma_s - 1)u_x u_y & (-i\gamma_s - 1)u_x u_z \\ i\beta\gamma_s u_y & (-i\gamma_s - 1)u_x u_y & 1 + (-i\gamma_s - 1)u_y^2 & (-i\gamma_s - 1)u_y u_z \\ i\beta\gamma_s u_z & (-i\gamma_s - 1)u_x u_z & (-i\gamma_s - 1)u_y u_z & 1 + (-i\gamma_s - 1)u_z^2 \end{bmatrix} \tag{2.33}$$

A simple case that illustrates the nature of the right-handed superluminal transformation is to assume a relative velocity in the x direction. Then eq. 2.33 becomes

$$\Lambda_R(\omega, \mathbf{u} = (1,0,0)) = \begin{bmatrix} -i\gamma_s & i\beta\gamma_s & 0 & 0 \\ i\beta\gamma_s & -i\gamma_s & 0 & 0 \\ 0 & 0 & 1 & 0 \\ 0 & 0 & 0 & 1 \end{bmatrix} \tag{2.34}$$

implementing the coordinate transformation:

[62] We call these transformations right-handed because they lead eventually to an alternate Standard Model with right-handed SU(2) doublets and left-handed SU(2) singlets. This alternate right-handed Standard Model does not appear to correspond to experimental reality.

[63] We note the singularities at $\beta = \pm 1$ or $\omega = \pm\infty$. As a result we have a branch cut in the complex ω-plane consisting of the entire real ω axis. Therefore three left-handed boosts are not equivalent to a right-handed boost but rather appear on another Riemann sheet.

$$X' = E_R(\omega, \mathbf{u})X = i\Lambda_R(\omega, \mathbf{u})X$$

or

$$\begin{aligned} t' &= \gamma_s(t - \beta x) \\ x' &= \gamma_s(x - \beta t) \\ y' &= iy \\ z' &= iz \end{aligned} \tag{2.35}$$

Comparing eq. 2.34 with eq. 2.15 for a left-handed superluminal boost we see that

$$PT\Lambda_L(\omega, \mathbf{u} = (1,0,0)) = \begin{bmatrix} -i\gamma_s & i\beta\gamma_s & 0 & 0 \\ i\beta\gamma_s & -i\gamma_s & 0 & 0 \\ 0 & 0 & -1 & 0 \\ 0 & 0 & 0 & -1 \end{bmatrix} \tag{2.36}$$

where P is the parity operator and T is the time reversal operator. If we now apply a spatial rotation $\mathcal{R}$ of π radians around the x axis then we obtain

$$\mathcal{R}PT\Lambda_L(\omega, \mathbf{u} = (1,0,0))\mathcal{R}^{-1} = \begin{bmatrix} -i\gamma_s & i\beta\gamma_s & 0 & 0 \\ i\beta\gamma_s & -i\gamma_s & 0 & 0 \\ 0 & 0 & 1 & 0 \\ 0 & 0 & 0 & 1 \end{bmatrix} \tag{2.37}$$

$$= \Lambda_R(\omega, \mathbf{u} = (1,0,0))$$

by eq. 2.34. Since P and T commute with spatial rotations we find

$$\Lambda_R(\omega, \mathbf{u} = (1,0,0)) = PT\mathcal{R}\Lambda_L(\omega, \mathbf{u} = (1,0,0))\mathcal{R}^{-1} \tag{2.38}$$

or, more generally, performing additional spatial rotations:

$$\Lambda_R(\omega, \mathbf{u}) = PT\mathcal{R}_u\mathcal{R}\mathcal{R}_w\Lambda_L(\omega, \mathbf{w})\mathcal{R}_w^{-1}\mathcal{R}^{-1}\mathcal{R}_u^{-1} \tag{2.39}$$

or,

$$\Lambda_R(\omega, \mathbf{u}) = PT\mathcal{R}_{tot}\Lambda_L(\omega, \mathbf{w})\mathcal{R}_{tot}^{-1} \tag{2.40}$$

where **u** and **w** are unit vectors. Alternately,

$$\Lambda_L(\omega, \mathbf{w}) = PT\mathcal{R}_{tot}^{-1}\Lambda_R(\omega, \mathbf{u})\mathcal{R}_{tot} \tag{2.41a}$$

or

$$\Lambda_L(\omega, \mathbf{w}) = PT\Lambda_R(\omega, \mathbf{u}') \tag{2.41b}$$

for some unit vector **u'**.

Thus we have shown that PT can be used to relate left-handed and right-handed boosts in a one-to-one fashion. *The appearance of the parity operator P takes on great significance when we derive features of the Standard Model. The appearance of left-handed doublets and right-handed singlets stems directly from the implicit parity dependence of the left-handed extended Lorentz group.*

For the right-handed boost of eq. 2.34 the addition rule for the x-component of velocity can be computed for infinitesimal displacements in space and time:

$$\begin{aligned} v_x' = \Delta x' / \Delta t' &= (\Delta x\, \gamma_s - \Delta t\, \beta\gamma_s)/(\Delta t\, \gamma_s - \Delta x\, \beta\gamma_s) \\ &= (v_x - \beta)/(1 - \beta v_x) \end{aligned} \tag{2.42}$$

in the limit $\Delta t \rightarrow 0$ where the x component of a particle's velocity in the unprimed frame is $v_x = \Delta x/\Delta t$. Note if $v_x = 1$ then $v_x' = 1$. *Thus right-handed superluminal transformations also preserve the constancy of the speed of light in all reference frames.*

2.4 Inhomogeneous Left-Handed Extended Lorentz Group

The *Left-Handed Extended Lorentz group* consists of the elements of the Lorentz group plus left-handed superluminal transformations including pure boosts and combinations of boosts and spatial rotations. Thus homogeneous left-handed superluminal transformations have the general form:

$$\Lambda_L(\mathbf{v}, \boldsymbol{\theta}) = \exp[i\omega_L \hat{\mathbf{u}}\cdot\mathbf{K} + i\boldsymbol{\theta}\cdot\mathbf{J}] \tag{2.43}$$

where $\omega_L' = \omega + i\pi/2$, $\boldsymbol{\theta}$ is the angular vector, and $\mathbf{J}$ is the angular momentum operator vector. Inhomogeneous left-handed superluminal transformations, which include displacements, can be expressed as

$$\Lambda_L(\mathbf{v}, \boldsymbol{\theta}, \mathbf{d}) = \exp[i\omega_L \hat{\mathbf{u}}\cdot\mathbf{K} + i\boldsymbol{\theta}\cdot\mathbf{J} - i\mathbf{d}\cdot\mathbf{P}] \tag{2.44}$$

where **P** is the momentum operator vector and **d** is a displacement vector.

We have verified the group structure of the left-handed extended Lorentz group is indeed a group in the preceding section (eqs. 2.23 – 2.29).

We also note

$$\det \Lambda_L(\omega, \mathbf{u}) = \pm 1 \tag{2.45a}$$

by eq. 2.7.

The ordinary Lorentz group is divided into four disjoint subgroups that are often denoted:

$$L_+^\uparrow: \quad \det \Lambda(\omega, \mathbf{u}) = +1; \;\; \operatorname{sgn} \Lambda(\omega, \mathbf{u})^0{}_0 = +1$$

$$L_{-}^{\uparrow}: \quad \det \Lambda(\omega, \mathbf{u}) = -1; \;\; \text{sgn}\, \Lambda(\omega, \mathbf{u})^0{}_0 = +1$$

$$L_{+}^{\downarrow}: \quad \det \Lambda(\omega, \mathbf{u}) = +1; \;\; \text{sgn}\, \Lambda(\omega, \mathbf{u})^0{}_0 = -1 \qquad (2.45b)$$

$$L_{-}^{\downarrow}: \quad \det \Lambda(\omega, \mathbf{u}) = -1; \;\; \text{sgn}\, \Lambda(\omega, \mathbf{u})^0{}_0 = -1$$

where $\text{sgn}\, \Lambda(\omega, \mathbf{u})^0{}_0$ is the sign of the 00 component of the $\Lambda(\omega, \mathbf{u})$ matrix. The various subgroups are related by the discrete transformations of parity P and time reversal T:

$$L_{+}^{\uparrow} \xrightarrow{P} L_{-}^{\uparrow}$$

$$L_{+}^{\uparrow} \xrightarrow{PT} L_{+}^{\downarrow}$$

$$L_{+}^{\uparrow} \xrightarrow{T} L_{-}^{\downarrow}$$

The left-handed superluminal transformations are disjoint in a somewhat different way. By eq. 2.45a the determinants are ±1. However the 0-0 matrix element of eq. 2.7 gives

$$\Lambda_L{}^0{}_0{}^2 - \Sigma_i\, (\Lambda_L{}^i{}_0)^2 = 1 \qquad (2.46)$$

The representation of superluminal boosts (eq. 2.14) shows that each factor in eq. 2.46 is imaginary. Thus eq. 2.46 implies

$$\Sigma_i\, |\Lambda_L{}^i{}_0|^2 \geq 1 \qquad (2.47)$$

$$|\Lambda_L{}^0{}_0| \geq 0 \qquad \text{(not } \geq 1 \text{ as in eq. 2.45b)} \qquad (2.48)$$

where || indicates absolute value. Thus the magnitude of $\Lambda_L{}^0{}_0$ does not have a gap. Therefore left-handed superluminal transformations can be divided into two categories:

$${}_L L_{+}: \quad \det \Lambda_L(\omega, \mathbf{u}) = +1$$

$${}_L L_{-}: \quad \det \Lambda_L(\omega, \mathbf{u}) = -1 \qquad (2.49)$$

Under a PT transformation a left-handed superluminal transformation becomes a right handed superluminal transformation (eq. 2.41b).

Again the various disjoint pieces are related by the discrete transformations of parity P and time reversal T:

$$_L L_+ \xrightarrow{P} {}_L L_-$$

$$_L L_+ \xrightarrow{T} {}_L L_- \tag{2.50}$$

2.5 Inhomogeneous Right-Handed Extended Lorentz Group

The inhomogeneous right-handed extended Lorentz group[64] consists of the Lorentz group plus right-handed superluminal transformations that have the form:

$$\Lambda_R(\mathbf{v}, \boldsymbol{\theta}, \mathbf{d}) = \exp[i\omega_R\hat{\mathbf{u}}\cdot\mathbf{K} + i\boldsymbol{\theta}\cdot\mathbf{J} - i\mathbf{d}\cdot\mathbf{P}] \tag{2.51}$$

in general where $\omega_R = \omega - i\pi/2$.

2.6 The Complex Lorentz Group L(C)

The Complex Lorentz group consists of all complex transformations of a 4-dimensional complex space-time that satisfies eq. 2.1 when the transformations are expressed as 4 by 4 matrices composed of complex numbers. At first glance the complex Lorentz group would not appear to be relevant to our universe since our universe seems to be described by real space-time coordinates. However as Streater (1964) points out,[65] "It is essential in the proof of the PCT theorem." Thus the complex Lorentz group has a deep role in Quantum Field Theory. The real homogeneous Lorentz group has six parameters – three boost parameters and three rotation parameters. The homogeneous complex Lorentz group has 12 generators and 12 parameters – six boost parameters and six rotation parameters.

SL(2,C)⊗SL(2,C) Formulation of Complex Lorentz Group

The complex Lorentz group can be realized in matrix form as a map from the proper orthochronous Lorentz group $L_+^\uparrow$ to SL(2,C):

$$\Lambda(A) \rightarrow A \tag{2.52}$$

where

$$A = A^0 I + \mathbf{A}\cdot\boldsymbol{\tau} \tag{2.53}$$

with $\boldsymbol{\tau}$ a 3-vector of the 2 × 2 Pauli matrices. The four coeffcients A^0 and $\mathbf{A}$ are *complex* numbers. A has determinant one. The Lorentz transformation

[64] Since γ_s has branch points at $v = \pm 1$ which corresponds to $\omega = \pm\infty$ for the left-handed and right-handed groups there is a cut along the real ω axis between $-\infty$ and $+\infty$. Therefore the product of three left-handed Lorentz transformations does not yield a right-handed transformation (as might be supposed from eqs. 2.43 and 2.51) but rather a left-handed transformation on the second sheet. A transformation with $\omega + 3i\pi/2$ is not equivalent to any with $\omega - i\pi/2$.

[65] Streater (1964) p. 13.

$$x' = \Lambda(A)x \tag{2.54}$$

corresponds to the transformation

$$x'^0 I + \mathbf{x'}\cdot\boldsymbol{\tau} = A(x^0 I + \mathbf{x}\cdot\boldsymbol{\tau})A^* \tag{2.55}$$

where * denotes complex conjugation.

The proper complex Lorentz group is the direct product of SL(2, C) groups representing the proper Lorentz group: SL(2,C)⊗SL(2,C). The direct product is represented by the set of all pairs of 2 × 2 complex matrices of determinant one that satisfy the multiplication rule:

$$(A, B)(C, D) = (AC, BD) \tag{2.56}$$

To each pair of 2 × 2 matrices there corresponds a 4 × 4 complex matrix Λ(A, B). These complex matrices satisfy

$$\Lambda(A, B)\Lambda(C, D) = \Lambda(AC, BD) \tag{2.57}$$

The coordinate transformations of this complex Lorentz group representation has the form

$$x'^0 I + \mathbf{x'}\cdot\boldsymbol{\tau} = A(x^0 I + \mathbf{x}\cdot\boldsymbol{\tau})B^T \tag{2.58}$$

where the superscript T represents transpose. If we express A and B in terms of 2 × 2 matrices:

$$A = A^0 I + \mathbf{A}\cdot\boldsymbol{\tau} \quad \text{and} \quad B = B^0 I + \mathbf{B}\cdot\boldsymbol{\tau} \tag{2.59}$$

we must use complex values for the parameters A^0, **A**, B^0, and **B**.

2.7 A Different Complex Lorentz Group, L_C

There is another possible formulation of a complex homogeneous Lorentz group based on the complexification[66] of the 4 × 4 matrix proper, orthochronous representation of boosts, with matrices $\Lambda(\omega, \hat{\mathbf{u}})$, discussed earlier together with complex spatial rotations, the parity transformation and the time reversal transformation and products thereof. We will denote this group as homogeneous L_C since its matrices contain complex entries. L_C has 6 complex parameters – three complex boost parameters and three complex rotation parameters just as the standard formulation of

---

[66] Just as in the case of the SL(2,C)⊗SL(2,C) representation where the parameters of both A and B (eq. 2.59) are complex.

the homogeneous complex Lorentz group SL(2,C)⊗SL(2,C). *L_C only has three boost operators and three rotation operators.*

It can be shown that the transformations of L_C satisfy the defining relation of the Lorentz group, eq. 2.1 since the algebraic proof of eq. 2.1 does not depend on the reality of the parameters. Thus the group L_C is a subgroup of the complex Lorentz group L(C).

In this subsection we will describe some of the basic features 4 × 4 matrix representation of L_C for use in the next chapter in the definition of a set of four spin ½ wave equations, and wave functions, that correspond to the four species of leptons and quarks of each generation.

The group elements of (homogeneous) L_C can be expressed as

$$\Lambda_C = \exp[i(\omega_r\hat{\mathbf{u}}_r + i\omega_i\hat{\mathbf{u}}_i)\cdot\mathbf{K} + i\boldsymbol{\theta}_c\cdot\mathbf{J}] \tag{2.60}$$

where the vector $\boldsymbol{\theta}_c$ is a complex 3-vector, $\omega_r \geq 0$ and $\omega_i \geq 0$ are real numbers, and $\hat{\mathbf{u}}_r$ and $\hat{\mathbf{u}}_i$ are real normalized 3-vectors such that $\hat{\mathbf{u}}_r\cdot\hat{\mathbf{u}}_r = 1 = \hat{\mathbf{u}}_i\cdot\hat{\mathbf{u}}_i$. The generators of the homogeneous group are **K**, and **J** just as for the real Lorentz group.

Matrix Representation of L_C Group Boost Transformations

We shall now turn our attention to L_C boosts because they will be crucial in the determination of the equations of motion of various types of spin ½ particles in chapter 3. From this analysis we will be able to reestablish the previous results of this chapter from a slightly deeper, more physical, viewpoint as well as to develop a more general picture.

An L_C boost can be expressed in the form

$$\Lambda_C(\mathbf{v}_c) = \exp[i\omega\hat{\mathbf{w}}\cdot\mathbf{K}] \tag{2.61}$$

where

$$\omega = (\omega_r^2 - \omega_i^2 + 2i\omega_r\omega_i\,\hat{\mathbf{u}}_r\cdot\hat{\mathbf{u}}_i)^{1/2} \tag{2.62}$$

and

$$\hat{\mathbf{w}} = (\omega_r\hat{\mathbf{u}}_r + i\omega_i\hat{\mathbf{u}}_i)/\omega \tag{2.63}$$

Since $\hat{\mathbf{u}}_r\cdot\hat{\mathbf{u}}_r = 1 = \hat{\mathbf{u}}_i\cdot\hat{\mathbf{u}}_i$

$$\hat{\mathbf{w}}\cdot\hat{\mathbf{w}} = 1 \tag{2.64a}$$

and the complex relative velocity is

$$\mathbf{v}_c = \hat{\mathbf{w}}\tanh(\omega) \tag{2.64b}$$

in analogy with eqs. 2.3 and 2.9.

Having placed L_C boost transformations in the form of eq. 2.3 we can take advantage of the form of real proper orthchronous Lorentz boost transformations, eq. 2.4, and analytically continue to complex ω and complex unit vectors $\hat{\mathbf{w}}$ provided eq.

2.64 is satisfied. The resulting complex generalization will be the matrix form of proper L_C boosts:

$$\Lambda_C(\mathbf{v_c}) = \exp[i\omega\hat{\mathbf{w}}\cdot\mathbf{K}] \equiv \Lambda_C(\omega, \hat{\mathbf{w}})$$

$$= \begin{bmatrix} \cosh(\omega) & -\sinh(\omega)\hat{w}_x & -\sinh(\omega)\hat{w}_y & -\sinh(\omega)\hat{w}_z \\ -\sinh(\omega)\hat{w}_x & 1 + (\cosh(\omega) - 1)\hat{w}_x^2 & (\cosh(\omega) - 1)\hat{w}_x\hat{w}_y & (\cosh(\omega) - 1)\hat{w}_x\hat{w}_z \\ -\sinh(\omega)\hat{w}_y & (\cosh(\omega) - 1)\hat{w}_x\hat{w}_y & 1 + (\cosh(\omega) - 1)\hat{w}_y^2 & (\cosh(\omega) - 1)\hat{w}_y\hat{w}_z \\ -\sinh(\omega)\hat{w}_z & (\cosh(\omega) - 1)\hat{w}_x\hat{w}_z & (\cosh(\omega) - 1)\hat{w}_y\hat{w}_z & 1 + (\cosh(\omega) - 1)\hat{w}_z^2 \end{bmatrix}$$

(2.65)

Since analytic continuations are unique, the above form for $\Lambda_C(\mathbf{v_c})$ is well-defined and unique. It spans the complete set of proper L_C boosts.

Connection to the Extended Lorentz Groups

The form of Left-handed and Right-handed Extended Lorentz group boosts, eqs. 2.8, 2.14 and 2.33, suggests we re-express eq. 2.57 in a different form to bring out the connection of L_C boosts with Left-handed and Right-handed Extended Lorentz group boosts.

Left-handed Extended Lorentz Group

If we let

$$\hat{\mathbf{u}}_\mathbf{i} = \hat{\mathbf{u}}_\mathbf{r} \equiv \hat{\mathbf{u}} \tag{2.66}$$

so that the vector $\hat{\mathbf{u}}_\mathbf{i}$ is parallel to $\hat{\mathbf{u}}_\mathbf{r}$, and

$$\omega_i = \pi/2 \tag{2.67}$$

then $\Lambda_C(\mathbf{v_c})$ becomes a Left-handed Extended Lorentz boost:

$$\Lambda_C(\mathbf{v_c}) = \Lambda_L(\omega_r, \mathbf{u}) \tag{2.68}$$

by eq. 2.28.

Right-handed Extended Lorentz Group

If we let

$$\hat{\mathbf{u}}_\mathbf{i} = -\hat{\mathbf{u}}_\mathbf{r} \equiv -\hat{\mathbf{u}} \tag{2.69}$$

so that the vector $\hat{\mathbf{u}}_\mathbf{i}$ is anti-parallel to $\hat{\mathbf{u}}_\mathbf{r}$, and

$$\omega_i = -\pi/2 \tag{2.70}$$

then $\Lambda_C(\mathbf{v_c})$ becomes a Right-handed Extended Lorentz boost:

$$\Lambda_C(\mathbf{v_c}) = \Lambda_R(\omega_r\,,\, \mathbf{u}) \tag{2.71}$$

by eq. 2.33.

Difference between the Extended Lorentz Groups Reduced to Parallelism of $\hat{\mathbf{u}}_r$ and $\hat{\mathbf{u}}_i$

Since the Left-handed Extended Lorentz group leads to the Standard Model's left-handed features, it seems that the parallel case $\hat{\mathbf{u}}_i = \hat{\mathbf{u}}_r \equiv \hat{\mathbf{u}}$ is favored by Nature.[67] To some extent this concept of parallel vectors $\hat{\mathbf{u}}_i$ and $\hat{\mathbf{u}}_r$, which leads to the Left-handed Extended Lorentz group, is more intuitively satisfying then the anti-parallel case that leads to the Right-handed Extended Lorentz group. However, a deeper reason for Nature's choice remains to be found.

[67] It is possible that parity violation might disappear at ultra-high energies. Then we would view the parity symmetric theory as broken to the left-handed Standard Model currently established by experiment with right-handed parts at higher energy.

3. Free Spin ½ Particles – Leptons & Quarks

3.0 Chapter Overview

In this chapter we begin by developing dynamical equations for spin ½ particles based on the Extended Lorentz groups, and L_C, described in chapter 2. These spin ½ particles are conventional Dirac particles (Majorana particles are also allowed), spin ½ tachyons, and "color" versions of both types totalling four species. We will identify leptons and quarks with these fields.

3.1 Introduction

Tachyons are particles that move faster than the speed of light. As we saw in chapter 0 tachyons exist inside Black Holes, and within current theories – particularly SuperString theories.

Attempts to create canonical tachyon quantum field theories began in the 1960's. These attempts were made within the framework of the Lorentz group and, consequently, were limited to spin 0 theories since there are no finite dimensional representations of the Lorentz group for negative m^2 except for the one-dimensional representation. None of these attempts, or attempts since then, succeeded in creating a canonically quantized spin 0 tachyon quantum field theory.[68]

In this chapter we will formulate a free spin ½ tachyon Quantum Field theory. We choose to develop a spin ½ tachyon theory first because spin ½ particles (quarks and leptons) play an extraordinary role in the Standard Model. In chapter 4 we will consider bosonic tachyons.

We will develop our spin ½ tachyon theory from the "ground up" by applying a Left-Handed Extended Lorentz boost to the Dirac equation, and the Dirac spinor wave function, for a particle at rest. This procedure will give a tachyon spinor wave function, and the momentum space tachyon equation equivalent of the Dirac equation. Then we will obtain the coordinate space tachyon Dirac equation, define a Lagrangian, and proceed to create a canonical quantum field theory for spin ½ tachyons.

3.2 A Method for Deriving the Dirac Equation

In this section we will review a method of obtaining the Dirac equation by Lorentz boosts of the spinor wave function of a particle at rest. In the case of a Lorentz transformation the 4 x 4 matrix form of a Lorentz transformation of the Dirac matrices is

$$S^{-1}(\Lambda(v))\gamma^{\nu}S(\Lambda(v)) = \Lambda^{\nu}{}_{\mu}(v)\gamma^{\mu} \tag{3.1}$$

where $S(\Lambda(v))$ is

$$S(\Lambda(v)) = \exp(-i\omega\sigma_{0i}v_i/(2|\mathbf{v}|)) = \exp(-\omega\gamma^0\boldsymbol{\gamma}\cdot\mathbf{v}/(2|\mathbf{v}|))$$

---

[68] Blaha (2006), the first edition of this book.

$$= \cosh(\omega/2)I + \sinh(\omega/2)\gamma^0\boldsymbol{\gamma}\cdot\mathbf{p}/|\mathbf{p}| \qquad (3.2)$$

with $\omega = \text{arctanh}(|\mathbf{v}|)$, $\cosh(\omega/2) = [(E+m)/(2m)]^{1/2}$ and $\sinh(\omega/2) = |\mathbf{p}|[2m(E+m)]^{-1/2}$. Also

$$S^{-1}(\Lambda(v)) = \gamma^0 S^\dagger(\Lambda(v))\gamma^0 = \exp(\omega\gamma^0\boldsymbol{\gamma}\cdot\mathbf{v}/(2|\mathbf{v}|))$$

$$= \cosh(\omega/2)I - \sinh(\omega/2)\gamma^0\boldsymbol{\gamma}\cdot\mathbf{p}/|\mathbf{p}| \qquad (3.3)$$

A generic positive energy plane wave solution of the Dirac equation for a particle at rest with rest energy m is

$$\psi(x) = e^{-imt}w(0) \qquad (3.4)$$

with w(0) a four component spinor column vector. It satisfies the momentum space Dirac equation for a particle at rest:

$$(m\gamma^0 - m)e^{-imt}w(0) = 0 \qquad (3.5)$$

If we now apply $S(\Lambda(v))$ we find

$$0 = S(\Lambda(v))(m\gamma^0 - m)e^{-imt}w(0) = [mS(\Lambda(v))\gamma^0 S^{-1}(\Lambda(v)) - m]S(\Lambda(v))w(0)$$

A straightforward evaluation shows

$$mS(\Lambda(v))\gamma^0 S^{-1}(\Lambda(v)) = g_{\mu\nu}p^\mu\gamma^\nu = \not{p} \qquad (3.6)$$

where $p^0 = (p^2 + m^2)^{1/2}$, $\mathbf{p} = \gamma m\mathbf{v}$, and $p = |\mathbf{p}|$. In addition

$$S(\Lambda(v))w(0) = w(p) \qquad (3.7)$$

is a positive energy Dirac spinor. Therefore the Dirac equation in momentum space has the form:

$$(\not{p} - m)e^{-ip\cdot x}w(p) = 0 \qquad (3.8)$$

where the exponential factor, mt, is also boosted to $p\cdot x$. Eq. 3.8 implies the free, coordinate space Dirac equation:

$$(i\gamma^\mu\partial/\partial x^\mu - m)\psi(x) = 0 \qquad (3.9)$$

3.3 Derivation of the Tachyonic Dirac Equation

The Left-handed Extended Lorentz boost has the form:

$$\Lambda_L(\omega, \mathbf{u}) = \Lambda(\omega + i\pi/2, \mathbf{u}) = \exp[i\omega_L \hat{\mathbf{u}}\cdot\mathbf{K}] \qquad (3.10)$$

where $\omega_L = \omega + i\pi/2$ and

$$\cosh(\omega_L) = i\sinh(\omega) = -\gamma = i\gamma_s$$
$$\sinh(\omega_L) = i\cosh(\omega) = -\beta\gamma = i\beta\gamma_s \qquad (2.11)$$

with, $\beta = v > 1$, $\gamma_s = (\beta^2 - 1)^{-\frac{1}{2}}$, and $\boldsymbol{\omega \geq 0}$. Thus

$$\sinh(\omega) = \gamma_s$$
$$\cosh(\omega) = \beta\gamma_s \qquad (2.12)$$

The corresponding spinor transformation is:

$$S_L(\Lambda_L(\omega, \mathbf{u})) = \exp(-i\omega_L\sigma_{0i}v_i/(2|\mathbf{v}|)) = \exp(-\omega_L\gamma^0\boldsymbol{\gamma}\cdot\mathbf{v}/(2|\mathbf{v}|))$$
$$= \cosh(\omega_L/2)I + \sinh(\omega_L/2)\gamma^0\boldsymbol{\gamma}\cdot\mathbf{p}/|\mathbf{p}| \qquad (3.11)$$

The inverse transformation is

$$S_L^{-1}(\Lambda_L(\omega, \mathbf{u})) = \gamma^2\gamma^0 K^{-1} S_L^\dagger K \gamma^0\gamma^2 = \gamma^2\gamma^0 S_L^T \gamma^0\gamma^2 = \exp(\omega_L\gamma^0\boldsymbol{\gamma}\cdot\mathbf{v}/(2|\mathbf{v}|))$$
$$= \cosh(\omega_L/2)I - \sinh(\omega_L/2)\gamma^0\boldsymbol{\gamma}\cdot\mathbf{p}/|\mathbf{p}| \qquad (3.12)$$

where the superscript T denotes the transpose and K is the complex conjugation operator (that also appears in the time-reversal operator). Note that S_L is not unitary just as the equivalent spinor Lorentz transformation $S(\Lambda(v))$ is not unitary.

We can now apply a left-handed superluminal transformation to the generic positive energy plane wave solution of the Dirac equation for a particle of mass m at rest. The result is

$$0 = S_L(\Lambda_L(\omega, \mathbf{u}))(m\gamma^0 - m)e^{-imt}w(0)$$
$$= [mS_L\gamma^0 S_L^{-1} - m]e^{-imt}S_L w(0)$$

where $S_L = S_L(\Lambda_L(\omega, \mathbf{u}))$. After some algebra

$$mS_L\gamma^0 S_L^{-1} = m[\cosh(\omega_L)\gamma^0 - \sinh(\omega_L)\boldsymbol{\gamma}\cdot\mathbf{p}/|\mathbf{p}|]$$
$$= i\gamma^0 E - i\boldsymbol{\gamma}\cdot\mathbf{p} = i\not{p} \qquad (3.13)$$

using eqs. 2.11 and the tachyon energy and momentum expressions

$$\mathbf{p} = m\mathbf{v}\gamma_s \qquad\qquad E = m\gamma_s \tag{3.14}$$

Also

$$S_L w(0) = w_T(p') \tag{3.15}$$

is a tachyon spinor. See Appendix 3-A (at the end of this chapter) for a discussion of tachyon spinors.

The momentum space tachyonic Dirac equation is

$$(i\not{p} - m)e^{ip\cdot x}w_T(p) = 0 \tag{3.16}$$

where $p\cdot x = Et - \mathbf{p\cdot x}$ after performing a corresponding left-handed superluminal coordinate transformation in the exponential factor based on eq. 2.10d. Thus the positive energy wave is transformed into a negative energy wave by the superluminal transformation.

If we apply $i\not{p}$ to we find the tachyon mass condition is satisfied

$$-E^2 + \mathbf{p}^2 = m^2 \tag{3.17}$$

Transforming back to coordinate space we obtain the *tachyonic Dirac equation*:

$$(\gamma^\mu\partial/\partial x^\mu - m)\psi_T(x) = 0 \tag{3.18}$$

The “missing” factor of i in the first term of eq. 3.18 requires the Lagrangian to be different from the conventional Dirac Lagrangian in order for the Lagrangian to be real. The simplest, physically acceptable, free spin ½ tachyon Lagrangian density is:

$$\mathcal{L}_T = \psi_T^{\ S}(\gamma^\mu\partial/\partial x^\mu - m)\psi_T(x) \tag{3.19}$$

where

$$\psi_T^{\ S} = \psi_T^{\ \dagger}\, i\gamma^0\gamma^5 \tag{3.20}$$

The corresponding action is

$$I = \int d^4x\mathcal{L}_T \tag{3.21}$$

Appendix 3-B proves I is real. The Hamiltonian density is

$$\mathcal{H} = \pi_T\dot{\psi}_T - \mathcal{L} = i\psi_T^{\ \dagger}\gamma^5(\boldsymbol{\alpha}\cdot\nabla + \beta m)\psi_T = -i\psi_T^{\ \dagger}\gamma^5\dot{\psi}_T \tag{3.22}$$

using the tachyon Dirac equation to obtain the last equality. The reader will note that the tachyon Hamiltonian is Hermitian by explicit calculation up to an irrelevant total spatial divergence.

Probability Conservation Law

The tachyon Dirac equation implies a probability conservation law:

$$\partial\rho_5/\partial t = \nabla\cdot\mathbf{j}_5 \tag{3.23}$$

where

$$\rho_5 = \psi_T^{\dagger}\gamma^5\psi_T \qquad\qquad \mathbf{j}_5 = \psi_T^{\dagger}\gamma^5\boldsymbol{\alpha}\psi_T \tag{3.24}$$

We are thus led to define the conserved axial charge Q_5

$$Q_5 = \int d^3x\, \psi_T^{\dagger}\gamma^5\psi_T \tag{3.25}$$

Energy-Momentum Tensor

The tachyon energy-momentum tensor is

$$\mathfrak{I}_{T\mu\nu} = -\, g_{\mu\nu}\, \mathfrak{L}_T + \partial\mathfrak{L}_T/\partial(\partial\psi_T/\partial x_\mu)\, \partial\psi_T/\partial x^\nu \tag{3.26}$$

$$= i\psi_T^{\dagger}\gamma^0\gamma^5\gamma_\mu\partial\psi_T/\partial x^\nu \tag{3.27}$$

and thus the conserved energy and momentum are

$$P^0 = H = \int d^3x\, \mathfrak{I}_T^{00} = i\int d^3x\psi_T^{\dagger}\gamma^5(\boldsymbol{\alpha}\cdot\nabla + \beta m)\psi_T \tag{3.28}$$

and

$$P^I = \int D^3x\, \mathfrak{I}_T^{0I} = -\, I\int D^3x\, \Psi_T^{\dagger}\Gamma^5\partial\Psi_T/\partial X_I \tag{3.29}$$

Both the energy and momentum differ significantly from the corresponding quantities for conventional Dirac fields.

3.4 Tachyon Canonical Quantization

Having defined a suitable tachyon Lagrangian we can now proceed to its canonical quantization. The conjugate momentum can be calculated from the Lagrangian density eq. 3.19:

$$\pi_{Ta} = \partial\, \mathfrak{L}_T/\partial\dot{\psi}_{Ta} \equiv \partial\mathfrak{L}_T/\partial(\partial\psi_{Ta}/\partial t) = -i(\psi_T^{\dagger}\gamma^5)_a \tag{3.30}$$

The resulting non-zero, canonical anti-commutation relations are

$$\{\pi_{\mathrm{T}a}(\mathrm{x}),\ \psi_{\mathrm{Tb}}(\mathrm{x}')\} = \mathrm{i}\,\delta_{\mathrm{ab}}\,\delta^3(\mathrm{x}-\mathrm{x}')$$

or

$$\{\psi_{\mathrm{T}}{}^{\dagger}{}_{a}(\mathrm{x}),\ \psi_{\mathrm{Tb}}(\mathrm{x}')\} = -\,[\gamma^5]_{\mathrm{ab}}\,\delta^3(\mathrm{x}-\mathrm{x}') \tag{3.31}$$

At this point we might attempt to complete the canonical quantization procedure in the conventional manner by Fourier expanding the field and specifying anti-commutation relations for the Fourier component amplitudes. However the incompleteness of the set of plane waves, which are limited by the restriction $|\mathbf{p}| \geq \mathrm{m}$, causes the anti-commutator of the fields *not* to yield a $\delta^3(\mathrm{x}-\mathrm{x}')$. Thus the conventional approach fails to yield the required anti-commutation relations.[69]

Other approaches: 1) decompose the tachyon field into left-handed and right-handed parts and then second quantize each part; and 2) second quantize in light-front coordinates ($\mathrm{x}^{\pm} = (\mathrm{x}^0 \pm \mathrm{x}^3)/\sqrt{2}$). These approaches also both fail.[70]

The only approach that does succeed[71] is to decompose the tachyon field into left-handed and right-handed parts and then second quantize in light-front coordinates. We follow that procedure in the following subsections.

Separation into Left-Handed and Right-Handed Fields

We will use a transformed set of Dirac matrices to develop our left-handed and right-handed tachyon formulations:

$$\gamma^0 = \begin{bmatrix} 0 & -\mathrm{I} \\ -\mathrm{I} & 0 \end{bmatrix} \qquad \gamma^i = \begin{bmatrix} 0 & \sigma_i \\ -\sigma_i & 0 \end{bmatrix} \qquad \gamma^5 = \begin{bmatrix} \mathrm{I} & 0 \\ 0 & -\mathrm{I} \end{bmatrix} \tag{3.32}$$

which are obtained from the usual Dirac matrices by applying the unitary transformation $\mathrm{U} = 2^{-\frac{1}{2}}(\mathrm{I} + \gamma^5\gamma^0)$. I is the 4×4 identity matrix. The γ^5 chirality operator's eigenvalues define handedness: +1 corresponds to right-handed; and –1 corresponds to left-handed:

$$\gamma^5\psi_{\mathrm{L}} = -\,\psi_{\mathrm{L}} \qquad\qquad \gamma^5\psi_{\mathrm{R}} = \psi_{\mathrm{R}} \tag{3.33}$$

Consequently, we can define left-handed and right-handed tachyon fields with the projection operators:

$$\mathrm{C}^{\pm} = \tfrac{1}{2}(\mathrm{I} \pm \gamma^5)$$

---

[69] See G. Feinberg, Phys. Rev. **159**, 1089 (1967) for example.

[70] See the first edition Blaha (2006) where these possibilities were considered and found to fail.

[71] Blaha (2006) discusses this case in detail.

$$C^+ + C^- = I \qquad (3.34)$$
$$C^{\pm\,2} = C^\pm$$
$$C^+C^- = 0$$

with the result

$$\psi_{TL} = C^-\psi_T \qquad (3.35)$$
$$\psi_{TR} = C^+\psi_T$$

We can calculate the commutation relations of the left-handed and right-handed tachyon fields from eq. 3.31 by pre-multiplying and post-multiplying by ½(1 – γ^5) and ½(1 + γ^5). The results are:

$$\{\psi_{TLa}{}^\dagger(x), \psi_{TLb}(x')\} = ½(1-\gamma^5)_{ab}\,\delta^3(x-x') \qquad (3.36)$$
$$\{\psi_{TRa}{}^\dagger(x), \psi_{TRb}(x')\} = -½(1+\gamma^5)_{ab}\,\delta^3(x-x') \qquad (3.37)$$
$$\{\psi_{TLa}{}^\dagger(x), \psi_{TRb}(x')\} = \{\psi_{TRa}{}^\dagger(x), \psi_{TLb}(x')\} = 0 \qquad (3.38)$$

The Lagrangian density of eq. 3.19 decomposes into left-handed and right-handed parts:

$$\mathcal{L}_T = \psi_{TL}{}^\dagger\gamma^0 i\gamma^\mu\partial_\mu\psi_{TL} - \psi_{TR}{}^\dagger\gamma^0 i\gamma^\mu\partial_\mu\psi_{TR} - im[\psi_{TR}{}^\dagger\gamma^0\psi_{TL} - \psi_{TL}{}^\dagger\gamma^0\psi_{TR}] \qquad (3.39)$$

Further Separation into + and – Light-Front Fields

There have been many studies of light-front (infinite momentum frame) physics in the past forty years.[72] Light-front coordinates cannot be obtained by a Lorentz transformation, or by a superluminal transformation, from a standard set of coordinate system variables even in a limiting sense. Instead they are a defined set of variables that have been used to develop quantum field theories that have been shown to be equivalent to quantum field theories based on conventional coordinates. In particular, light-front quantum field theories have been shown to yield fully Lorentz covariant S matrix elements that are the same as S matrix elements calculated in the conventional way.

Light-front variables can be defined by:

$$x^\pm = (x^0 \pm x^3)/\surd 2 \qquad (3.40)$$
$$\partial/\partial x^\pm \equiv \partial^\mp \equiv (\partial/\partial x^0 \pm \partial/\partial x^3)/\surd 2$$

with the "transverse" coordinate variables, x^1 and x^2, unchanged.

---

[72] L. Susskind, Phys. Rev. **165**, 1535 (1968); K. Bardakci and M. B. Halpern Phys. Rev. **176**, 1686 (1968), S. Weinberg, Phys. Rev. **150**, 1313 (1966); J. Kogut and D. Soper, Phys. Rev. **D1**, 2901 (1970); J. D. Bjorken, J. Kogut, and D. Soper, Phys. Rev. **D3**, 1382 (1971); R. A. Neville and F. Rohrlich, Nuov. Cim. **A1**, 625 (1971); F. Rohrlich, Acta Phys Austr. Suppl. **8**, 277 (1971); S-J Chang, R. Root, and T-M Yan, Phys. Rev. **D7**, 1133 (1973); S-J Chang, and T-M Yan, Phys. Rev. **D7**, 1147 (1973); T-M Yan, Phys. Rev. **D7**, 1761 (1973); T-M Yan, Phys. Rev. **D7**, 1780 (1973); C. Thorn, Phys. Rev. **D19**, 639 (1979); and references therein.

The inner product of two 4-vectors has the form

$$x{\cdot}y = x^+y^- + y^+x^- - x^1y^1 - x^2y^2 \tag{3.41}$$

and the light-front definition of Dirac matrices is:

$$\gamma^\pm = (\gamma^0 \pm \gamma^3)/\sqrt{2} \tag{3.42}$$

with transverse matrices γ^1 and γ^2 defined as usual. Note the useful identity:

$$\gamma^{\pm\,2} = 0$$

We define "+" and "–" tachyon fields with the projection operators:

$$R^\pm = ½(I \pm \gamma^0\gamma^3) \tag{3.43}$$

Left-handed, ± light-front fields: $\quad \psi_{TL}{}^\pm = R^\pm C^-\psi_T$

Right-handed, ± light-front fields: $\quad \psi_{TR}{}^\pm = R^\pm C^+\psi_T$ (3.44)

Now if we transform to light-front variables and fields as above we obtain the light-front free tachyon Lagrangian:

$$\begin{aligned}\mathcal{L}_T = {} & 2^{½}\psi_{TL}{}^{+\dagger}i\partial^-\psi_{TL}{}^+ + 2^{½}\psi_{TL}{}^{-\dagger}i\partial^+\psi_{TL}{}^- - \psi_{TL}{}^{+\dagger}\gamma^0 i\gamma^j\partial^j\psi_{TL}{}^- - \psi_{TL}{}^{-\dagger}\gamma^0 i\gamma^j\partial^j\psi_{TL}{}^+ - \\ & - 2^{½}\psi_{TR}{}^{+\dagger}i\partial^-\psi_{TR}{}^+ - 2^{½}\psi_{TR}{}^{-\dagger}i\partial^+\psi_{TR}{}^- + \psi_{TR}{}^{+\dagger}\gamma^0 i\gamma^j\partial^j\psi_{TR}{}^- + \psi_{TR}{}^{-\dagger}\gamma^0 i\gamma^j\partial^j\psi_{TR}{}^+ - \\ & - im[\psi_{TR}{}^{+\dagger}\gamma^0\psi_{TL}{}^- - \psi_{TL}{}^{+\dagger}\gamma^0\psi_{TR}{}^- + \psi_{TR}{}^{-\dagger}\gamma^0\psi_{TL}{}^+ - \psi_{TL}{}^{-\dagger}\gamma^0\psi_{TR}{}^+]\end{aligned} \tag{3.45}$$

with implied sums over j = 1,2. In contrast to the light-front tachyon Lagrangian we note the corresponding light-front Dirac fermion Lagrangian is

$$\begin{aligned}\mathcal{L}_D = {} & 2^{½}\psi_L{}^{+\dagger}i\partial^-\psi_L{}^+ + 2^{½}\psi_L{}^{-\dagger}i\partial^+\psi_L{}^- - \psi_L{}^{+\dagger}\gamma^0 i\gamma^j\partial^j\psi_L{}^- - \psi_L{}^{-\dagger}\gamma^0 i\gamma^j\partial^j\psi_L{}^+ - \\ & + 2^{½}\psi_R{}^{+\dagger}i\partial^-\psi_R{}^+ + 2^{½}\psi_R{}^{-\dagger}i\partial^+\psi_R{}^- - \psi_R{}^{+\dagger}\gamma^0 i\gamma^j\partial^j\psi_R{}^- - \psi_R{}^{-\dagger}\gamma^0 i\gamma^j\partial^j\psi_R{}^+ - \\ & - im[\psi_R{}^{+\dagger}\gamma^0\psi_L{}^- + \psi_L{}^{+\dagger}\gamma^0\psi_R{}^- + \psi_R{}^{-\dagger}\gamma^0\psi_L{}^+ + \psi_L{}^{-\dagger}\gamma^0\psi_R{}^+]\end{aligned} \tag{3.46}$$

The difference in signs between eqs. 3.45 and 3.46 will turn out to be a crucial factor in the derivation of features of the Standard Model later.

Returning to the tachyon Lagrangian eq. 3.45 we obtain equations of motion through the standard variational techniques:

$$
\begin{aligned}
2^{1/2} i\partial^- \psi_{TL}{}^+ - \gamma^0 i\gamma^j \partial^j \psi_{TL}{}^- + im\gamma^0 \psi_{TR}{}^- &= 0 \\
2^{1/2} i\partial^- \psi_{TR}{}^+ - \gamma^0 i\gamma^j \partial^j \psi_{TR}{}^- + im\gamma^0 \psi_{TL}{}^- &= 0 \\
2^{1/2} i\partial^+ \psi_{TL}{}^- - \gamma^0 i\gamma^j \partial^j \psi_{TL}{}^+ + im\gamma^0 \psi_{TR}{}^+ &= 0 \\
2^{1/2} i\partial^+ \psi_{TR}{}^- - \gamma^0 i\gamma^j \partial^j \psi_{TR}{}^+ + im\gamma^0 \psi_{TL}{}^+ &= 0
\end{aligned} \tag{3.47}
$$

Eqs. 3.47 show that $\psi_{TL}{}^-$ and $\psi_{TR}{}^-$ are dependent fields that are functions of $\psi_{TL}{}^+$ and $\psi_{TR}{}^+$ on the light-front where x^+ equals a constant. They can be expressed in an integral form as well. (The independent fields $\psi_{TL}{}^+$ and $\psi_{TR}{}^+$ play a fundamental role in tachyon theory and are used to define "in" and "out" tachyon states in perturbation theory.)

The conjugate momenta implied by eq. 3.45 are

$$
\begin{aligned}
\pi_{TL}{}^+ &= \partial\mathcal{L}/\partial(\partial^- \psi_{TL}{}^+) = 2^{1/2} i\psi_{TL}{}^{+\dagger} && (3.48)\\
\pi_{TL}{}^- &= \partial\mathcal{L}/\partial(\partial^- \psi_{TL}{}^-) = 0 \\
\pi_{TR}{}^+ &= \partial\mathcal{L}/\partial(\partial^- \psi_{TR}{}^+) = -2^{1/2} i\psi_{TR}{}^{+\dagger} && (3.49)\\
\pi_{TR}{}^- &= \partial\mathcal{L}/\partial(\partial^- \psi_{TR}{}^-) = 0
\end{aligned}
$$

Quantization on surfaces of constant x^+ (light-front surfaces) has been shown to support satisfactory formulations of Quantum Electrodynamics and other quantum field theories. Thus x^+ plays the role of the "time" variable in light-front quantized theories. So we will define canonical equal x^+ anti-commutation relations for spin ½ tachyons.

The resulting canonical equal-light-front ($x^+ = y^+$) anti-commutation relations of the independent fields are:

$$\{\psi_{TL}{}^{+\dagger}{}_a(x), \psi_{TL}{}^+{}_b(y)\} = 2^{-1}[C^- R^+]_{ab}\, \delta(x^- - y^-)\delta^2(x - y) \tag{3.50}$$

$$\{\psi_{TR}{}^{+\dagger}{}_a(x), \psi_{TR}{}^+{}_b(y)\} = -2^{-1}[C^+ R^+]_{ab}\, \delta(x^- - y^-)\delta^2(x - y) \tag{3.51}$$

$$\{\psi_{TL}{}^{+\dagger}{}_a(x), \psi_{TR}{}^+{}_b(y)\} = \{\psi_{TR}{}^{+\dagger}{}_a(x), \psi_{TL}{}^+{}_b(y)\} = 0 \tag{3.52}$$

$$\{\psi_{TL}{}^+{}_a(x), \psi_{TR}{}^+{}_b(y)\} = \{\psi_{TR}{}^{+\dagger}{}_a(x), \psi_{TL}{}^{+\dagger}{}_b(y)\} = 0 \tag{3.53}$$

where the factors of 2^{-1} are the result of the $2^{1/2}$ factor in eqs. 3.48 and 3.49, and the factor of $2^{-1/2}$ in the definition of x^- in eq. 3.40.

If we compare eqs. 3.50 and 3.51 with the corresponding anti-commutation relations of conventional Dirac quantum fields:

$$\{\psi_L{}^{+\dagger}{}_a(x), \psi_L{}^{+}{}_b(y)\} = 2^{-1}[C^-R^+]_{ab}\, \delta(x^- - y^-)\delta^2(x - y) \tag{3.54}$$
$$\{\psi_R{}^{+\dagger}{}_a(x), \psi_R{}^{+}{}_b(y)\} = 2^{-1}[C^+R^+]_{ab}\, \delta(x^- - y^-)\delta^2(x - y) \tag{3.55}$$

we see that the right-handed tachyon anti-commutation relation (eq. 3.51) has a minus sign relative to the corresponding right-handed conventional anti-commutation relation (eq. 3.55). The right-handed tachyon anti-commutation relation (eq. 3.51) with its minus sign will require compensating minus signs in its creation and annihilation Fourier component operators' anti-commutation relations.

The sign differences between the Lagrangian terms in eqs. 3.47 and 3.48 ultimately lead to parity violating features in the Standard Model Lagrangian and thus resolve the long-standing question: Why parity violation? Answer: Nature chooses the Left-handed Extended Lorentz group. Thus the source of parity violation, and much of the form of the Standard Model, is in superluminal physics.

Left-Handed Tachyons

The free, "+" light-front, left-handed tachyon wave function Fourier expansion is:

$$\psi_{TL}{}^{+}(x) = \sum_{\pm s} \int d^2p dp^+ N_{TL}{}^{+}(p)\theta(p^+)[b_{TL}{}^{+}(p, s)u_{TL}{}^{+}(p, s)e^{-ip.x} + \\ + d_{TL}{}^{+\dagger}(p, s)v_{TL}{}^{+}(p, s)e^{+ip.x}] \tag{3.56}$$

and its Hermitian conjugate is

$$\psi_{TL}{}^{+\dagger}(x) = \sum_{\pm s} \int d^2p dp^+ N_{TL}{}^{+}(p)\theta(p^+)\, [b_{TL}{}^{+\dagger}(p, s)u_{TL}{}^{+\dagger}(p,s)e^{+ip.x} + \\ + d_{TL}{}^{+}(p, s)v_{TL}{}^{+\dagger}(p, s)e^{-ip.x}] \tag{3.57}$$

where † indicates Hermitian conjugate, where

$$N_{TL}{}^{+}(p) = [2m|\mathbf{p}|/((2\pi)^3(p^+(p^+ - p^-) + p_\perp{}^2))]^{1/2} \tag{3.57a}$$

where the anti-commutation relations of the Fourier coefficient operators are

$$\begin{aligned}
&\{b_{TL}{}^{+}(q,s), b_{TL}{}^{+\dagger}(p,s')\} = \delta_{ss'}\delta^2(\mathbf{q} - \mathbf{p})\delta(q^+ - p^+) \\
&\{d_{TL}{}^{+}(q,s), d_{TL}{}^{+\dagger}(p,s')\} = \delta_{ss'}\delta^2(\mathbf{q} - \mathbf{p})\delta(q^+ - p^+) \\
&\{b_{TL}{}^{+}(q,s), b_{TL}{}^{+}(p,s')\} = \{d_{TL}{}^{+}(q,s), d_{TL}{}^{+}(p,s')\} = 0 \\
&\{b_{TL}{}^{+\dagger}(q,s), b_{TL}{}^{+\dagger}(p,s')\} = \{d_{TL}{}^{+\dagger}(q,s), d_{TL}{}^{+\dagger}(p,s')\} = 0 \\
&\{b_{TL}{}^{+}(q,s), d_{TL}{}^{+\dagger}(p,s')\} = \{d_{TL}{}^{+}(q,s), b_{TL}{}^{+\dagger}(p,s')\} = 0 \\
&\{b_{TL}{}^{+\dagger}(q,s), d_{TL}{}^{+\dagger}(p,s')\} = \{d_{TL}{}^{+}(q,s), b_{TL}{}^{+}(p,s')\} = 0
\end{aligned} \tag{3.58}$$

and where the spinors are

$$
\begin{aligned}
u_{TL}^{+}(p, s) &= C^- R^+ S_L(\Lambda_L(\mathbf{p}))w^1(0)\\
u_{TL}^{+}(p, -s) &= C^- R^+ S_L(\Lambda_L(\mathbf{p}))w^2(0)\\
v_{TL}^{+}(p, s) &= C^- R^+ S_L(\Lambda_L(\mathbf{p}))w^3(0)\\
v_{TL}^{+}(p, -s) &= C^- R^+ S_L(\Lambda_L(\mathbf{p}))w^4(0)
\end{aligned}
\tag{3.59}
$$

$$
\begin{aligned}
u_{TL}^{+\dagger}(p, s) &= w^{1T}(0)S_L^{\dagger}(\Lambda_L(\mathbf{p}))R^+C^-\\
u_{TL}^{+\dagger}(p, -s) &= w^{2T}(0)S_L^{\dagger}(\Lambda_L(\mathbf{p}))R^+C^-\\
v_{TL}^{+\dagger}(p, s) &= w^{3T}(0)S_L^{\dagger}(\Lambda_L(\mathbf{p}))R^+C^-\\
v_{TL}^{+\dagger}(p, -s) &= w^{4T}(0)S_L^{\dagger}(\Lambda_L(\mathbf{p}))R^+C^-
\end{aligned}
$$

where the superscript "T" indicates the transpose. (These spinors are described in Appendix 3-A.)

The canonical left-handed, light-front anti-commutation relation (eq. 3.50) follows from eqs. 3.56 – 3.59:

$$
\{\psi_{TL}^{+}{}_a(x), \psi_{TL}^{+\dagger}{}_b(y)\} = \sum_{\pm s,s'} \int d^2p dp^+ \int d^2p' dp'^+ \, N_{TL}^{+}(p) N_{TL}^{+}(p') \theta(p^+) \theta(p'^+) \cdot
$$

$$
\cdot[\{b_{TL}^{+\dagger}(p',s'), b_{TL}^{+}(p,s)\} u_{TL}^{+}{}_a(p,s) u_{TL}^{+\dagger}{}_b(p',s') e^{+ip'\cdot y - ip\cdot x} +
$$

$$
+ \{d_{TL}^{+}(p',s'), d_{TL}^{+\dagger}(p,s)\} v_{TL}^{+}{}_a(p,s) v_{TL}^{+\dagger}{}_b(p',s') e^{-ip'\cdot y + ip\cdot x}]
$$

$$
= \sum_{\pm s} \int d^2p dp^+ \, N_{TL}^{+2}(p) \theta(p^+) [u_{TL}^{+}{}_a(p,s) u_{TL}^{+\dagger}{}_b(p,s) e^{+ip\cdot(y-x)} +
$$

$$
+ v_{TL}^{+}{}_a(p,s) v_{TL}^{+\dagger}{}_b(p,s) e^{-ip\cdot(y-x)}]
$$

$$
= -i\int d^2p dp^+ \theta(p^+) N_{TL}^{+2}(p)(2m|\mathbf{p}|)^{-1}\{[C^-R^+(i\not{p} - m)\gamma\cdot\mathbf{p}R^+C^-]_{ab} e^{+ip\cdot(y-x)} +
$$

$$
+ [C^-R^+(i\not{p} + m)\gamma\cdot\mathbf{p}R^+C^-]_{ab} e^{-ip\cdot(y-x)}\}
$$

$$
= -i\int d^2p_\perp \int_0^\infty dp^+ N_{TL}^{+2}(p)\{[C^-R^+(ip^+(p^+ - p^-) + ip_\perp{}^2 - mp_\perp\cdot\gamma_\perp)C^-]_{ab} \cdot
$$

$$\cdot e^{+ip^+(y^- - x^-) - ip_\perp\cdot(y_\perp - x_\perp)} -$$

$$- [C^- R^+(-ip^+(p^+ - p^-) - ip_\perp^2 - mp_\perp\cdot\gamma_\perp)C^-]_{ab} e^{-ip^+(y^- - x^-) + ip_\perp\cdot(y_\perp - x_\perp)}\}/(2m|\mathbf{p}|)$$

$$= \int d^2p_\perp \int_{-\infty}^{\infty} dp^+ N_{TL}^{+2}(p)[C^- R^+(p^+(p^+ - p^-) + p_\perp^2)]_{ab}\cdot$$
$$\cdot e^{+ip^+(y^- - x^-) - ip_\perp\cdot(y_\perp - x_\perp)}/(2m|\mathbf{p}|)$$

upon letting $p^+ \to -p^+$ and $\mathbf{p}_\perp \to -\mathbf{p}_\perp$ in the second term after using $N_{TL}^{+2}(p)(p^+(p^+ - p^-) + p_\perp^2) = 1$. The result

$$= ½\int d^2p_\perp \int_{-\infty}^{\infty} dp^+ (2\pi)^{-3}[C^- R^+]_{ab} e^{+ip^+(y^- - x^-) - ip_\perp\cdot(y_\perp - x_\perp)}$$

$$= 2^{-1}[C^- R^+]_{ab}\,\delta(y^- - x^-)\delta^2(\mathbf{y} - \mathbf{x}) \qquad (3.60)$$

Therefore we have left-handed, light-front quantized tachyons with canonical commutation relations and localized tachyons. As a result we have a canonical tachyon Quantum Field Theory.

Right-Handed Tachyons

The case of right-handed tachyons is similar to the left-handed case with only two differences: a minus sign in the creation and annihilation operator anti-commutation relations, and the use of right-handed projection operators. The right-handed tachyon wave function light-front Fourier expansion is:

$$\psi_{TR}^+(x) = \sum_{\pm s}\int d^2p\,dp^+ N_{TR}^+(p)\theta(p^+)[b_{TR}^+(p, s)u_{TR}^+(p, s)e^{-ip.x} +$$
$$+ d_{TR}^{+\dagger}(p, s)v_{TR}^+(p, s)e^{+ip.x}] \qquad (3.61)$$

and its Hermitian conjugate is

$$\psi_{TR}^{+\dagger}(x) = \sum_{\pm s}\int d^2p\,dp^+ N_{TR}^+(p)\theta(p^+)\,[b_{TR}^{+\dagger}(p, s)u_{TR}^{+\dagger}(p, s)e^{+ip.x} +$$
$$+ d_{TR}^+(p, s)v_{TR}^{+\dagger}(p, s)e^{-ip.x}] \qquad (3.62)$$

where $N_{TR}^+(p) = N_{TL}^+(p)$, where the anti-commutation relations of the Fourier coefficient operators are

$$\{b_{TR}^{\ +}(q,s),\ b_{TR}^{\ +\dagger}(p,s')\} = -\delta_{ss'}\delta^2(\mathbf{q} - \mathbf{p})\delta(q^+ - p^+) \qquad (3.63)$$
$$\{d_{TR}^{\ +}(q,s),\ d_{TR}^{\ +\dagger}(p,s')\} = -\delta_{ss'}\delta^2(\mathbf{q} - \mathbf{p})\delta(q^+ - p^+)$$
$$\{b_{TR}^{\ +}(q,s),\ b_{TR}^{\ +}(p,s')\} = \{d_{TR}^{\ +}(q,s),\ d_{TR}^{\ +}(p,s')\} = 0$$
$$\{b_{TR}^{\ +\dagger}(q,s),\ b_{TR}^{\ +\dagger}(p,s')\} = \{d_{TR}^{\ +\dagger}(q,s),\ d_{TR}^{\ +\dagger}(p,s')\} = 0$$
$$\{b_{TR}^{\ +}(q,s),\ d_{TR}^{\ +\dagger}(p,s')\} = \{d_{TR}^{\ +}(q,s),\ b_{TR}^{\ +\dagger}(p,s')\} = 0$$
$$\{b_{TR}^{\ +\dagger}(q,s),\ d_{TR}^{\ +\dagger}(p,s')\} = \{d_{TR}^{\ +}(q,s),\ b_{TR}^{\ +}(p,s')\} = 0$$

and where the spinors are

$$u_{TR}^{\ +}(p,\ s) = C^+R^+u_T(p,s) \qquad (3.64)$$
$$v_{TR}^{\ +}(p,\ s) = C^+R^+v_T(p,s) \qquad (3.65)$$

by Appendix 3-A (eq. 3-A.7).

The right-handed anti-commutation relation (eq. 3.51) with the minus sign follows in particular because of the minus signs in eqs. 3.63.

Interpretation of Tachyon Creation and Annihilation Operators

To properly discuss the physical interpretation of tachyon creation and annihilation operators we must first determine the Hamiltonian and momentum operators in terms of creation and annihilation operators.

The energy-momentum tensor density is the symmetrized version of

$$\mathfrak{T}^{\mu\nu} = \sum_i \partial\mathfrak{L}/\partial(\partial\chi_i/\partial x_\mu)\ \partial\chi_i/\partial x_\nu - g^{\mu\nu}\mathfrak{L} \qquad (3.66)$$

where the sum over i is over the fields. The light-front Hamiltonian is

$$H \equiv P^- = T^{+-} = \int dx^- d^2x\,\mathfrak{T}^{+-} \qquad (3.67)$$

and the "momenta" are

$$P^+ = T^{++} = \int dx^- d^2x\,\mathfrak{T}^{++} \qquad (3.68)$$
$$P^i = T^{+i} = \int dx^- d^2x\,\mathfrak{T}^{+i} \qquad (3.69)$$

for i = 1,2.

The light-front, left-handed and right-handed tachyon Lagrangian $\mathfrak{L}_T$ is eq. 3.45 and its equations of motion are eqs. 3.47. They imply

$$H = i2^{-1/2}\int dx^- d^2x\ [\psi_{TL}^{\ +\dagger}\partial^-\psi_{TL}^{\ +} - \partial^-\psi_{TL}^{\ +\dagger}\psi_{TL}^{\ +} + \psi_{TL}^{\ -\dagger}\partial^+\psi_{TL}^{\ -} - \partial^+\psi_{TL}^{\ -\dagger}\psi_{TL}^{\ -} -$$

$$- \psi_{TR}^{+\dagger}\partial^{-}\psi_{TR}^{+} + \partial^{-}\psi_{TR}^{+\dagger}\psi_{TR}^{+} - \psi_{TR}^{-\dagger}\partial^{+}\psi_{TR}^{-} + \partial^{+}\psi_{TR}^{-\dagger}\psi_{TR}^{-} + \text{mass terms}] \tag{3.70}$$

After substituting for the various fields we find the *independent fields* (which create the in and out particle states) have the Hamiltonian terms:

$$H = \sum_{\pm s} \int d^2p dp^+ \, p^- [b_{TL}^{+\dagger}(p,s) b_{TL}^{+}(p,s) - d_{TL}^{+}(p,s) d_{TL}^{+\dagger}(p,s) - \\ - b_{TR}^{+\dagger}(p,s) b_{TR}^{+}(p,s) + d_{TR}^{+}(p,s) d_{TR}^{+\dagger}(p,s)] \tag{3.71}$$

$$= \sum_{\pm s} \int d^2p dp^+ \, p^- [b_{TL}^{+\dagger}(p,s) b_{TL}^{+}(p,s) + d_{TL}^{+\dagger}(p,s) d_{TL}^{+}(p,s) - \\ - b_{TR}^{+\dagger}(p,s) b_{TR}^{+}(p,s) - d_{TR}^{+\dagger}(p,s) d_{TR}^{+}(p,s)] \tag{3.72}$$

up to the usual infinite constants due to left-handed operator rearrangement and right-handed operator rearrangement that are discarded. Eq. 3.72 is the basis for our particle interpretation of tachyon creation and annihilation operators based on Dirac's hole theory. Dirac hole theory as applied in light-front coordinates assumes all negative p^- ("energy") states are filled.

Left-Handed Tachyon Creation and Annihilation Operators

1. We identify $b_{TL}^{+\dagger}(p,s)$ and $d_{TL}^{+}(p,s)$ as creation operators for left-handed tachyons. $b_{TL}^{+\dagger}(p,s)$ creates a positive p^- ("energy") state and $d_{TL}^{+}(p,s)$ creates a negative p^- ("energy") state.

2. $b_{TL}^{+}(p,s)$ and $d_{TL}^{+\dagger}(p,s)$ are the corresponding annihilation operators for left-handed tachyons. $b_{TL}^{+}(p,s)$ annihilates a positive p^- ("energy") state and $d_{TL}^{+\dagger}(p,s)$ annihilates a negative p^- ("energy") state.

3. We assume Dirac hole theory holds for the left-handed tachyon vacuum with all negative energy states filled. There is no tachyon energy gap as there is for Dirac fermions. There is also the problem that the left-handed tachyon vacuum is not invariant under ordinary Lorentz transformations or Superluminal transformations. *However if we confine ourselves to light-front coordinates for computations no ambiguity can result and the Lorentz covariant quantities that we calculate, such as the S matrix, are well-defined.*

4. Using tachyon hole theory we identify $b_{TL}^{+}(p,s)$ and $d_{TL}^{+\dagger}(p,s)$ as annihilation operators for left-handed tachyons. $b_{TL}^{+}(p,s)$ annihilates a positive p^- ("energy")

state and $d_{TL}^{++}(p,s)$ annihilates a negative p^- ("energy") state – thus creating a hole in the tachyon sea that we view as the creation of a positive p^- ("energy"), left-handed antitachyon. $d_{TL}^{+}(p,s)$ annihilates a positive p^- ("energy"), left-handed antitachyon.

Right-Handed Tachyons

The anti-commutation relations of right-handed tachyon creation and annihilation operators (eqs. 3.63) and the right-handed Hamiltonian terms have the "wrong" sign compared to corresponding Dirac operators and left-handed tachyon operators. This situation is completely analogous to the situation of time-like photons in the covariant formulation of quantum Electrodynamics.[73] In the case of time-like photons it was possible to introduce an indefinite metric (Gupta-Bleuler formulation), and then to use the subsidiary condition $\partial A^\nu/\partial x^\nu = 0$ to reduce the dynamics of QED to the transverse components. Thus the time-like photons were intermediate artifacts needed to have a manifestly covariant formulation while QED observables depended solely on the transverse components of the electromagnetic field.

In the present case of free tachyons, and in leptonic ElectroWeak Theory there is no evident "subsidiary condition" to eliminate the right-handed tachyon fields. But since the only manner in which the right-handed leptonic tachyon fields[74] interact is through mass terms, which can be easily 'integrated out", right-handed leptonic tachyon fields are removed from the observable part of the leptonic ElectroWeak Theory by their "lack of interaction" with left-handed fields.

In the case of quark ElectroWeak Theory right-handed tachyon quark fields have charge (–1/3) and thus experience an electromagnetic interaction as well as a Z interaction. However, since quarks are totally confined, right-handed tachyon quarks will not be able to continuously emit photons or Z's due to energy conservation and their confinement to bound states of fixed positive energy. Starting in section 2.6, when we consider complex Lorentz group boosts, we will suggest that quarks may not consist of Dirac particles or tachyons of the type considered up to this point in this chapter. Rather they may be variants on Dirac particles and tachyons satisfying different dynamical equations. However, the preceding comments on quarks would still apply.

Thus right-handed tachyons are analogous to time-like photons – necessary theoretically but prevented from causing a negative energy disaster by the forms of their interactions. We discuss this subject in more detail in the following chapters.

3.5 Tachyon Feynman Propagator

In this section we develop the light-front propagator for tachyons. We begin with a subsection describing the light-front propagators of Dirac fields.

Dirac Field Light-Front Propagators

The light-front Feynman propagator for the ψ^+ field of a Dirac fermion is

---

[73] Bogoliubov (1959) pp. 130-136.

[74] The tachyon fields are provisionally assumed to be neutrino fields in the leptonic sector, and d, s and b quarks in the quark sector.

$$iS^{+}{}_{F}(x,y)\gamma^{0} = \theta(x^{+} - y^{+})\langle 0|\psi^{+}(x)\psi^{+\dagger}(y)|0\rangle - \theta(y^{+} - x^{+})\langle 0|\psi^{+\dagger}(y)\psi^{+}(x)|0\rangle \tag{3.73}$$

and does not contain a non-covariant piece due to the projection operators:

$$\begin{aligned} iS^{+}{}_{F}(x,y) &= \int d^2p dp^{+}\theta(p^{+})[1/(2(2\pi)^3p^{+})]\{\theta(x^{+}-y^{+})[R^{+}(\not p + m)R^{-}]\, e^{-ip\cdot(x-y)} + \\ &\quad + \theta(y^{+}-x^{+})[R^{+}(-\not p + m)R^{-}]e^{+ip\cdot(x-y)}\} \\ &= R^{+}iS_F(x,y)R^{-} \end{aligned} \tag{3.74}$$

where $S_F(x,y)$ is the usual Feynman propagator.

The light-front Feynman propagator for a *left-handed* Dirac field ψ^{+} is

$$\begin{aligned} iS^{+}{}_{LF}(x,y) &= \int d^2p dp^{+}\theta(p^{+})[1/(2(2\pi)^3p^{+})]\{\theta(x^{+}-y^{+})[C^{-}R^{+}(\not p+m)R^{-}C^{-}]e^{-ip\cdot(x-y)} + \\ &\quad + \theta(y^{+}-x^{+})[C^{-}R^{+}(-\not p + m)R^{-}C^{-}]e^{+ip\cdot(x-y)}\} \\ &= C^{-}R^{+}iS_F(x,y)R^{-}C^{-} \end{aligned} \tag{3.75}$$

Tachyon Field Feynman Propagator

Turning now to tachyons, the light-front Feynman propagator for the left-handed ψ_{TL}^{+} *tachyon* field is (using the Fourier expansion of the left-handed tachyon field eqs. 3.56 and 3.57):

$$\begin{aligned} iS^{+}{}_{TLF}(x,y) &= \theta(x^{+} - y^{+})\langle 0|\psi_{TL}{}^{+}(x)\psi_{TL}{}^{+\dagger}(y)\gamma^{0}|0\rangle - \\ &\quad - \theta(y^{+} - x^{+})\langle 0|\psi_{TL}{}^{+\dagger}(y)\gamma^{0}\psi_{TL}{}^{+}(x)|0\rangle \\ &= -i\int d^2p dp^{+}\theta(p^{+})N_{TL}{}^{+2}(2m|\mathbf{p}|)^{-1}C^{-}R^{+}\{\theta(x^{+}-y^{+})[(i\not p - m)\boldsymbol{\gamma}\cdot\mathbf{p}]e^{-ip\cdot(x-y)} + \\ &\quad + \theta(y^{+}-x^{+})[(i\not p + m)\boldsymbol{\gamma}\cdot\mathbf{p}]e^{+ip\cdot(x-y)}\}R^{+}C^{-}\gamma^{0} \end{aligned}$$

If we define the on-shell momentum variable $p_0^{-} = (p_0{}^1p_0{}^1 + p_0{}^2p_0{}^2 - m^2)/(2p_0{}^{+})$, $p_0{}^{+} = p^{+}$, $p_0{}^j = p^j$ (for j = 1, 2), $p_{\perp 0}{}^2 = p_0{}^jp_0{}^j$ and $\not p_0 = p_0\cdot\gamma$ then the above equation can be rewritten as

$$\begin{aligned} &= -iC^{-}R^{+}\int d^4p[32\pi^4(p_0{}^{+}(p_0{}^{+} - p_0{}^{-}) + p_{0\perp}{}^2)]^{-1}e^{-ip\cdot(x-y)}\,\cdot \\ &\quad \cdot\{\theta(p^{+})(i\not p_0 - m)\boldsymbol{\gamma}\cdot\mathbf{p}_0]/[p^{-} - p_0{}^{-} + i\varepsilon] + \\ &\quad + \theta(-p^{+})(i\not p_0 + m)\boldsymbol{\gamma}\cdot\mathbf{p}_0]/[p^{-} + p_0{}^{-} - i\varepsilon]\}R^{+}C^{-}\gamma^{0} \end{aligned}$$

$$= -\tfrac{1}{2}\, i\int d^4p(2\pi)^{-4}[C^-R^+(i\not{p} - m)\gamma\cdot\mathbf{p}R^+C^-\gamma^0]e^{-ip\cdot(x-y)}\cdot$$
$$\cdot[(p^2 + m^2 + i\varepsilon)(p^+(p^+ - p^-) + p_\perp^{\,2}))]^{-1}$$

and using $C^-R^+(i\not{p} - m)\gamma\cdot\mathbf{p}R^+C^- = i\, C^-R^+(p^+(p^+ - p^-) + p_\perp^{\,2})$ we find

$$= \tfrac{1}{2}C^-R^+\gamma^0\int d^4p(2\pi)^{-4}\, p^+e^{-ip\cdot(x-y)}/(p^2 + m^2 + i\varepsilon) \qquad (3.76)$$

Similarly the light-front Feynman propagator for the right-handed $\psi_{TR}^{\ +}$ tachyon field is

$$iS^+_{\ TRF}(x,y) = \theta(x^+ - y^+)<0|\psi_{TR}^{\ +}(x)\psi_{TR}^{\ +\dagger}(y)\gamma^0|0> -$$
$$- \theta(y^+ - x^+)<0|\psi_{TR}^{\ +\dagger}(y)\gamma^0\psi_{TR}^{\ +}(x)|0>$$

$$= -\tfrac{1}{2}C^+R^+\gamma^0\int d^4p(2\pi)^{-4}\, p^+e^{-ip\cdot(x-y)}/(p^2 + m^2 + i\varepsilon) \qquad (3.77)$$

where the relative minus sign between eqs. 3.76 and 3.77 is due to the relative minus signs of the Fouier component operator anti-commutation relations in eq. 3.58 and 3.63. Thus we find *tachyon* pole terms in the tachyon propagator as one would expect.

3.6 Complex Lorentz Group L_C Spin ½ Fermion Equations and Wave Functions

In this section we consider L_C group boosts and use them to develop a wider set of dynamical equations for free spin ½ fermions.[75] In chapter 2 we defined L_C boosts with

$$\Lambda_C(\mathbf{v_c}) = \exp[i\omega\hat{\mathbf{w}}\cdot\mathbf{K}] \qquad (2.61)$$
$$\omega = (\omega_r^{\,2} - \omega_i^{\,2} + 2i\omega_r\omega_i\, \hat{\mathbf{u}}_r\cdot\hat{\mathbf{u}}_i)^{1/2} \qquad (2.62)$$
$$\hat{\mathbf{w}} = (\omega_r\hat{\mathbf{u}}_r + i\omega_i\hat{\mathbf{u}}_i)/\omega \qquad (2.63)$$

$$\hat{\mathbf{w}}\cdot\hat{\mathbf{w}} = \hat{\mathbf{u}}_r\cdot\hat{\mathbf{u}}_r = \hat{\mathbf{u}}_i\cdot\hat{\mathbf{u}}_i = 1 \qquad (2.64a)$$
$$\mathbf{v_c} = \hat{\mathbf{w}}\tanh(\omega) \qquad (2.64b)$$

3.7 L_C Spinor "Normal" Lorentz Boosts & More Spin ½ Particle Types

Spinor boost transformations of extended Lorentz groups were used in sections 3.2 and 3.3 to develop the dynamical equations for Dirac fields and tachyon fields. In

---

[75] The complexon theory that we develop and use for quark dynamics in the Standard Model is not required. Our Standard Model could use Dirac fermion dynamics for the up-type quarks and tachyon dynamics for down-type quarks. We choose to use complexon dynamics for quarks because they have an internal SU(3)-like structure suggestive of color SU(3). More importantly, their spin dynamics is different and thus may resolve the differences between theory and experiment for the deep inelastic parton spin-dependent structure functions.

this section we will use L_C group spinor boosts to generate additional particle field dynamical equations.

The form of the L_C spinor boost transformation corresponding to the coordinate transformation eq. 2.61 is:

$$S_C(\omega, \mathbf{v_c}) = \exp(-i\omega\sigma_{0k}\hat{w}_k/2) = \exp(-\omega\gamma^0\boldsymbol{\gamma}\cdot\hat{\mathbf{w}}/2)$$

$$= \cosh(\omega/2)I + \sinh(\omega/2)\gamma^0\boldsymbol{\gamma}\cdot\hat{\mathbf{w}} \tag{3.78}$$

The inverse transformation is

$$S_C^{-1}(\omega, \mathbf{v_c}) = \gamma^2\gamma^0K^{-1}S_C{}^\dagger K\gamma^0\gamma^2 = \gamma^2\gamma^0S_C{}^T\gamma^0\gamma^2 = \exp(\omega\gamma^0\boldsymbol{\gamma}\cdot\hat{\mathbf{w}}/2)$$

$$= \cosh(\omega/2)I - \sinh(\omega/2)\gamma^0\boldsymbol{\gamma}\cdot\hat{\mathbf{w}} \tag{3.79}$$

where the superscript T denotes the transpose and K is the complex conjugation operator (that also appears in the time-reversal operator). Note that S_C is not unitary just as in previous cases considered in this chapter.

We now redo the development of spin ½ dynamical equations of motion of sections 3.2 and 3.3 in this more general case of complex ω and $\hat{\mathbf{w}}$. Again we apply a boost to a Dirac equation for a positive energy plane wave particle of mass m at rest:

$$0 = S_C(\omega, \mathbf{v_c}))(m\gamma^0 - m)e^{-imt}w(0)$$

$$= [mS_C\gamma^0S_C^{-1} - m]e^{-imt}S_Cw(0) \tag{3.80}$$

where $S_C = S_C(\omega, \hat{\mathbf{w}})$. After some algebra

$$mS_C\gamma^0S_C^{-1} = m[\cosh(\omega)\gamma^0 - \sinh(\omega)\boldsymbol{\gamma}\cdot\hat{\mathbf{w}}] \tag{3.81}$$

Case 1: Parallel Real and Imaginary Relative Vector

If the real and imaginary relative vectors parts of $\hat{\mathbf{w}}$, namely $\hat{\mathbf{u}}_\mathbf{r}$ and $\hat{\mathbf{u}}_\mathbf{i}$, are parallel, then $\hat{\mathbf{u}}_\mathbf{r}\cdot\hat{\mathbf{u}}_\mathbf{i} = 1$ and

$$\omega = \omega_r + i\omega_i \tag{3.82}$$

Eq. 2.64b enables us to re-express eq. 3.81 as

$$mS_C\gamma^0S_C^{-1} = m[\cosh(\omega_r)\cos(\omega_i) + i\sinh(\omega_r)\sin(\omega_i)]\gamma^0 -$$

$$- m[\sinh(\omega_r)\cos(\omega_i) + i\cosh(\omega_r)\sin(\omega_i)]\gamma\cdot\hat{\mathbf{u}}_r \tag{3.83}$$

or

$$mS_C\gamma^0 S_C^{-1} = \cos(\omega_i)\gamma\cdot p_r + i\sin(\omega_i)\gamma\cdot p_i \tag{3.84}$$

where

$$p_r^0 = m\cosh(\omega_r) \qquad p_i^0 = m\sinh(\omega_r) \tag{3.85}$$

and

$$\mathbf{p}_r = m\hat{\mathbf{u}}_r\sinh(\omega_r) \qquad \mathbf{p}_i = m\hat{\mathbf{u}}_r\cosh(\omega_r) \tag{3.86}$$

If $\omega_i = 0$, then we recover the momentum space Dirac equation eq. 3.8. If $\omega_i = \pi/2$, then we obtain the left-handed momentum space tachyon equation eq. 3.16. Since the range of ω_i is $[0, \infty>$ (due to the cut along the real ω-plane axis) eq. 3.84 corresponds to the results of the Left-Handed Extended Lorentz group development discussed earlier.

Case 2: Anti-Parallel Real and Imaginary Relative Vector

If the real and imaginary relative vectors parts of $\hat{\mathbf{w}}$, $\hat{\mathbf{u}}_r$ and $\hat{\mathbf{u}}_i$, are anti-parallel $\hat{\mathbf{u}}_r = -\hat{\mathbf{u}}_i$, then $\hat{\mathbf{u}}_r\cdot\hat{\mathbf{u}}_i = -1$ and

$$\omega - \omega_r - i\omega_i \tag{3.87}$$

We can then express eq. 3.81 as

$$mS_C\gamma^0 S_C^{-1} = m[\cosh(\omega_r)\cos(\omega_i) - i\sinh(\omega_r)\sin(\omega_i)]\gamma^0 -$$

$$- m[\sinh(\omega_r)\cos(\omega_i) - i\cosh(\omega_r)\sin(\omega_i)]\gamma\cdot\hat{\mathbf{u}}_r \tag{3.88}$$

or

$$mS_C\gamma^0 S_C^{-1} = \cos(\omega_i)\gamma\cdot p_r - i\sin(\omega_i)\gamma\cdot p_i \tag{3.89}$$

where

$$p_r^0 = m\cosh(\omega_r) \qquad p_i^0 = m\sinh(\omega_r) \tag{3.90}$$

and

$$\mathbf{p}_r = m\hat{\mathbf{u}}_r\sinh(\omega_r) \qquad \mathbf{p}_i = m\hat{\mathbf{u}}_r\cosh(\omega_r) \tag{3.91}$$

If $\omega_i = 0$, then we again recover the momentum space Dirac equation eq. 3.8. If $\omega_i = \pi/2$, then we obtain the right-handed momentum space tachyon equation. (The range of ω_i is again $[0, \infty>$.)

Note: Since the matrix elements in the boost eq. 2.65 depend on $\gamma = (1 - \beta^2)^{-1/2}$ with a singularities at $\beta = \pm 1$, which in turn corresponds to $\omega = \pm\infty$, there is a branch cut along the ω axis in the complex ω-plane. Therefore we point out again the product of three Left-handed extended Lorentz transformations is *not* equivalent to a Right-handed extended Lorentz transformation.

Case 3: Complexons: A New Type of Particle with Perpendicular Real and Imaginary 3-Momenta

If the real and imaginary relative vectors parts of $\hat{\mathbf{w}}$, namely $\hat{\mathbf{u}}_r$ and $\hat{\mathbf{u}}_i$, are perpendicular, $\hat{\mathbf{u}}_r \cdot \hat{\mathbf{u}}_i = 0$, then by eq. 2.62

$$\omega = (\omega_r^{\,2} - \omega_i^{\,2})^{1/2} \tag{3.92}$$

Thus ω is either pure real ($\omega_r \geq \omega_i$) or pure imaginary ($\omega_r < \omega_i$).

The momentum space equation generated by the corresponding L_C spinor boost is

$$\{m \cosh(\omega)\gamma^0 - m \sinh(\omega)\boldsymbol{\gamma}\cdot(\omega_r\hat{\mathbf{u}}_r + i\omega_i\hat{\mathbf{u}}_i)/\omega - m\}e^{-ip\cdot x}w_c(p) = 0 \tag{3.93}$$

Defining the momentum 4-vector

$$p = (p^0, \mathbf{p}) \tag{3.94}$$

where

$$p^0 = m \cosh(\omega) \qquad\qquad \mathbf{p} = \mathbf{p}_r + i\mathbf{p}_i \tag{3.95}$$

with

$$\mathbf{p}_r = m\omega_r\hat{\mathbf{u}}_r \sinh(\omega)/\omega \qquad\qquad \mathbf{p}_i = m\omega_i\hat{\mathbf{u}}_i \sinh(\omega)/\omega \tag{3.96a}$$

$$\mathbf{p}_r\cdot\mathbf{p}_i = 0 \tag{3.96b}$$

then eq. 3.93 becomes a positive energy Dirac-like equation

$$[p\cdot\gamma - m]e^{-ip\cdot x}w_c(p) = 0$$

or, explicitly,

$$[p^0\gamma^0 - (\mathbf{p}_r + i\mathbf{p}_i)\cdot\boldsymbol{\gamma} - m]e^{-ip\cdot x}w_c(p) = 0 \tag{3.97}$$

with a complex 3-momentum **p** and the 4-momentum mass shell condition:

$$p^2 = p^{0\,2} - \mathbf{p}_r\cdot\mathbf{p}_r + \mathbf{p}_i\cdot\mathbf{p}_i = m^2 \tag{3.98}$$

Note

$$|\mathbf{v}| = |\mathbf{p}|/p^0 = [(\mathbf{p}_r + i\mathbf{p}_i)\cdot(\mathbf{p}_r + i\mathbf{p}_i)]^{1/2}/p^0 = \tanh(\omega) \tag{3.99}$$

and so the Lorentz factor

$$\gamma = \cosh(\omega) \tag{3.100}$$

Eq. 3.97 is the momentum space equivalent of the wave equation

$$[i\gamma^0\partial/\partial t + i\gamma\cdot(\nabla_r + i\nabla_i) - m]\psi_C(t, \mathbf{x}_r, \mathbf{x}_i) = 0 \qquad (3.101)$$

where $\mathbf{x} = \mathbf{x}_r - i\mathbf{x}_i$, and where the grad operators ∇_r and ∇_i are with respect to $\mathbf{x}_r$ and $\mathbf{x}_i$ respectively. Since $\hat{\mathbf{u}}_r\cdot\hat{\mathbf{u}}_i = 0$, which in turn implies eq. 3.96b, we see that there is a subsidiary condition on the wave function

$$\nabla_r\cdot\nabla_i\ \psi_C(t, \mathbf{x}_r, \mathbf{x}_i) = 0 \qquad (3.102a)$$

We will call the particles satisfying eqs. 3.101 and 3.102a *complexons*. In addition eq. 3.96b implies the anti-commutation relation

$$\{\gamma\cdot\mathbf{p}_r, \gamma\cdot\mathbf{p}_i\} = 0 \qquad (3.102b)$$

which in turn implies

$$\gamma\cdot\nabla_r\gamma\cdot\nabla_i\psi_C(t, \mathbf{x}_r, \mathbf{x}_i) = \gamma\cdot\nabla_i\gamma\cdot\nabla_r\psi_C(t, \mathbf{x}_r, \mathbf{x}_i) = 0 \qquad (3.102c)$$

also holds. We note that eq. 3.101 is covariant under the real Lorentz group and eq. 3.102 can be easily put into covariant form since the difference of these 4-vectors squared (which is real Lorentz group invariant): $[\gamma^0\partial/\partial t + \gamma\cdot(\nabla_r + i\nabla_i)]^2 - [\gamma^0\partial/\partial t + i\gamma\cdot(\nabla_r - i\nabla_i)]^2 = 4\nabla_r\cdot\nabla_i$.

Before considering a Lagrangian formulation and the Fourier operator representation of $\psi_C(t, \mathbf{x}_r, \mathbf{x}_i)$ we will define the spinors and associated real and imaginary spin operators.

The spinor generated from a spin up Dirac spinor at rest by the complex Lorentz boost eqs. 3.78 is

$$w_c(p) = S_C(p)w(0) = [\cosh(\omega/2)I + \sinh(\omega/2)\gamma^0\gamma\cdot\hat{\mathbf{w}}]w(0) \qquad (3.103)$$

Following a procedure similar to Appendix 3-A (which the reader may wish to examine first) we define four spinors for Dirac particles at rest:

$$w^k(0) = \begin{bmatrix} \delta_{1k} \\ \delta_{2k} \\ \delta_{3k} \\ \delta_{4k} \end{bmatrix} \qquad (3\text{-A}.2)$$

where Kronecker deltas appear in the brackets. Then by applying eq. 3.103 to the spinors defined by eq. 3-A.2 we find the L_C spinors

$$S_C w^k(0) = w_{Cr}{}^k(p) + i w_{Ci}{}^k(p) \tag{3.104}$$

where

$$S_{Cr} = \cosh(\omega/2)I + (\omega_r/\omega)\sinh(\omega/2)\gamma^0\boldsymbol{\gamma}\cdot\hat{\mathbf{u}}_r$$

$$= [(m + E)/(2m)]^{1/2}I + [m(m + E)]^{-1/2}\gamma^0\boldsymbol{\gamma}\cdot\mathbf{p}_r = aI + b\gamma^0\boldsymbol{\gamma}\cdot\mathbf{p}_r \tag{3.105}$$

by eqs. 3.95 and 3.96. Thus the "real" spinors $w_{Cr}{}^k(p)$ are the columns of

$$S_{Cr} = \begin{array}{cccc} \underline{w_{Cr}{}^1(p)} & \underline{w_{Cr}{}^2(p)} & \underline{w_{Cr}{}^3(p)} & \underline{w_{Cr}{}^4(p)} \end{array}$$
$$\left[\begin{array}{cccc} a & 0 & bp_{r\,z} & bp_{r-} \\ 0 & a & bp_{r+} & -bp_{r\,z} \\ bp_{r\,z} & bp_{r-} & a & 0 \\ bp_{r+} & -bp_{r\,z} & 0 & a \end{array}\right] \tag{3.106}$$

where $p_{r\pm} = p_{r\,x} \pm ip_{r\,y}$. The "imaginary" spinors are the columns of

$$S_{Ci} = (\omega_i/\omega)\sinh(\omega/2)\gamma^0\boldsymbol{\gamma}\cdot\hat{\mathbf{u}}_i = [m(m + E)]^{-1/2}\gamma^0\boldsymbol{\gamma}\cdot\mathbf{p}_i = b\gamma^0\boldsymbol{\gamma}\cdot\mathbf{p}_i \tag{3.107}$$

$$S_{Ci} = \begin{array}{cccc} \underline{w_{Ci}{}^1(p)} & \underline{w_{Ci}{}^2(p)} & \underline{w_{Ci}{}^3(p)} & \underline{w_{Ci}{}^4(p)} \end{array}$$
$$\left[\begin{array}{cccc} 0 & 0 & bp_{i\,z} & bp_{i-} \\ 0 & 0 & bp_{i+} & -bp_{i\,z} \\ bp_{i\,z} & bp_{i-} & 0 & 0 \\ bp_{i+} & -bp_{i\,z} & 0 & 0 \end{array}\right] \tag{3.108}$$

where $p_{i\pm} = p_{i\,x} \pm ip_{i\,y}$.

Eqs. 3.101 through 3.108 imply that the wave function solution of eq. 3.101, subject to the subsidiary condition eq. 102a, is[76, 77]

[76] Note that when $|\mathbf{p}_i| \geq |\mathbf{p}_r|$ (for imaginary $\omega = (\omega_r{}^2 - \omega_i{}^2)^{1/2}$) the 3-momentum becomes imaginary $\mathbf{p}\cdot\mathbf{p} < 0$. However, since we will be identifying confined quarks with this type of particle – much modified by a confining color quark interaction – the issue of an imaginary 3-momentum in the hypothetical free quark case becomes moot. We note the energy gap between positive and negative energy states disappears so E = 0 is possible. Thus real Lorentz transformations can mix positive and negative energy states. The solution is to do all calculations in the light-front frame as we do for tachyons. Then the mixing issue is resolved. In the present case we second quantize on the "time-front" for illustrative purposes.

$$\psi_C(x_r, x_i) = \sum_{\pm s} \int d^3p_r d^3p_i\, N_C(p)\delta(\mathbf{p}_r\cdot\mathbf{p}_i/m^2)[b_C(p,s)u_C(p, s)e^{-i(p.x + p^*.x^*)/2} + \\ + d_C^{\dagger}(p,s)v_C(p, s)e^{+i(p.x + p^*.x^*)/2}] \quad (3.109)$$

where $\mathbf{p} = \mathbf{p_r} + i\mathbf{p_i}$ (eq. 3.95), $\mathbf{x} = \mathbf{x_r} - i\mathbf{x_i}$, $p\cdot x = p^0x^0 - \mathbf{p}\cdot\mathbf{x}$, and where we use

$$(p\cdot x + p^*\cdot x^*)/2 = p^0x^0 - \mathbf{p_r}\cdot\mathbf{x_r} - \mathbf{p_i}\cdot\mathbf{x_i} \quad (3.109a)$$

in the exponentials in order to avoid divergences that would appear in the calculation of the equal-time commutator, the Feynman propagator and other quantities of interest after second quantization. Note that

$$(\nabla_r + i\nabla_i)e^{-i(p.x + p^*.x^*)/2} = i(\mathbf{p_r} + i\mathbf{p_i})e^{-i(p.x + p^*.x^*)/2} \quad (3.109b)$$

and

$$(\nabla_r + i\nabla_i)e^{-ip^*.x^*} = 0 \quad (3.109c)$$

for all p.

The wave function's conjugate (the Hermitian conjugate modified by letting $\mathbf{x_i} \rightarrow -\mathbf{x_i}$ in addition to Hermitian conjugation) is

$$\psi_C^{\dagger}(x) = \psi_C^{\dagger}(x_r, -x_i) = \sum_{\pm s} \int d^3p_r d^3p_i\, \delta(\mathbf{p}_r\cdot\mathbf{p}_i/m^2)N_C(p^*)\cdot \\ \cdot[b_C^{\dagger}(p^*,s)u_C^{\dagger}(p^*,s)e^{+i(p.x^* + p^*.x)/2} + d_C(p^*,s)v_C^{\dagger}(p^*,s)e^{-i(p.x^* + p^*.x)/2}] \quad (3.110)$$

where $\mathbf{p} = \mathbf{p_r} + i\mathbf{p_i}$ (eq. 3.95), $\mathbf{x} = \mathbf{x_r} - i\mathbf{x_i}$, $p\cdot x = p^0x^0 - \mathbf{p}\cdot\mathbf{x}$, and † indicates Hermitian Hermitian conjugation.

The spinors are

$$\begin{aligned} u_C(p, s) &= S_C(p)w^1(0) \\ u_C(p, -s) &= S_C(p)w^2(0) \\ v_C(p, s) &= S_C(p)w^3(0) \\ v_C(p, -s) &= S_C(p)w^4(0) \quad (3.110a) \\ u_C^{\dagger}(p^*, s) &= w^{1T}(0)S_C^{\dagger}(p^*)= w^{1T}(0)S_C(p) \\ u_C^{\dagger}(p^*, -s) &= w^{2T}(0)S_C^{\dagger}(p^*) = w^{2T}(0)S_C(p) \\ v_C^{\dagger}(p^*, s) &= w^{3T}(0)S_C^{\dagger}(p^*) = w^{3T}(0)S_C(p) \\ v_C^{\dagger}(p^*, -s) &= w^{4T}(0)S_C^{\dagger}(p^*) = w^{4T}(0)S_C(p) \end{aligned}$$

with the superscript "T" indicating the transpose. Note that

---

[77] We scale $\mathbf{p}_r\cdot\mathbf{p}_i$ with m^2 in the delta function for convenience. In the case of a zero mass particle some other scale could be used.

$$S_C^{\dagger}(p^*) = [S_C(p^*)]^{\dagger} = S_C(p) \qquad (3.110b)$$

The normalization factor $N_C(p)$ is

$$N_C(p) = [2m/((2\pi)^6 p^0)]^{1/2} \qquad (3.110c)$$

Since $\mathbf{p_r} = \mathbf{p_i} = 0$ in the particle rest frame prior to the complex group boost, the boosted particle spin 4-vector s^μ satisfies

$$s^\mu p_r^{\ \mu} = s^\mu p_i^{\ \mu} = 0 \qquad (3.111)$$

Note that s^μ is itself complex[78] and, if the spin points in the z-direction prior to the complex boost (eq. 2.57), then the boosted s^μ has the form

$$s^\mu = (-\sinh(\omega)\hat{w}_z\, , (0,0,1) + (\cosh(\omega) - 1)\hat{w}_z\hat{\mathbf{w}}) \qquad (3.112)$$

with $\hat{\mathbf{w}}$ defined by eq. 2.63: $\hat{\mathbf{w}} = (\omega_r\hat{\mathbf{u}}_r + i\omega_i\hat{\mathbf{u}}_i)/\omega = \mathbf{p}/(m\sinh(\omega))$ using eq. 3.96a.

A Global SU(3) Symmetry Revealed

Before proceding to consider the second quantization of this case, we will consider a global SU(3) symmetry implicit in eqs. 3.101, 3.102 and the solution 3.109. The defining property of the group SU(3) is that it preserves the invariance of inner products of complex 3-vectors of the form:

$$u^*\cdot v = u^{1*}v^1 + u^{2*}v^2 + u^{3*}v^3 \qquad (3.113)$$

If we examine the dynamical equation eq. 3.101 we see that the differential operator is invariant under an SU(3) transformation U (using $\nabla_c = (\nabla_c^*)^* = \mathbf{D}_c^*$)

$$[i\gamma^0\partial/\partial t + i\mathbf{D}_c^*\cdot\boldsymbol{\gamma} - m] = [i\gamma^0\partial/\partial t + i\mathbf{D}_c'^*\cdot\boldsymbol{\gamma}' - m] \qquad (3.114)$$

where

$$\mathbf{D}_c^* = \nabla_c = \nabla_r + i\nabla_i$$

and

$$\gamma'^a = U^{ab}\gamma'^b$$
$$D_c'^{*a} = D_c'^{*b}U^{\dagger ab}$$

---

[78] This feature of partons, which is not present in ordinary Dirac particles, might be the source of the discrepancies between theory and experiment in deep inelastic parton spin physics which is based on conventional real parton spins.

where U is a global SU(3) transformation and $U^\dagger = U^{-1}$. By theorem[79] all 4 × 4 γ matrices such as **γ'** are equivalent up to a unitary transformation V. Thus $V^\dagger\boldsymbol{\gamma}'V = \boldsymbol{\gamma}$ and eq. 3.114 is equivalent to

$$[i\gamma^0\partial/\partial t + i\mathbf{D}_c{}^*\cdot\boldsymbol{\gamma} - m] = [i\gamma^0\partial/\partial t + i\mathbf{D}_c{}'^*\cdot\boldsymbol{\gamma} - m] \quad (3.115)$$

$$= [i\gamma^0\partial/\partial t + i\nabla_c{}'\cdot\boldsymbol{\gamma} - m]$$

where $\nabla_c{}'_a = U^{ab}\nabla_{cb}$ showing that eq. 3.101 is invariant under an SU(3) transformation if

$$\psi_C(t, \mathbf{x}) = \psi_C(t, U\mathbf{x}) = \psi_C{}'(t, \mathbf{x}') \quad (3.116)$$

where $\psi_C(t, \mathbf{x}) \equiv \psi_C(t, \mathbf{x}_r, \mathbf{x}_i)$.

The subsidiary condition eq. 3.102a can be seen to transform as

$$\nabla_r\cdot\nabla_i\, \psi_C(t, \mathbf{x}) = \nabla_r{}^*\cdot\nabla_i\, \psi_C(t, \mathbf{x}) = \nabla_r{}'^*\cdot\nabla_i{}'\psi_C{}'(t, \mathbf{x}') = 0 \quad (3.117)$$

under an SU(3) rotation. The invariance of the orthogonality condition is preserved.

The wave function (eq. 3.109) transforms in the following way under the SU(3) transformation U. If we define

$$q^{*\mu} = (q^0, \mathbf{q}^*) = (p^0, \mathbf{p}_r + i\mathbf{p}_i) = (p^0, \mathbf{p}) = p^\mu \quad (3.118)$$

then eq. 3.109 can be rewritten in an almost[80] manifestly invariant form under a SU(3) transformation:

$$\psi_C(x) = \sum_{\pm s} \int d^3q_r d^3q_i\, N_C(p^0)\delta(\mathbf{q}_r{}^*\cdot\mathbf{q}_i/m^2)[b_C(q^*,s)u_C(q^*,s)e^{-i(q^*.x + q.x^*)/2} +$$
$$+ d_C^\dagger(q^*,s)v_C(q^*,s)e^{+i(q^*.x + q.x^*)/2}] \quad (3.119)$$

subject to an examination of the transformation properties of the Fourier coefficients and spinors. Note both terms in each exponential are separately invariant under global SU(3). (Note also $\mathbf{q}_r{}^* = \mathbf{q}_r$ since $\mathbf{q}_r$ is real.)

From the form of S_C (eq. 3.78) it is clear that an argument similar to that for the dynamical equations (eqs. 3.114 – 3.115) shows S_C is invariant under an SU(3)

---

[79] R. H. Good, Rev. Mod. Phys., **27**, 187 (1955).

[80] The δ-function in eq. 3.119, $\delta(\mathbf{q}_r{}^*\cdot\mathbf{q}_i/m^2)$, is not invariant under a global SU(3) transformation rather it is a constraint that breaks the SU(3) symmetry to O(3)×O(2) rather like a δ-function term in the path integral formalism of gauge theories "breaks" the symmetry and fixes the gauge. The O(3)×O(2) symmetry can be visualized by mapping (conceptually) the $\mathbf{q}_r$ and $\mathbf{q}_i$ to 3-space and considering the perpendicular vector **q** to the pair as rotating under O(3) transformations, and $\mathbf{q}_r$ and $\mathbf{q}_i$ rotating in the plane perpendicular to **q** under O(2) transformations.

transformation and thus the spinors defined by eqs. 3.110a are also invariant under SU(3) transformations. The Fourier coefficients, if second quantized in a direct generalization of the usual way, have covariant anti-commutation relations under an SU(3) transformation. For example

$$\{b_C(q,s), b_C^\dagger(q'^*,s')\} = \delta_{ss'}\delta^3(q_r - q'_{r'})\delta^3(q_i - q'_{i'}) \qquad (3.120)$$

Under an SU(3) transformation, z = Uq and z' = Uq', the right side of eq. 3.120 transforms to

$$\delta^3(q_r - q'_{r'})\delta^3(q_i - q'_{i'}) \rightarrow \delta^3(z_r - z'_{r'})\delta^3(z_i - z'_{i'})/|\partial(q)/\partial(z)|$$
$$= \delta^3(z_r - z'_{r'})\delta^3(z_i - z'_{i'}) \qquad (3.121)$$

where

$$|\partial(q)/\partial(z)| = |\partial(q_r^{\,1},q_r^{\,2},q_r^{\,3},q_i^{\,1}, q_i^{\,2}, q_i^{\,3})/\partial(z_r^{\,1},z_r^{\,2},z_r^{\,3},z_i^{\,1}, z_i^{\,2}, z_i^{\,3})| = 1 \qquad (3.122)$$

is the Jacobian of the transformation U. Thus the Fourier coefficients transform trivially under SU(3). For example,

$$b_C(q^*,s) \rightarrow b_C(z^*,s) \qquad (3.123)$$

Since the integrand transforms as

$$\int d^3q_r d^3q_i \rightarrow \int d^3z_r d^3z_i\, |\partial(q)/\partial(z)| = \int d^3z_r d^3z_i \qquad (3.124)$$

the wave function $\psi_C(t, \mathbf{x})$ transforms as an SU(3) scalar (eq. 3.116) up to an inessential unitary transformation V of γ matrices: $\psi_C(t, \mathbf{x}) \rightarrow V\psi_C(t, \mathbf{x})$.[81]

Global SU(3) Spin ½ Complexon Fields

Having uncovered an SU(3) symmetry in the scalar field equations of Case 3A the generalization of the scalar field equations to the 3 representation of SU(3) is direct:

$$\psi_C^{\,a}(x) = \sum_{\pm s} \int d^3p_r d^3p_i\, N_C(p)\delta(\mathbf{p}_r\cdot\mathbf{p}_i/m^2)[b_C(p,a,s)u_C^{\,a}(p, s)e^{-i(p.x + p^*.x^*)/2} +$$
$$+ d_C^{\,\dagger}(p,a,s)v_C^{\,a}(p, s)e^{+i(p.x + p^*.x^*)/2}] \qquad (3.125a)$$

---

[81] The spinors $u_C(q^*,s)$ and $v_C(q^*,s)$ are unchanged up to a unitary transformation of the γ matrices ($V^\dagger\gamma'V = \gamma$). Thus the term $(U\mathbf{w})^*\cdot\gamma' = \mathbf{w}^*\cdot V\gamma V^\dagger \equiv \mathbf{w}^*\cdot\gamma$ in the expressions for the $u_C(q^*,s)$ and $v_C(q^*,s)$ spinors.

for a = 1, 2, 3 with $u_C{}^a(p, s)$ and $v_C{}^a(p, s)$ being the product a spinor of type eq. 3.110a and a 3 element column vector c^a with b^{th} element

$$c^a(b) = \delta^{ab} \tag{3.125b}$$

Under a global SU(3) transformation U the 3 complexon wave functions transform as

$$\psi_C{}'^a(x) = U^{ab}\psi_C{}^b(x) \tag{3.126}$$

In a subsequent chapter we will extend the global SU(3) symmetry described in these subsections to color local SU(3) upon the introduction of the Yang-Mills color gluon interaction and detach it from the global SU(3) coordinate transformation feature found in this section.

Lagrangian Formulation and Second Quantization of Complexons

In this subsection we will outline the canonical quantization of SU(3) singlet complexons with the quantum field equation

$$[i\gamma^0\partial/\partial t + i\gamma\cdot(\nabla_\mathbf{r} + i\nabla_\mathbf{i}) - m]\psi_C(t, \mathbf{x_r}, \mathbf{x_i}) = 0 \tag{3.101}$$

and subsidiary condition

$$\nabla_\mathbf{r}\cdot\nabla_\mathbf{i}\, \psi_C(t, \mathbf{x_r}, \mathbf{x_i}) = 0 \tag{3.102a}$$

We begin with the Lagrangian density

$$\mathcal{L} = \bar{\psi}_C(i\gamma^\mu D_\mu - m)\psi_C(x) \tag{3.127}$$

where $\bar{\psi}_C = \psi_C{}^\dagger\gamma^0$:

$$\psi_C{}^\dagger = [\psi_C(\mathbf{x_r}, \mathbf{x_i})]^\dagger \big|_{\mathbf{X_i} = -\mathbf{X_i}} \tag{3.128}$$

$$\begin{aligned} D_0 &= \partial/\partial x^0 \\ D_k &= \partial/\partial x^k + i\,\partial/\partial x_i{}^k \end{aligned} \tag{3.129}$$

with $x^k = x_r{}^k$ for k = 1,2, 3. The invariant action (under real Lorentz transformations) is

$$I = \int d^7x\mathcal{L} \tag{3.130}$$

It is easy to show that the action is real

$$I^* = I \tag{3.131}$$

in a manner similar to the case considered in Appendix 3-A due to the form of $\psi_C^\dagger$ in eq. 3.128. (One has to change the integration over $\mathbf{x_i}$ to $-\mathbf{x_i}$ after taking the complex conjugate of I and performing manipulations similar to those in Appendix 3-A.)

The conjugate momentum is

$$\pi_{Ca} = \partial\mathcal{L}/\partial\dot{\psi}_{Ca} \equiv \partial\mathcal{L}/\partial(\partial\psi_{Ca}/\partial x^0) = i\psi_{C\,a}^{\dagger} \tag{3.132}$$

where a is a spinor index. It yields the non-zero anti-commutation relation

$$\{\psi_{C\,a}^{\dagger}(x), \psi_{Cb}(y)\} = \delta_{ab}\, \delta^3(x_r - y_r)\delta^3(x_i - y_i) \tag{3.133a}$$

However we will see that the constraint eq. 3.102a is required. So the correct anti-commutator turns out to be

$$\{\psi_{C\,a}^{\dagger}(x), \psi_{Cb}(y)\} = -\delta_{ab}\delta'(\nabla_r\cdot\nabla_i/m^2)[\delta^3(x_r - y_r)\delta^3(x_i - y_i)] \tag{3.133b}$$

where all ∇_r and ∇_i are ∇ derivatives with respect to x, and where $\delta'(\nabla_r\cdot\nabla_i)$ is the derivative of a delta function with the argument being differential operators such as those in eq. 3.102a. The minus sign is due to the presence of a *derivative* of a delta-function and not an issue.

The Hamiltonian density is

$$\mathcal{H} = \pi_C\dot{\psi}_C - \mathcal{L} = \psi_C^{\dagger}(-i\boldsymbol{\alpha}\cdot\mathbf{D} + \beta m)\psi_C \tag{3.134}$$

and the (unsymmetrized) energy-momentum tensor is

$$\mathcal{T}_{\mu\nu} = -\, g_{\mu\nu}\,\mathcal{L} + \partial\mathcal{L}/\partial(D^{\mu}\psi_C)D_{\nu}\psi_C \tag{3.135}$$

The conserved energy and momentum are

$$P^0 = H = \int d^3x_r d^3x_i\, \mathcal{T}^{00} = \int d^3x_r d^3x_i\, \mathcal{H} \tag{3.136}$$

and

$$P^I = \int D^3X_R D^3X_I\, \mathcal{T}^{0I}$$

(3.137)

We now proceed to establish the canonical anti-commutation relations eq. 3.133. First, the second quantization of the complexon field (eqs. 3.109 and 3.110) uses the Fourier coefficient anti-commutation relations eq. 3.120 (suitably rewritten):

$$\begin{aligned}
&\{b_C(p,s),\, b_C^\dagger(p'^*,s')\} = \delta_{ss'}\delta^3(\mathbf{p}_r - \mathbf{p}'_{r'})\delta^3(\mathbf{p}_i + \mathbf{p}'_{i'}) \\
&\{d_C(p,s),\, d_C^\dagger(p'^*,s')\} = \delta_{ss'}\delta^3(\mathbf{p}_r - \mathbf{p}'_{r'})\delta^3(\mathbf{p}_i + \mathbf{p}'_{i'}) \\
&\{b_C(p,s),\, b_C(p'^*,s')\} = \{d_C(p,s),\, d_C(p'^*,s')\} = 0 \\
&\{b_C^\dagger(p,s),\, b_C^\dagger(p'^*,s')\} = \{d_C^\dagger(p,s),\, d_C^\dagger(p'^*,s')\} = 0 \qquad (3.138) \\
&\{b_C(p,s),\, d_C^\dagger(p'^*,s')\} = \{d_C(p,s),\, b_C^\dagger(p'^*,s')\} = 0 \\
&\{b_C^\dagger(p,s),\, d_C^\dagger(p'^*,s')\} = \{d_C(p,s),\, b_C(p'^*,s')\} = 0
\end{aligned}$$

The delta-function arguments $\delta^3(\mathbf{p}_i + \mathbf{p}'_{i'})$ above have a positive sign in order to obtain $\delta^3(\mathbf{x}_i - \mathbf{y}_i)$ in the field anti-commutator eq. 3.133b.

The spinors, eq. 3.110a, satisfy

$$\begin{aligned}
&\sum_{\pm s} u_\alpha(p,s)\bar{u}_\beta(p^*,s) = (2m)^{-1}(\not{p} + m)_{\alpha\beta} \qquad (3.139) \\
&\sum_{\pm s} v_\alpha(p,s)\bar{v}_\beta(p^*,s) = (2m)^{-1}(\not{p} - m)_{\alpha\beta}
\end{aligned}$$

by eqs. 3.78, 3.81, 3.85 and 3.103 remembering

$$\bar{u}_C(p^*,s) = w^{1T}(0)S_C(p)\gamma^0 = w^{1T}(0)[\cosh(\omega/2)I + \sinh(\omega/2)\gamma^0\boldsymbol{\gamma}\cdot\hat{\mathbf{w}}]\gamma^0 \qquad (3.140)$$

by eqs. 3.110 since $\hat{\mathbf{w}}^{**} = \hat{\mathbf{w}}$.

We will now evaluate the equal-time anti-commutation relation using eqs. 3.109 and 3.110:

$$\begin{aligned}
\{\psi_{C\,\alpha}^\dagger(x),\, \psi_{Cb}(y)\} = &\sum_{\pm s,\, s'} \int d^3p_r d^3p_i\, d^3p'_r d^3p'_i\, \delta(\mathbf{p}_r\cdot\mathbf{p}_i/m^2)\delta(\mathbf{p}'_r\cdot\mathbf{p}'_i/m^2)\, N_C(p')N_C(p)\cdot \\
&[\{b_C^\dagger(p^*,s)u_{Ca}^\dagger(p^*,s)e^{+i(p.x^* + p^*.x)/2},\, b_C(p',s')u_{Cb}(p',s')e^{-i(p'.y + p'^*.y^*)/2}\} + \\
&+ \{d_C(p^*,s)v_{Ca}^\dagger(p^*,s)e^{-i(p.x^* + p^*.x)/2},\, d_C^\dagger(p',s')v_{Cb}(p',s')e^{+i(p'.y + p'^*.y^*)/2}\}]
\end{aligned}$$

$$\begin{aligned}
= \int d^3p_r d^3p_i\, N_C^2(p)[\delta(\mathbf{p}_r\cdot\mathbf{p}_i/m^2)]^2[((\not{p} + m)\gamma^0)_{ba} e^{+i(p.x^* + p^*.x)/2 - i(p^*.y + p.y^*)/2} + \\
+((\not{p} - m)\gamma^0)_{ba} e^{-i(p.x^* + p^*.x)/2 + i(p^*.y + p.y^*)/2}]/(2m)
\end{aligned}$$

Next we use eq. 3.110c and the identity

$$[\delta(x - y)]^2 = -\tfrac{1}{2}\,\delta'(x - y) \equiv -\tfrac{1}{2}\, d\delta(x - y)/dx \qquad (3.141)$$

which can be derived from the step function identity $\theta(x - y) = [\theta(x - y)]^2$ to obtain

$$\{\psi_{C}{}^{\dagger}{}_{a}(x),\psi_{Cb}(y)\} = -\tfrac{1}{2}\int d^3p_r d^3p_i N_C{}^2(p)\delta'(\mathbf{p}_r\cdot\mathbf{p}_i/m^2)[((\not{p}+m)\gamma^0)_{ba}e^{-i\mathbf{p}_r\cdot(\mathbf{x}_r-\mathbf{y}_r)+i\mathbf{p}_i\cdot(\mathbf{x}_i-\mathbf{y}_i)} + ((\not{p}-m)\gamma^0)_{ba}e^{+i\mathbf{p}_r\cdot(\mathbf{x}_r-\mathbf{y}_r)-i\mathbf{p}_i\cdot(\mathbf{x}_i-\mathbf{y}_i)}]/(2m)$$

$$= -\tfrac{1}{2}\delta_{ba}\int d^3p_r d^3p_i N_C{}^2(p)\delta'(\mathbf{p}_r\cdot\mathbf{p}_i/m^2)p^0 e^{-i\mathbf{p}_r\cdot(\mathbf{x}_r-\mathbf{y}_r)+i\mathbf{p}_i\cdot(\mathbf{x}_i-\mathbf{y}_i)}/m$$

$$= -\delta_{ab}\,\delta'(\nabla_r\cdot\nabla_i/m^2)[\delta^3(x_r-y_r)\delta^3(x_i-y_i)] \qquad (3.142)$$

The grad operators, ∇_r and ∇_i, are derivatives are with respect to x of the Dirac delta functions. The factor[82] $\delta'(\nabla_r\cdot\nabla_i)$ expresses the orthogonality constraint in coordinate space on the momenta eq. 3.102a. It is analogous to the transversality constraint on the electromagnetic vector potential commutator:

$$[\pi_A{}^j(x), A_k(y)] = -i\,\delta^{tr}{}_{jk}(x-y) \qquad (3.143)$$

$$\delta^{tr}{}_{jk}(x-y) = (\delta_{jk} - \partial_j\partial_k/\nabla^2)\,\delta^3(x-y) \qquad (3.144)$$

where $\partial_k = \partial/\partial x_k$.

Complexon Feynman Propagator

The complexon Feynman propagator for ψ_C is[83]

$$iS_C(x, y) = \theta(x^0 - y^0)\langle 0|\psi_C(x)\psi_C{}^{\dagger}(y)\gamma^0|0\rangle - \theta(y^0 - x^0)\langle 0|\psi_C{}^{\dagger}(y)\gamma^0\psi_C(x)|0\rangle \qquad (3.145)$$

$$= \int d^3p_r d^3p_i N_C{}^2(p)[\delta(\mathbf{p}_r\cdot\mathbf{p}_i/m^2)]^2\{\theta(x^0 - y^0)(\not{p} + m)e^{-i(p^*\cdot(x-y) + p\cdot(x^*-y^*))/2} - \theta(y^0 - x^0)(\not{p} - m)e^{+i(p^*\cdot(x-y) + p\cdot(x^*-y^*))/2}\}/(2m)$$

$$= -(4\pi)^{-1}\int dp^0 d^3p_r d^3p_i (2\pi)^{-6}\delta'(\mathbf{p}_r\cdot\mathbf{p}_i/m^2)(\not{p}+m)e^{-i(p^*\cdot(x-y) + p\cdot(x^*-y^*))/2}/(p^2-m^2+i\varepsilon)$$

$$= -\tfrac{1}{2}\int dp^0 d^3p_r d^3p_i\, \delta'(\mathbf{p}_r\cdot\mathbf{p}_i/m^2)(\not{p} + m)(2\pi)^{-7}\cdot$$

---

[82] A derivative of a delta function containing grad operators.

[83] The reader, upon seeing the additional integrations d^3p_i might suspect that they would ultimately lead to divergence issues in perturbation theory calculations. However the $\delta'(\mathbf{p}_r\cdot\mathbf{p}_i/m^2)$ term compensates in part for the additional integrations by four powers of momentum since $\delta'(\mathbf{p}_r\cdot\mathbf{p}_i/m^2) = (|\mathbf{p}_r||\mathbf{p}_i|/m^2)^{-2}\delta'(\cos\theta_{ri})$ where θ_{ri} is the angle between the momenta. As a result only 2 fermion and 3 fermion loop integrations would potentially have difficulties if one uses the conventional approach to perturbation theory. If one uses the approach of Blaha (2003) and (2005a) then there are no divergences.

$$\cdot\exp[-ip^0(x^0 - y^0) + i\mathbf{p_r}\cdot(\mathbf{x_r} - \mathbf{y_r}) - i\mathbf{p_i}\cdot(\mathbf{x_i} - \mathbf{y_i})]/(p^2 - m^2 + i\varepsilon) \qquad (3.146)$$

The integral can be written in the form:

$$I=\int dp^0 d^3p_r d^3p_i \delta'(\mathbf{p_r}\cdot\mathbf{p_i}/m^2)(\not{p}+m)\exp[-ip^0(x^0-y^0)+i\mathbf{p_r}\cdot(\mathbf{x_r}-\mathbf{y_r})-i\mathbf{p_i}\cdot(\mathbf{x_i}-\mathbf{y_i})]/(p^2 - m^2 + i\varepsilon)$$

$$=\int d^4p_r dM^2\delta'(\nabla_r\cdot\nabla_I/m^2)(p^0\gamma^0-(\mathbf{p_r}-\nabla_i)\cdot\gamma+m)\exp[-ip^0(x^0-y^0)+i\mathbf{p_r}\cdot(\mathbf{x_r}-\mathbf{y_r})]\cdot$$
$$\cdot J(\mathbf{x_i} - \mathbf{y_i}, M^2)/(p_r^2-M^2 + i\varepsilon) \qquad (3.147)$$

where $p_r^2 = p^{0\,2} - \mathbf{p_r}\cdot \mathbf{p_r}$ and

$$J(\mathbf{x_i} - \mathbf{y_i}, M^2) = (2\pi)^{-3}\int d^3p_i\,\delta(M^2 + \mathbf{p_i}^2 - m^2)\exp[-i\mathbf{p_i}\cdot(\mathbf{x_i}- \mathbf{y_i})] \quad (3.148)$$

$$= (2\pi)^{-2}|\mathbf{x_i}- \mathbf{y_i}|^{-1}\theta(m^2 - M^2)\sin((m^2 - M^2)^{1/2}|\mathbf{x_i}- \mathbf{y_i}|)$$

The complexon Feynman propagator eq. 3.146 can be rearranged into the form of a spectral integral:

$$iS_C(x, y) = -\int dM\,(i\gamma^0\partial/\partial x^0-i(\nabla_r-i\nabla_i)\cdot\gamma + m)\delta'(\nabla_r\cdot\nabla_i/m^2)\cdot$$
$$\cdot J(\mathbf{x_i} - \mathbf{y_i}, M^2)\Delta_F(x - y, M) \qquad (3.149)$$

where

$$\Delta_F(x - y, M) = (2\pi)^{-4}\int d^4p_r\,\exp[-ip^0(x^0 - y^0) + i\mathbf{p_r}\cdot(\mathbf{x_r} - \mathbf{y_r})]/(p_r^2 - M^2 + i\varepsilon)$$
$$(3.150)$$

Case 4: Left-handed Tachyon Complexons

In this case $\hat{\mathbf{u}}_\mathbf{r}\cdot\hat{\mathbf{u}}_\mathbf{i} = 0$ again. However we add an imaginary term to ω to obtain a manifest Left-handed L_C boost[84]

$$\Lambda_{CL}(\mathbf{v_c}) = \exp[i(\omega + i\pi/2)\hat{\mathbf{w}}\cdot\mathbf{K}] \qquad (3.151)$$

where ω remains

$$\omega = (\omega_r^2 - \omega_i^2)^{1/2} \qquad (2.62)$$

and

$$\hat{\mathbf{w}} = (\omega_r\hat{\mathbf{u}}_\mathbf{r} + i\omega_i\hat{\mathbf{u}}_\mathbf{i})/\omega \qquad (2.63)$$
$$\hat{\mathbf{w}}\cdot\hat{\mathbf{w}} = \hat{\mathbf{u}}_\mathbf{r}\cdot\hat{\mathbf{u}}_\mathbf{r} = \hat{\mathbf{u}}_\mathbf{i}\cdot\hat{\mathbf{u}}_\mathbf{i} = 1 \qquad (2.64a)$$
$$\mathbf{v_c} = \hat{\mathbf{w}}\tanh(\omega + i\pi/2) = \hat{\mathbf{w}}\,\mathrm{cotanh}(\omega) \qquad (3.152)$$

[84] The reader can readily verify the form is consistent with an L_C boost transformation.

Letting $\omega_L = \omega + i\pi/2$ we find, as before,

$$\cosh(\omega_L) = i\sinh(\omega) = -\gamma = i\,\gamma_s$$
$$\sinh(\omega_L) = i\cosh(\omega) = -\beta\gamma = i\beta\gamma_s \tag{2.11}$$

with, $\beta = v_c = |\mathbf{v_c}| > 1$, $\gamma_s = (\beta^2 - 1)^{-1/2}$, and

$$\sinh(\omega) = \gamma_s$$
$$\cosh(\omega) = \beta\gamma_s \tag{2.12}$$

Thus we denote $\Lambda_{CL}(\mathbf{v_c})$ by

$$\Lambda_{CL}(\mathbf{v_c}) \equiv \Lambda_{CL}(\omega, \hat{\mathbf{w}}) \tag{3.153}$$

The corresponding spinor boost transformation is:

$$S_{CL}(\Lambda_{CL}(\omega, \hat{\mathbf{w}})) = \exp(-i\omega_L\sigma_{0i}\hat{w}_i/2) = \exp(-\omega_L\gamma^0\boldsymbol{\gamma}\cdot\hat{\mathbf{w}}/2)$$
$$= \cosh(\omega_L/2)I + \sinh(\omega_L/2)\gamma^0\boldsymbol{\gamma}\cdot\hat{\mathbf{w}} \tag{3.154}$$

The momentum space equation generated by $S_{CL}(\Lambda_{CL}(\omega, \hat{\mathbf{w}}))$ is

$$\{m\cosh(\omega_L)\gamma^0 - m\sinh(\omega_L)\boldsymbol{\gamma}\cdot(\omega_r\hat{\mathbf{u}}_r + i\omega_i\hat{\mathbf{u}}_i)/\omega - m\}e^{+ip.x}w_{cL}(p) = 0 \tag{3.155}$$

or

$$\{im\sinh(\omega)\gamma^0 - im\cosh(\omega)\boldsymbol{\gamma}\cdot(\omega_r\hat{\mathbf{u}}_r + i\omega_i\hat{\mathbf{u}}_i)/\omega - m\}e^{+ip.x}w_{cL}(p) = 0 \tag{3.156}$$

where $p\cdot x = Et - \mathbf{p}\cdot\mathbf{x}$ after performing a corresponding left-handed superluminal coordinate transformation in the exponential factor based on eq. 2.10d. Thus the positive energy wave is transformed into a negative energy wave by the transformation.

The momentum 4-vector is defined by

$$p = (p^0, \mathbf{p}) \tag{3.157}$$

where

$$p^0 = m\sinh(\omega) \qquad \mathbf{p} = \mathbf{p}_r + i\mathbf{p}_i \tag{3.158}$$

with

$$\mathbf{p_r} = m\omega_r\hat{\mathbf{u}}_r \cosh(\omega)/\omega \qquad \mathbf{p_i} = m\omega_i\hat{\mathbf{u}}_i \cosh(\omega)/\omega \tag{3.159}$$

and

$$\mathbf{p_r}\cdot\mathbf{p_i} = 0 \tag{3.160}$$

then eq. 3.156 becomes the complexon tachyon equation

$$[ip\cdot\gamma - m]e^{+ip.x}w_{cL}(p) = 0 \tag{3.161}$$

with a complex 3-momentum **p** and the tachyon 4-momentum mass shell condition:[85]

$$p^2 - p^{\mathbf{0}\,2} \quad \mathbf{p_r}^2 + \mathbf{p_i}^2 - \quad m^2 \tag{3.162}$$

Eq. 3.161 is the momentum space equivalent of the wave equation

$$[\gamma^{\mathbf{0}}\partial/\partial t + \boldsymbol{\gamma}\cdot(\nabla_r + i\nabla_i) - m]\psi_{CL}(t, \mathbf{x_r}, \mathbf{x_i}) = 0 \tag{3.163a}$$

or

$$[\gamma\cdot\nabla - m]\psi_{CL}(t, \mathbf{x_r}, \mathbf{x_i}) = 0 \tag{3.163b}$$

with the subsidiary condition on the wave function

$$\nabla_r\cdot\nabla_i\, \psi_{CL}(t, \mathbf{x_r}, \mathbf{x_i}) = 0 \tag{3.164}$$

also holds. We note that eq. 3.163 is covariant under the real Lorentz group and eq. 3.164 can be easily put into (real Lorentz group) covariant form.

Before considering a Lagrangian formulation and the Fourier operator representation of $\psi_{CL}(t, \mathbf{x_r}, \mathbf{x_i})$ we will define the tachyon spinors, and its associated real and imaginary spin operators.

The spinor generated from a spin up Dirac spinor at rest by the L_C spinor boost eqs. 3.154 is

$$w_{cL}(p) = S_{CL}w(0) = [\cosh(\omega_L/2)I + \sinh(\omega_L/2)\gamma^0\boldsymbol{\gamma}\cdot\hat{\mathbf{w}}]w(0) \tag{3.165}$$

Following a procedure similar to Appendix 3-A (which the reader may wish to examine first) we define four spinors for Dirac particles at rest with eq. 3-A.2. Then by applying a boost to these rest spinors we find the L_C tachyon spinors:

$$S_{CL}w^k(0) = w_{cL}{}^k(p) \tag{3.166}$$

---

[85] Note that the presence of the $\mathbf{p_i}^2$ term does not change the tachyon requirement that $\mathbf{p_r}^2 \geq m^2$ as seen in the previous cases.

and from these tachyon spinors we generalize to tachyon spinors $u_{CL}(p, s)$ and $v_{CL}(p, s)$ in a manner similar to eqs. 3.110a of the previous case.

Eqs. 3.161 through 3.164 imply that the wave function solution of eq. 3.161, subject to the subsidiary condition eq. 3.164, has the form

$$\psi_{CL}(x) = \sum_{\pm s} \int_{\mathbf{p}_r^2 \geq m^2} d^3p_r d^3p_i \, N_{CL}(p)\delta(\mathbf{p}_r\cdot\mathbf{p}_i/m^2)[b_{CL}(p,s)u_{CL}(p, s)e^{-i(p\cdot x + p^*\cdot x^*)/2} + d_{CL}^{\dagger}(p,s)v_{CL}(p, s)e^{+i(p\cdot x + p^*\cdot x^*)/2}] \quad (3.167)$$

where $\mathbf{p} = \mathbf{p}_r + i\mathbf{p}_i$, $\mathbf{x} = \mathbf{x}_r - i\mathbf{x}_i$, $p\cdot x = p^0x^0 - \mathbf{p}\cdot\mathbf{x}$, and $b_{CL}(p, s)$ and $d_{CL}(p,s)$ are tachyon Fourier coefficients.

Global SU(3) Symmetry

We can show that there is also a global SU(3) symmetry present here as shown in the previous case. The demonstration is similar to that of eqs. 3.113 – 3.126.

Light-Front Quantization of Tachyonic Complexons

Because of the momentum constraint $\mathbf{p}_r^2 \geq m^2$ the set of solutions of the form of eq. 3.167 is incomplete and the result of second quantization will not be an equal time anti-commutator expression consisting of derivatives of delta functions (eq. 3.142) but rather an analogue to previous unsuccessful attempts to create a second quantized tachyon theory.[86]

Therefore we will use light-front coordinates, and left and right handed field operators (as previously) to obtain a successful second quantization of this new type of tachyon.

The "missing" factor of i in the first term of eq. 3.163b requires the Lagrangian to be different from the conventional Dirac Lagrangian in order for the Lagrangian to be real. The simplest, physically acceptable, free spin ½ tachyon Lagrangian density for ψ_{CL} is:

$$\mathcal{L}_{CL} = \psi_{CL}{}^{C}(x)(\gamma\cdot\nabla - m)\psi_{CL}(x) \quad (3.168)$$

where

$$\psi_{CL}{}^{C}(x) = [\psi_{CL}(x)]^{\dagger}\big|_{\mathbf{x}_i = -\mathbf{x}_i} \, i\gamma^0\gamma^5 \quad (3.169)$$

similar to eq. 3.128. In words, eq. 3.169 states take the Hermitian conjugate of $\psi_{CL}(x)$; change $\mathbf{x}_i$ to $-\mathbf{x}_i$; and then post-multiply by the indicated factors.

The free complexon invariant action (under real Lorentz transformations) is

[86] Such as G. Feinberg, Phys. Rev. **159**, 1089 (1967).

$$I = \int d^7x \mathcal{L}_{CL} \tag{3.170}$$

The action can be shown to be real

$$I^* = I \tag{3.171}$$

in a manner similar to the case considered in Appendix 3-A. The tachyonic complexon's energy-momentum tensor is

$$\mathfrak{T}_{CL\mu\nu} = -\, g_{\mu\nu}\, \mathcal{L}_{CL} + \partial\mathcal{L}_{CL}/\partial(D^{\mu}\psi_{CL})\, D_{\nu}\psi_{CL} \tag{3.172}$$

$$= i\psi_{CL}{}^{C}\gamma^{0}\gamma^{5}\gamma_{\mu}D_{\nu}\psi_{CL}$$

where

$$D_0 = \partial/\partial x^0$$
$$D_k = \partial/\partial x^k + i\, \partial/\partial x_i{}^k \tag{3.129}$$

and thus the conserved energy and momentum are

$$P^0 = H = \int d^3x_r d^3x_i\, \mathfrak{T}_{CL}{}^{00} = i\int d^3x_r d^3x_i \psi_{CL}{}^{C}\gamma^5(\boldsymbol{\alpha}\cdot\mathbf{D} + \beta m)\psi_{CL} \tag{3.173}$$

and

$$P^k = \int d^3x_r d^3x_i\, \mathfrak{T}_{CL}{}^{0k} = -\, i\int d^3x_r d^3x_i\, \psi_{CL}{}^{C}\gamma^5 D^k\psi_{CL} \tag{3.174}$$

Having defined a suitable tachyon Lagrangian we can now proceed to its canonical quantization. The conjugate momentum can be calculated from the Lagrangian density eq. 3.168:

$$\pi_{CLa} = \partial\mathcal{L}_{CL}/\partial\dot{\psi}_{CLa} \equiv \partial\mathcal{L}_{CL}/\partial(\partial\psi_{CLa}/\partial t) = -i([\psi_{CL}(x)]^{\dagger}|_{\mathbf{x}_i = -\mathbf{x}_i}\gamma^5)_a \tag{3.175}$$

The resulting non-zero, canonical anti-commutation relations are presumably

$$\{\pi_{CL_a}(x), \psi_{CLb}(y)\} = i\, \delta_{ab}\, \delta^3(x_r - y_r)\delta^3(x_i - y_i)$$

based on locality in both real and imaginary coordinates, or

$$\{\psi_{CL}{}^{\dagger}{}_{a}(x)|_{\mathbf{x}_i = -\mathbf{x}_i}, \psi_{Tb}(y)\} = -\, [\gamma^5]_{ab}\, \delta^3(x_r - y_r)\delta^3(x_i - y_i) \tag{3.176}$$

At this point we might attempt to complete the canonical quantization procedure in the conventional manner by Fourier expanding the field and specifying anti-commutation relations for the Fourier component amplitudes. However the incompleteness of the set of plane waves, which are limited by the restriction $\mathbf{p_r}^2 \geq m^2$, causes the equal time anti-commutator of the fields *not* to yield a δ-functions.

Therefore we turn to the previous successful approach to tachyon quantization[87] and decompose the tachyonic complexon field into left-handed and right-handed parts and then second quantize in light-front coordinates.

Separation into Left-Handed and Right-Handed Fields

As before we will use a transformed set of Dirac matrices (eq. 3.32) to develop our left-handed and right-handed tachyon formulations. The γ^5 chirality operator's eigenvalues define handedness: +1 corresponds to right-handed; and –1 corresponds to left-handed:

$$\gamma^5\psi_{CLL} = -\psi_{CLL} \qquad\qquad \gamma^5\psi_{CLR} = \psi_{CLR} \tag{3.177}$$

As before we define left-handed and right-handed tachyon fields with the projection operators:

$$\begin{aligned} C^{\pm} &= \tfrac{1}{2}(I \pm \gamma^5) \\ C^{+} + C^{-} &= I \\ C^{\pm\,2} &= C^{\pm} \\ C^{+}C^{-} &= 0 \end{aligned} \tag{3.34}$$

with the result

$$\begin{aligned} \psi_{CLL} &= C^{-}\psi_{CL} \\ \psi_{CLR} &= C^{+}\psi_{CL} \end{aligned} \tag{3.178}$$

We can calculate the commutation relations of the left-handed and right-handed tachyonic complexon fields from eq. 3.176 by pre-multiplying and post-multiplying by $\tfrac{1}{2}(1 - \gamma^5)$ and $\tfrac{1}{2}(1 + \gamma^5)$. The results are:

$$\{\psi_{CLLa}{}^{\dagger}(x)\big|_{\mathbf{x_i} = -\mathbf{x_i}},\ \psi_{CLLb}(y)\} = C^{-}{}_{ab}\,\delta^6(x - y) \tag{3.179}$$

$$\{\psi_{CLRa}{}^{\dagger}(x)\big|_{\mathbf{x_i} = -\mathbf{x_i}},\ \psi_{CLRb}(y)\} = -C^{+}{}_{ab}\,\delta^6(x - y) \tag{3.180}$$

$$\{\psi_{CLLa}{}^{\dagger}(x)\big|_{\mathbf{x_i} = -\mathbf{x_i}},\ \psi_{CLRb}(y)\} = \{\psi_{CLRa}{}^{\dagger}(x)\big|_{\mathbf{x_i} = -\mathbf{x_i}},\ \psi_{CLLb}(x')\} = 0 \tag{3.181}$$

where

[87] Blaha (2006) discusses this case in detail.

$$\delta^6(x - y) = \delta^3(x_r - y_r)\delta^3(x_i - y_i) \tag{3.182}$$

The Lagrangian density of eq. 3.168 decomposes into left-handed and right-handed parts: (The change $\mathbf{x_i}$ to $-\mathbf{x_i}$ will be understood in $\psi_{CLL}{}^{\dagger}(x)$ and $\psi_{CLR}{}^{\dagger}(x)$ in the following.)

$$\mathcal{L}_{CL} = \psi_{CLL}{}^{\dagger}\gamma^0 i\gamma^\mu\partial_\mu\psi_{CLL} - \psi_{CLR}{}^{\dagger}\gamma^0 i\gamma^\mu\partial_\mu\psi_{CLR} - im[\psi_{CLR}{}^{\dagger}\gamma^0\psi_{CLL} - \psi_{CLL}{}^{\dagger}\gamma^0\psi_{CLR}] \tag{3.183}$$

Further Separation into + and – Light-Front Complexon Fields

As previously we now use light-front coordinates and quantization to obtain a successful second quantization of this form of tachyon field. Light-front variables, in the present case where we have to contend with complex 3-vectors, are defined by the real coordinates and derivatives:

$$x^{\pm} = (x^0 \pm x_r{}^3)/\sqrt{2}$$
$$\partial/\partial x^{\pm} \equiv \partial^{\mp} \equiv (\partial/\partial x^0 \pm \partial/\partial x_r{}^3)/\sqrt{2} \tag{3.184}$$

with the "transverse" real coordinate variables, $x_r{}^1$ and $x_r{}^2$, and imaginary coordinate variables $x_i{}^1$, $x_i{}^2$, and $x_i{}^3$.

The inner product of two 4-vectors has the form

$$x\cdot y = x^{+}y^{-} + y^{+}x^{-} + i[y_i{}^3(x^+ - x^-) + x_i{}^3(y^+ - y^-)]/\sqrt{2} + x_i{}^3y_i{}^3 - (\mathbf{x_{r\perp}} - i\mathbf{x_{i\perp}})\cdot(\mathbf{y_{r\perp}} \;\; i\mathbf{y_{i\perp}}) \tag{3.185}$$

with

$$\mathbf{x_{r\perp}} = (x_r{}^1, x_r{}^2) \qquad \mathbf{x_{i\perp}} = (x_i{}^1, x_i{}^2) \tag{3.186}$$
$$\mathbf{y_{r\perp}} = (y_r{}^1, y_r{}^2) \qquad \mathbf{y_{i\perp}} = (y_i{}^1, y_i{}^2)$$

where $x = (x^0, \mathbf{x} = \mathbf{x_r} - i\mathbf{x_i})$ and $y = (y^0, \mathbf{y} = \mathbf{y_r} - i\mathbf{y_i})$. Momenta are always defined as $p = (p^0, \mathbf{p} = \mathbf{p_r} + i\mathbf{p_i})$.

The light-front definition of Dirac matrices is:

$$\gamma^{\pm} = (\gamma^0 \pm \gamma^3)/\sqrt{2} \tag{3.42}$$

with transverse matrices γ^1 and γ^2 defined as usual. Note:

$$\gamma^{\pm\,2} = 0$$

We define "+" and "–" tachyon fields with the projection operators:

$$R^{\pm} = \tfrac{1}{2}(I \pm \gamma^0\gamma^3) \tag{3.43}$$

Left-handed, ± light-front fields: $\psi_{CLL}{}^{\pm} = R^{\pm}C^{-}\psi_{CL}$ (3.187)
Right-handed, ± light-front fields: $\psi_{CLR}{}^{\pm} = R^{\pm}C^{+}\psi_{CL}$

Transforming to light-front variables and fields as above we obtain the light-front free tachyon Lagrangian:

$$
\begin{aligned}
\mathcal{L}_{CL} = {} & 2^{1/2}\psi_{CLL}{}^{+\dagger}i\partial^{-}\psi_{CLL}{}^{+} + 2^{1/2}\psi_{CLL}{}^{-\dagger}i\partial^{+}\psi_{CLL}{}^{-} - \psi_{CLL}{}^{+\dagger}\gamma^{0}[i\gamma_{\perp}\cdot\nabla_{r\perp} - \gamma\cdot\nabla_{i}]\psi_{CLL}{}^{-} - \\
& - \psi_{CLL}{}^{-\dagger}\gamma^{0}[i\gamma_{\perp}\cdot\nabla_{r\perp} - \gamma\cdot\nabla_{i}]\psi_{CLL}{}^{+} - 2^{1/2}\psi_{CLR}{}^{+\dagger}i\partial^{-}\psi_{CLR}{}^{+} - 2^{1/2}\psi_{CLR}{}^{-\dagger}i\partial^{+}\psi_{CLR}{}^{-} + \\
& + \psi_{CLR}{}^{+\dagger}\gamma^{0}[i\gamma_{\perp}\cdot\nabla_{r\perp} - \gamma\cdot\nabla_{i}]\psi_{CLR}{}^{-} + \psi_{CLR}{}^{-\dagger}\gamma^{0}[i\gamma_{\perp}\cdot\nabla_{r\perp} - \gamma\cdot\nabla_{i}]\psi_{CLR}{}^{+} - \\
& - im[\psi_{CLR}{}^{+\dagger}\gamma^{0}\psi_{CLL}{}^{-} - \psi_{CLL}{}^{+\dagger}\gamma^{0}\psi_{CLR}{}^{-} + \psi_{CLR}{}^{-\dagger}\gamma^{0}\psi_{CLL}{}^{+} - \psi_{CLL}{}^{-\dagger}\gamma^{0}\psi_{CLR}{}^{+}]
\end{aligned} \tag{3.188}
$$

(Note the similarity to eq. 3.45 in the previous tachyon case.) Again the difference in signs between the left-handed and right-handed terms will be a crucial factor in the derivation of the left-handed features of the Standard Model.

Eq. 3.188 generates the equations of motion:

$$
\begin{aligned}
2^{1/2}i\partial^{-}\psi_{CLL}{}^{+} - \gamma^{0}[i\gamma_{\perp}\cdot\nabla_{r\perp} - \gamma\cdot\nabla_{i}]\psi_{CLL}{}^{-} + im\gamma^{0}\psi_{CLR}{}^{-} &= 0 \\
2^{1/2}i\partial^{-}\psi_{CLR}{}^{+} - \gamma^{0}[i\gamma_{\perp}\cdot\nabla_{r\perp} - \gamma\cdot\nabla_{i}]\psi_{CLR}{}^{-} + im\gamma^{0}\psi_{CLL}{}^{-} &= 0 \\
2^{1/2}i\partial^{+}\psi_{CLL}{}^{-} - \gamma^{0}[i\gamma_{\perp}\cdot\nabla_{r\perp} - \gamma\cdot\nabla_{i}]\psi_{CLL}{}^{+} + im\gamma^{0}\psi_{CLR}{}^{+} &= 0 \\
2^{1/2}i\partial^{+}\psi_{CLR}{}^{-} - \gamma^{0}[i\gamma_{\perp}\cdot\nabla_{r\perp} - \gamma\cdot\nabla_{i}]\psi_{CLR}{}^{+} + im\gamma^{0}\psi_{CLL}{}^{+} &= 0
\end{aligned} \tag{3.189}
$$

Eqs. 3.189 show that $\psi_{CLL}{}^{-}$ and $\psi_{CLR}{}^{-}$ are dependent fields that are functions of $\psi_{CLL}{}^{+}$ and $\psi_{CLR}{}^{+}$ on the light-front where x^{+} equals a constant. They can be expressed in an integral form as well. (The independent fields $\psi_{CLL}{}^{+}$ and $\psi_{CLR}{}^{+}$ play a fundamental role in tachyonic complexon theory and are used to define "in" and "out" tachyon states in perturbation theory.)

The conjugate momenta implied by eq. 3.188 are

$$
\begin{aligned}
\pi_{CLL}{}^{+} &= \partial\mathcal{L}/\partial(\partial^{-}\psi_{CLL}{}^{+}) = 2^{1/2}i\psi_{CLL}{}^{+\dagger} \\
\pi_{CLL}{}^{-} &= \partial\mathcal{L}/\partial(\partial^{-}\psi_{CLL}{}^{-}) = 0
\end{aligned} \tag{3.190}
$$

$$
\begin{aligned}
\pi_{CLR}{}^{+} &= \partial\mathcal{L}/\partial(\partial^{-}\psi_{CLR}{}^{+}) = -2^{1/2}i\psi_{CLR}{}^{+\dagger} \\
\pi_{CLR}{}^{-} &= \partial\mathcal{L}/\partial(\partial^{-}\psi_{CLR}{}^{-}) = 0
\end{aligned} \tag{3.191}
$$

x^{+} plays the role of the "time" variable in light-front quantized theories. So we define canonical equal x^{+} anti-commutation relations for spin ½ tachyonic complexons also.

The canonical equal-light-front ($x^+ = y^+$) anti-commutation relations of the independent fields would normally be:

$$\{\psi_{CLL}{}^{+\dagger}{}_a(x), \psi_{CLL}{}^{+}{}_b(y)\} = 2^{-1}[C^-R^+]_{ab}\delta(x^- - y^-)\delta^2(x_r - y_r)\delta^3(x_I - y_i) \quad (3.192)$$

$$\{\psi_{CLR}{}^{+\dagger}{}_a(x), \psi_{CLR}{}^{+}{}_b(y)\} = -2^{-1}[C^+R^+]_{ab}\,\delta(x^- - y^-)\delta^2(x_r - y_r)\delta^3(x_I - y_i) \quad (3.193)$$

$$\{\psi_{CLL}{}^{+}{}_a{}^{\dagger}(x), \psi_{CLR}{}^{+}{}_b(y)\} = \{\psi_{CLR}{}^{+}{}_a{}^{\dagger}(x), \psi_{CLL}{}^{+}{}_b(y)\} = 0 \quad (3.194)$$

$$\{\psi_{CLL}{}^{+}{}_a(x), \psi_{CLR}{}^{+}{}_b(y)\} = \{\psi_{CLR}{}^{+}{}_a{}^{\dagger}(x), \psi_{CLL}{}^{+\dagger}{}_b(y)\} = 0 \quad (3.195)$$

But as in the previous case they will be modified.

Again we see that the right-handed tachyon anti-commutation relation (eq. 3.193) has a minus sign relative to the corresponding conventional right-handed anti-commutation relation (eq. 3.55).

The sign differences between the left-handed and right-handed Lagrangian terms ultimately lead to parity violating features in the Standard Model Lagrangian.

Left-Handed Tachyonic Complexons

The free, "+" light-front, left-handed tachyonic complexon Fourier expansion is:

$$\psi_{CLL}{}^{+}(x_r, x_i) = \sum_{\pm s}\int d^2p_r dp^+ d^3p_i\, N_{CLL}{}^{+}(p)\theta(p^+)\delta((p_i^3(p^+ - p^-)/\sqrt{2} + \mathbf{p}_{r\perp}\cdot\mathbf{p}_{i\perp})/m^2)\cdot$$
$$\cdot[b_{CLL}{}^{+}(p, s)u_{CLL}{}^{+}(p, s)e^{-i(p.x + p^*.x^*)/2} + d_{CLL}{}^{+\dagger}(p, s)v_{CLL}{}^{+}(p, s)e^{+i(p.x + p^*.x^*)/2}] \quad (3.196)$$

(Compare to eq. 3.109.) Its Hermitian conjugate is

$$\psi_{CLL}{}^{+\dagger}(x_r, x_i) = \sum_{\pm s}\int d^2p_r dp^+ d^3p_i\, N_{CLL}{}^{+}(p)\theta(p^+)\delta((p_i^3(p^+ - p^-)/\sqrt{2} + \mathbf{p}_{r\perp}\cdot\mathbf{p}_{i\perp})/m^2)\cdot$$
$$\cdot[b_{CLL}{}^{\dagger}(p^*,s)u_{CLL}{}^{\dagger}(p^*,s)e^{+i(p^*.x + p.x^*)/2} + d_{CLL}(p^*,s)v_{CLL}{}^{\dagger}(p^*,s)e^{-i(p^*.x + p.x^*)/2}] \quad (3.197)$$

where $\mathbf{p} = \mathbf{p}_r + i\mathbf{p}_i$ (eq. 3.95), $\mathbf{x} = \mathbf{x}_r - i\mathbf{x}_i$, $p\cdot x = p^0x^0 - \mathbf{p}\cdot\mathbf{x}$, and † indicates Hermitian conjugate. (Compare to eq. 3.110.) The spinors are

$$u_{CLL}{}^{+}(p, s) = C^- R^+ S_{CL}w^1(0)$$
$$u_{CLL}{}^{+}(p, -s) = C^- R^+ S_{CL}w^2(0)$$

$$\begin{aligned}
v_{CLL}^{\;+}(p, s) &= C^- R^+ S_{CL} w^3(0) \\
v_{CLL}^{\;+}(p, -s) &= C^- R^+ S_{CL} w^4(0) \\
u_{CLL}^{\;+\dagger}(p^*, s) &= w^{1T}(0) S_{CL} R^+ C^- \\
u_{CLL}^{\;+\dagger}(p^*, -s) &= w^{2T}(0) S_{CL} R^+ C^- \\
v_{CLL}^{\;+\dagger}(p^*, s) &= w^{3T}(0) S_{CL} R^+ C^- \\
v_{CLL}^{\;+\dagger}(p^*, -s) &= w^{4T}(0) S_{CL} R^+ C^-
\end{aligned} \tag{3.198}$$

using eq. 3.166 where the superscript "T" indicates the transpose (These spinors are described in Appendix 3-A.) and

$$N_{CLL}^{\;+}(p) = (2\pi)^{-3}(2m/p^+)^{½} \tag{3.199}$$

The anti-commutation relations of the Fourier coefficient operators are

$$\begin{aligned}
\{b_{CLL}(p,s), b_{CLL}^{\dagger}(p'^*,s')\} &= 2^{-½}\delta_{ss'}\delta(p^+ - p'^+)\delta^2(\mathbf{p}_r - \mathbf{p'}_{r'})\delta^3(\mathbf{p}_i + \mathbf{p'}_{i'}) \\
\{d_{CLL}(p,s), d_{CLL}^{\dagger}(p'^*,s')\} &= 2^{-½}\delta_{ss'}\delta(p^+ - p'^+)\delta^2(\mathbf{p}_r - \mathbf{p'}_{r'})\delta^3(\mathbf{p}_i + \mathbf{p'}_{i'}) \\
\{b_{CLL}(p,s), b_{CLL}(p'^*,s')\} &= \{d_{CLL}(p,s), d_{CLL}(p'^*,s')\} = 0 \\
\{b_{CLL}^{\dagger}(p,s), b_{CLL}^{\dagger}(p'^*,s')\} &= \{d_{CLL}^{\dagger}(p,s), d_{CLL}^{\dagger}(p'^*,s')\} = 0 \\
\{b_{CLL}(p,s), d_{CLL}^{\dagger}(p'^*,s')\} &= \{d_{CLL}(p,s), b_{CLL}^{\dagger}(p'^*,s')\} = 0 \\
\{b_{CLL}^{\dagger}(p,s), d_{CLL}^{\dagger}(p'^*,s')\} &= \{d_{CLL}(p,s), b_{CLL}(p'^*,s')\} = 0
\end{aligned} \tag{3.200}$$

The delta-function arguments $\delta^3(\mathbf{p}_i + \mathbf{p'}_{i'})$ above have a positive sign as in eq. 3.138 in order to obtain $\delta^3(\mathbf{x}_i - \mathbf{y}_i)$ in the field anti-commutators.

The spinors, eq. 3.198, satisfy

$$\begin{aligned}
\sum_{\pm s} u_{CLL}^{\;+}{}_{\alpha}(p, s)\bar{u}_{CLL}^{\;+}{}_{\beta}(p^*, s) &= (2m)^{-1}[C^- R^+(i\not{p} + m)R^- C^+]_{\alpha\beta} \\
\sum_{\pm s} v_{CLL}^{\;+}{}_{\alpha}(p, s)\bar{v}_{CLL}^{\;+}{}_{\beta}(p^*, s) &= (2m)^{-1}[C^- R^+(i\not{p} - m)R^- C^+]_{\alpha\beta}
\end{aligned} \tag{3.201}$$

where $\bar{u}_{CLL}^{\;+} = u_{CLL}^{\;+\dagger}\gamma^0$ and $\bar{v}_{CLL}^{\;+} = v_{CLL}^{\;+\dagger}\gamma^0$.

We now evaluate the canonical left-handed, light-front anti-commutation relation (eq. 3.192):

$$\{\psi_{CLL}^{\;+}{}_{a}(x), \psi_{CLL}^{\;+\dagger}{}_{b}(y)\} = \sum_{\pm s,s'} \int d^3p_i d^2p dp^+ \int d^3p_i' d^2p' dp'^+ N_{CLL}^{\;+}(p)\, N_{CLL}^{\;+}(p')\cdot$$

$$\cdot\theta(p^+)\theta(p'^+)\delta((p_i^3(p^+ - p^-)/\surd 2 + \mathbf{p}_{r\perp}\cdot\mathbf{p}_{i\perp})/m^2)\,\delta((p_i'^3(p'^+ - p'^-)/\surd 2 + \mathbf{p'}_{r\perp}\cdot\mathbf{p'}_{i\perp})/m^2)\cdot$$

$$\cdot[\{b_{CLL}{}^{+\dagger}(p'^*,s'),b_{CLL}{}^{+}(p,s)\}u_{CLL}{}^{+}{}_{a}(p,s)u_{CLL}{}^{+\dagger}{}_{b}(p'^*,s')e^{+i(p'^*.y+p'.y^*)2-i(p.x+p^*.x^*)/2}+$$
$$+\{d_{CLL}{}^{+}(p'^*,s'),d_{CLL}{}^{+\dagger}(p,s)\}v_{CLL}{}^{+}{}_{a}(p,s)v_{CLL}{}^{+\dagger}{}_{b}(p'^*,s')e^{-i(p'^*.y+p'.y^*)/2+i(p.x+p^*.x^*)/2}]$$
$$=2^{-\frac{1}{2}}\sum_{\pm s}\int d^3p_id^2p_rdp^+\,[N_{CLL}{}^{+}(p)]^2\theta(p^+)[\delta((p_i{}^3(p^+-p^-)/\surd 2+\mathbf{p}_{r\perp}\cdot\mathbf{p}_{i\perp})/m^2)]^2$$
$$[u_{CLL}{}^{+}{}_{a}(p,s)u_{CLL}{}^{+}{}_{b}{}^{\dagger}(p^*,s)e^{+i(p^*.(y-x)+p.(y^*-x^*))/2}+$$
$$+v_{CLL}{}^{+}{}_{a}(p,s)v_{CLL}{}^{+\dagger}{}_{b}(p^*,s)e^{-i(p^*.(y-x)+p.(y^*-x^*))/2}]$$

and, using eq. 3.141,

$$=-2^{-3/2}\int d^3p_id^2pdp^+\theta(p^+)[N_{CLL}{}^{+}(p)]^2\delta'((p_i{}^3(p^+-p^-)/\surd 2+\mathbf{p}_{r\perp}\cdot\mathbf{p}_{i\perp})/m^2)(2m)^{-1}$$
$$\{[C^-R^+(i\not{p}+m)\gamma^0R^+C^-]_{ab}e^{+i(p^*.(y-x)+p.(y^*-x^*))/2}+$$
$$+[C^-R^+(i\not{p}-m)\gamma^0R^+C^-]_{ab}e^{-i(p^*.(y-x)+p.(y^*-x^*))/2}\}$$

$$=-(1/2)C^-R^+\delta_{ab}\int d^3p_i\,d^2p_\perp\int_0^\infty dp^+\,\delta'((p_i{}^3(p^+-p^-)/\surd 2+\mathbf{p}_\perp\cdot\mathbf{p}_{i\perp})/m^2)(2\pi)^{-6}\cdot$$
$$\cdot\{e^{+i\{p^+(y^--x^-)-\mathbf{p}_{r\perp}\cdot(\mathbf{y}_{r\perp}-\mathbf{x}_{r\perp})+\mathbf{p}_i.(\mathbf{y}_i-\mathbf{x}_i)\}}+e^{-i\{p^+(y^--x^-)-\mathbf{p}_{r\perp}\cdot(\mathbf{y}_{r\perp}-\mathbf{x}_{r\perp})+\mathbf{p}_i\cdot(\mathbf{y}_i-\mathbf{x}_i)\}}\}$$

$$=-C^-R^+\delta_{ab}(4\pi)^{-1}\int_0^\infty dp^+\delta'(\nabla_r\cdot\nabla_i/m^2)\delta^3(y_i-x_i)\,\delta^2(y_r-x_r)\{e^{+ip^+(y^--x^-)}+e^{-ip^+(y^--x^-)}\}$$

whereupon we revert back to the original form of the constraint: $\delta(\nabla_r\cdot\nabla_i/m^2)$

$$\{\psi_{CLL}{}^{+}{}_{a}(x),\,\psi_{CLL}{}^{+\dagger}{}_{b}(y)\}=-(1/2)C^-R^+\delta_{ab}\,\delta'(\nabla_r\cdot\nabla_i/m^2)\delta(y^--x^-)\delta^2(y_r-x_r)\delta^3(y_i-x_i) \tag{3.202}$$

The result is the left-handed, light-front equivalent of the earlier non-tachyon result eq. 3.142. Again the constraint is apparent in the anti-commutator. (The factor of 2 difference is due to light-front coordinate definitions.)

Therefore we have left-handed, light-front quantized tachyonic complexons with the equivalent of canonical anti-commutation relations, and with localized tachyonic complexons. As a result we have a canonical tachyonic complexon Quantum Field Theory

Left-handed Case 4 Tachyonic Complexon Feynman Propagator

The light-front Feynman propagator for the left-handed $\psi_{CLL}{}^{+}$ *tachyonic* complexon field is

$$iS^{+}_{CLLF}(x,y) = \theta(x^{+} - y^{+})<0|\psi_{CLL}{}^{+}(x)\psi_{CLL}{}^{+\dagger}(y)\gamma^{0}|0> - \\ - \theta(y^{+} - x^{+})<0|\psi_{CLL}{}^{+\dagger}(y)\gamma^{0}\psi_{CLL}{}^{+}(x)|0> \qquad (3.203)$$

$$= -\tfrac{1}{2}\int d^3p_i d^2p_r dp^{+}\theta(p^{+})N_{CLL}{}^{+2}\delta'((p_i{}^{3}(p^{+} - p^{-})/\sqrt{2} + \mathbf{p}_{r\perp}\cdot\mathbf{p}_{i\perp})/m^2)(2m)^{-1}C^{-}R^{+} \\ \{\theta(x^{+} - y^{+})[(i\not{p} + m)\gamma^{0}]e^{+i(p^*.(y-x)+p.(y^*-x^*))/2} + \\ + \theta(y^{+} - x^{+})[(i\not{p} - m)\gamma^{0}]e^{-i(p^*.(y-x)+p.(y^*-x^*))/2}\}R^{+}C^{-}\gamma^{0}$$

If we define the on-shell momentum variable $p_0{}^{-} = (p_{r0}{}^{1}p_{r0}{}^{1} + p_{r0}{}^{2}p_{r0}{}^{2} - \mathbf{p}_{i0}\cdot\mathbf{p}_{i0} - m^2)/(2p_0{}^{+})$, $p_0{}^{+} = p^{+}$, $p_{r0}{}^{j} = p_r{}^{j}$ (for j = 1, 2), $\mathbf{p}_{i0} = \mathbf{p}_i$, $p_{r\perp 0}{}^{2} = p_{r0}{}^{j}p_{r0}{}^{j}$ and $\not{p}_0 = p_0\cdot\gamma$ with $p_0 = (p^0, \mathbf{p}_{r0} + i\mathbf{p}_{r0})$ then the above equation can be rewritten as

$$= -\tfrac{1}{2}C^{-}R^{+}\int d^4p d^3p_i N_{CLL}{}^{+2}\delta'((p_{i0}{}^{3}(p_0{}^{+} - p_0{}^{-})/\sqrt{2} + \mathbf{p}_{r\perp 0}\cdot\mathbf{p}_{i\perp 0})/m^2)(4\pi m)^{-1}e^{+i(p^*.(y-x)+p.(y^*-x^*))/2}$$

$$\cdot\{\theta(p^{+})(i\not{p}_0 + m)\gamma^{0}]/[p^{-} - p_0{}^{-} + i\varepsilon] + \theta(-p^{+})(i\not{p}_0 - m)\gamma^{0}]/[p^{-} + p_0{}^{-} - i\varepsilon]\}R^{+}C\gamma^{0}$$

$$= -\tfrac{1}{2}\int d^4p_r d^3p_i N_{CLL}{}^{+2}\delta'((p_{i0}{}^{3}(p^{+} - p^{-})/\sqrt{2} + \mathbf{p}_{r\perp}\cdot\mathbf{p}_{i\perp})/m^2)(p^{+}/4\pi m)\, e^{+i(p^*.(y-x)+p.(y^*-x^*))/2}\,\cdot \\ \cdot[C^{-}R^{+}(i\not{p} + m)\gamma^{0}R^{+}C^{-}\gamma^{0}][(p^2 + m^2 + i\varepsilon)]^{-1}$$

with $p_r = (p^0, \mathbf{p}_r)$ and $p = (p^0, \mathbf{p}_r + i\mathbf{p}_r)$. Substituting for N_{CLL} and using $x\delta'(x) = -\delta(x)$ we obtain

$$= -\tfrac{1}{2}\int d^4p_r d^3p_i(2\pi)^{-7}\delta'(\mathbf{p}_r\cdot\mathbf{p}_i/m^2)\exp[ip^0(y^0 - x^0) - i\mathbf{p}_r\cdot(\mathbf{y}_r - \mathbf{x}_r) + i\mathbf{p}_i\cdot(\mathbf{y}_i - \mathbf{x}_i)]\cdot \\ \cdot[C^{-}R^{+}(i\not{p} + m)R^{-}C^{+}]/(p^2 + m^2 + i\varepsilon)$$

since $C^{-}R^{+}(i\not{p} + m)\gamma^{0}R^{+}C^{-}\gamma^{0} = C^{-}R^{+}(i\not{p} + m)R^{-}C^{+}$. The integral can be written:

$$I = \int d^4p_r d^3p_i\delta'(\mathbf{p}_r\cdot\mathbf{p}_i/m^2)C^{-}R^{+}(i\not{p} + m)R^{-}C^{+}\cdot \\ \cdot\exp[-ip^0(x^0 - y^0) + i\mathbf{p}_r\cdot(\mathbf{x}_r - \mathbf{y}_r) - i\mathbf{p}_i\cdot(\mathbf{x}_i - \mathbf{y}_i)]/(p^2 + m^2 + i\varepsilon)$$

$$= \int d^4p_r dM^2\delta'(\nabla_r\cdot\nabla_i/m^2)C^{-}R^{+}(ip^{\mathbf{0}}\gamma^{\mathbf{0}} - (\nabla_r - i\nabla_i)\cdot\boldsymbol{\gamma} + m)R^{-}C^{+}\cdot \\ \cdot\exp[-ip^0(x^0 - y^0) + i\mathbf{p}_r\cdot(\mathbf{x}_r - \mathbf{y}_r)]J_2(\mathbf{x}_i - \mathbf{y}_i, M^2)/(p_r{}^2 + M^2 + i\varepsilon)$$

where

$$J_2(\mathbf{x}_i - \mathbf{y}_i, M^2) = (2\pi)^{-3}\int d^3p_i\,\delta(M^2 - \mathbf{p}_i{}^2 - m^2)\exp[-i\mathbf{p}_i\cdot(\mathbf{x}_i - \mathbf{y}_i)] \qquad (3.204)$$

$$= (2\pi)^{-2}|\mathbf{x_i} - \mathbf{y_i}|^{-1}\theta(M^2 - m^2)\sin((M^2 - m^2)^{1/2}|\mathbf{x_i} - \mathbf{y_i}|)$$

This tachyonic complexon Feynman propagator can be rearranged into the form of a spectral integral:

$$iS^{+}{}_{CLLF}(x, y) = -\int dM\, C^{-}R^{+}(\gamma^{0}\partial/\partial x^{0} + (\nabla_{r} - i\nabla_{i})\cdot\gamma - m)R^{-}C^{+}\,\delta'(\nabla_{r}\cdot\nabla_{i}/m^{2})\cdot$$
$$\cdot J_2(\mathbf{x_i} - \mathbf{y_i}, M^2)\Delta_{FT}(x - y, M) \qquad (3.205)$$

with ∇_r and ∇_i derivatives with respect to $\mathbf{x_r}$ and $\mathbf{x_i}$ and where

$$\Delta_{FT}(x - y, M) = (2\pi)^{-4}\int d^4p_r \exp[-ip^0(x^0 - y^0) + i\mathbf{p_r}\cdot(\mathbf{x_r} - \mathbf{y_r})]/(p_r^{\,2} + M^2 + i\varepsilon) \qquad (3.206)$$

Right-Handed Tachyonic Complexons

The case of right-handed tachyonic complexons is similar to left-handed complexons with only one difference: a minus sign in the canonical right-handed equal-time commutation relations resulting in a minus sign in the creation and annihilation operator anti-commutation relations. The right-handed tachyonic complexon wave function (eq. 3.187) light-front Fourier expansion is:

$$\psi_{CLR}^{+}(x_r, x_i) = \sum_{\pm s}\int d^2p_r dp^{+} d^3p_i\, N_{CLR}^{+}(p)\theta(p^{+})\delta((p_i^{3}(p^{+} - p^{-})/\sqrt{2} + \mathbf{p_{r\perp}}\cdot\mathbf{p_{i\perp}})/m^2)\cdot$$
$$\cdot[b_{CLR}^{+}(p, s)u_{CLR}^{+}(p, s)e^{-i(p.x + p^*.x^*)/2} + d_{CLR}^{+\dagger}(p, s)v_{CLR}^{+}(p, s)e^{+i(p.x + p^*.x^*)/2}] \qquad (3.207)$$

where

$$N_{CLR}^{+}(p) = (2\pi)^{-3}(2m/p^{+})^{1/2} \qquad (3.208)$$

Its Hermitian conjugate is

$$\psi_{CLR}^{+\dagger}(x_r, x_i) = \sum_{\pm s}\int d^2p_r dp^{+} d^3p_i\, N_{CLR}^{+}(p)\theta(p^{+})\delta((p_i^{3}(p^{+} - p^{-})/\sqrt{2} + \mathbf{p_{r\perp}}\cdot\mathbf{p_{i\perp}})/m^2)\cdot$$
$$\cdot[b_{CLR}^{\dagger}(p^*,s)u_{CLR}^{\dagger}(p^*,s)e^{+i(p^*.x + p.x^*)/2} + d_{CLR}(p^*,s)v_{CLR}^{\dagger}(p^*,s)e^{-i(p^*.x + p.x^*)/2}] \qquad (3.209)$$

where $\mathbf{p} = \mathbf{p_r} + i\mathbf{p_i}$ (eq. 3.95), $\mathbf{x} = \mathbf{x_r} - i\mathbf{x_i}$, $p\cdot x = p^0x^0 - \mathbf{p}\cdot\mathbf{x}$, and † indicates Hermitian conjugate. (Compare to eq. 3.110.) The right-handed spinors are

$$u_{CLR}^{+}(p, s) = C^{+} R^{+} S_{CR} w^{1}(0)$$
$$u_{CLR}^{+}(p, -s) = C^{+} R^{+} S_{CR} w^{2}(0)$$

$$
\begin{aligned}
\overset{+}{v}_{CLR}(p, s) &= C^+ R^+ S_{CR} w^3(0) \\
\overset{+}{v}_{CLR}(p, -s) &= C^+ R^+ S_{CR} w^4(0) \\
\overset{+}{u}_{CLR}{}^{\dagger}(p^*, s) &= w^{1T}(0) S_{CR} R^+ C^+ \\
\overset{+}{u}_{CLR}{}^{\dagger}(p^*, -s) &= w^{2T}(0) S_{CR} R^+ C^+ \\
\overset{+}{v}_{CLR}{}^{\dagger}(p^*, s) &= w^{3T}(0) S_{CR} R^+ C^+ \\
\overset{+}{v}_{CLR}{}^{\dagger}(p^*, -s) &= w^{4T}(0) S_{CR} R^+ C^+
\end{aligned}
\tag{3.210}
$$

where the superscript "T" indicates the transpose. The anti-commutation relations of the Fourier coefficient operators are

$$
\begin{aligned}
\{b_{CLR}(p,s), b_{CLR}{}^{\dagger}(p'^*,s')\} &= -2^{-1/2}\delta_{ss'}\delta(p^+ - p'^+)\delta^2(\mathbf{p_r} - \mathbf{p'_{r'}})\delta^3(\mathbf{p_i} + \mathbf{p'_{i'}}) \\
\{d_{CLR}(p,s), d_{CLR}{}^{\dagger}(p'^*,s')\} &= -2^{-1/2}\delta_{ss'}\delta(p^+ - p'^+)\delta^2(\mathbf{p_r} - \mathbf{p'_{r'}})\delta^3(\mathbf{p_i} + \mathbf{p'_{i'}}) \\
\{b_{CLR}(p,s), b_{CLR}(p'^*,s')\} &= \{d_{CLR}(p,s), d_{CLR}(p'^*,s')\} = 0 \\
\{b_{CLR}{}^{\dagger}(p,s), b_{CLR}{}^{\dagger}(p'^*,s')\} &= \{d_{CLR}{}^{\dagger}(p,s), d_{CLR}{}^{\dagger}(p'^*,s')\} = 0 \\
\{b_{CLR}(p,s), d_{CLR}{}^{\dagger}(p'^*,s')\} &= \{d_{CLR}(p,s), b_{CLR}{}^{\dagger}(p'^*,s')\} = 0 \\
\{b_{CLR}{}^{\dagger}(p,s), d_{CLR}{}^{\dagger}(p'^*,s')\} &= \{d_{CLR}(p,s), b_{CRR}(p'^*,s')\} = 0
\end{aligned}
\tag{3.211}
$$

The spinors satisfy

$$
\sum_{\pm s} \overset{+}{u}_{CLR\,\alpha}(p, s)\overset{-\;+}{u}_{CLR\,\beta}(p^*, s) = (2m)^{-1}[C^+R^+(-i\not{p} + m)R^-C^-]_{\alpha\beta} \tag{3.212}
$$

$$
\sum_{\pm s} \overset{+}{v}_{CLR\,\alpha}(p, s)\overset{-\;+}{v}_{CLR\,\beta}(p^*, s) = (2m)^{-1}[C^+R^+(-i\not{p} - m)R^-C^-]_{\alpha\beta}
$$

where $\overset{-\;+}{u}_{CLR} = \overset{+}{u}_{CLR}{}^{\dagger}\gamma^0$ and $\overset{-\;+}{v}_{CLR} = \overset{+}{v}_{CLR}{}^{\dagger}\gamma^0$.

The right-handed anti-commutation relation with a minus sign follows in particular because of the minus signs in eqs. 3.211.

Case 4 Right-handed Tachyonic Complexon Feynman Propagator

The Feynman propagator for right-handed tachyonic complexons can be obtained from eqs. 3.205 and 3.206 by changing the parity projection operator and some numerator signs in the integral (basically p → –p) resulting in

$$
\begin{aligned}
i\overset{+}{S}_{CLRF}(x, y) = \int dM\, C^+R^+(\gamma^0\partial/\partial x^0 + (\nabla_r - i\nabla_i)\cdot\gamma - m)R^-C^-\,\delta'(\nabla_r\cdot\nabla_i/m^2)\cdot \\
\cdot J_2(\mathbf{x_i} - \mathbf{y_i}, M^2)\Delta_{FT}(x - y, M)
\end{aligned}
\tag{3.213}
$$

with $\nabla_r + i\nabla_i$ derivatives with respect to $\mathbf{x}_r$ and $\mathbf{x}_i$ and where

$$\Delta_{FT}(x - y, M) = (2\pi)^{-4}\int d^4p_r \exp[-ip^0(x^0 - y^0) + i\mathbf{p}_r\cdot(\mathbf{x}_r - \mathbf{y}_r)]/(p_r^{\ 2} + M^2 + i\varepsilon) \tag{3.214}$$

Other Cases? No

The four cases considered above are the only cases having symmetry under the real Lorentz group L and a single real energy (with a corresponding single real time parameter) that is independent of the direction of the boost thus preserving (real) spatial rotation invariance. The realness of the time variable survives the breakdown to conventional Lorentz invariance.

One might think that using the other type of spinor boost operator (Compare to eq. 3.154.)

$$S_{CR}(\Lambda_{CR}(\omega, \hat{\mathbf{w}})) = \exp(-i\omega_R\sigma_{0i}w_i/2) = \exp(-\omega_R\gamma^0\boldsymbol{\gamma}\cdot\hat{\mathbf{w}}/2) \tag{3.215}$$

$$= \cosh(\omega_R/2)I + \sinh(\omega_R/2)\gamma^0\boldsymbol{\gamma}\cdot\hat{\mathbf{w}}$$

where $\omega_R = \omega - i\pi/2$ might lead to more possible forms of spin ½ wave equations and particles. In fact it merely leads to the same particle types but with the role of the left-handed and right-handed fields reversed. The result would be a "right-handed" Standard Model contrary to experiment.

3.8 L_C Spinor Lorentz Boosts Generate 4 Types of Particles that suggest they are Analogues of Leptons and Color Quarks

In this chapter we have found four types of fermions using a complex form of Lorentz boosts, L_C boosts, that correspond in a natural way with the four general types of known fermions: charged leptons, neutrinos, up-type color quarks and down-type color quarks.[88]

Charged lepton fermions

The conventional Dirac equation and solutions.

Neutrinos

Simple tachyons with real energy and 3-momentum. Their free field equation is:

$$(\gamma^\mu\partial/\partial x^\mu - m)\psi_T(x) = 0 \tag{3.18}$$

and their left-handed $\psi_{TL}^{\ +}$ Feynman propagator is:

$$iS^+_{TLF}(x, y) = \tfrac{1}{2}C^-R^+\gamma^0\int d^4p(2\pi)^{-4}\, p^+ e^{-ip.(x-y)}/(p^2 + m^2 + i\varepsilon) \tag{3.76}$$

---

[88] We call each type of fermion a *species*. Each species has three known generations.

Similarly the light-front Feynman propagator for the right-handed ψ_{TR}^{+} tachyon field is

$$iS^{+}_{TRF}(x,y) = -\tfrac{1}{2}C^{+}R^{+}\gamma^{0}\int d^4p(2\pi)^{-4}\, p^{+}e^{-ip\cdot(x-y)}/(p^2 + m^2 + i\varepsilon) \quad (3.77)$$

Up-type Color Quarks

Up-type quarks are assumed[89] to be fermions with complex 3-momenta - complexons, and an internal color SU(3) symmetry, that are "normal" with $p^2 = m^2$. Their field equation with a color SU(3) index, denoted a, inserted is

$$[i\gamma^{0}\partial/\partial t + i\boldsymbol{\gamma}\cdot(\nabla_r + i\nabla_i) - m]\psi_C{}^{a}(t, \mathbf{x}_r, \mathbf{x}_i) = 0 \quad (3.101)$$

with the subsidiary condition

$$\nabla_r\cdot\nabla_i\, \psi_C{}^{a}(t, \mathbf{x}_r, \mathbf{x}_i) = 0 \quad (3.102a)$$

The free field solution is:

$$\psi_C{}^{a}(x) = \sum_{\pm s} \int d^3p_r d^3p_i\, N_C(p)\delta(\mathbf{p}_r\cdot\mathbf{p}_i/m^2)[b_C(p,a,s)u_C{}^{a}(p, s)e^{-i(p\cdot x + p^*\cdot x^*)/2} + $$
$$+ d_C{}^{\dagger}(p,a,s)v_C{}^{a}(p, s)e^{+i(p\cdot x + p^*\cdot x^*)/2}] \quad (3.125a)$$

The free Feynman propagator arranged into the form of a spectral integral is

$$iS_C{}^{ab}(x,y) = -\delta^{ab}\int dM\, (i\gamma^{0}\partial/\partial x^{0} - i(\nabla_r - i\nabla_i)\cdot\boldsymbol{\gamma} + m)\delta'(\nabla_r\cdot\nabla_i/m^2)\cdot$$
$$\cdot J(\mathbf{x}_i - \mathbf{y}_i, M^2)\Delta_F(x - y, M) \quad (3.149)$$

where

$$\Delta_F(x - y, M) = (2\pi)^{-4}\int d^4p_r\, \exp[-ip^{0}(x^{0} - y^{0}) + i\mathbf{p}_r\cdot(\mathbf{x}_r - \mathbf{y}_r)]/(p_r{}^2 - M^2 + i\varepsilon) \quad (3.150)$$

and

$$J(\mathbf{x}_i, M^2) = (2\pi)^{-3}\int d^3p_i\, \delta(M^2 + \mathbf{p}_i{}^2 - m^2)\exp[-i\mathbf{p}_i\cdot(\mathbf{x}_i - \mathbf{y}_i)] \quad (3.148)$$

$$= (2\pi)^{-2}|\mathbf{x}_i - \mathbf{y}_i|^{-1}\theta(m^2 - M^2)\sin((m^2 - M^2)^{1/2}|\mathbf{x}_i - \mathbf{y}_i|)$$

---

[89] The complexon theory that we develop and use for quark dynamics in the Standard Model is not required. Our Standard Model could use Dirac fermion dynamics for the up-type quarks and tachyon dynamics for down-type quarks. Then the (broken) Left-handed Extended Lorentz group would be the basic space-time group rather than L_C. We choose to use complexon dynamics for quarks because they have an internal SU(3)-like structure suggestive of color SU(3). More importantly, their spin dynamics is different and thus may resolve the differences between theory and experiment for the deep inelastic parton spin-dependent structure functions.

Down-type Color Quarks

Tachyonic complexons with complex 3-momenta, and an internal global SU(3) symmetry, that have mass shell condition $p^2 = -m^2$. Their field equation with a color SU(3) index, denoted a, inserted is

$$[\gamma^0\partial/\partial t + \boldsymbol{\gamma}\cdot(\nabla_r + i\nabla_i) - m]\psi_{CL}{}^a(t, \mathbf{x_r}, \mathbf{x_i}) = 0 \qquad (3.163a)$$

with the subsidiary condition on the wave function

$$\nabla_r\cdot\nabla_i\, \psi_{CL}{}^a(t, \mathbf{x_r}, \mathbf{x_i}) = 0 \qquad (3.164)$$

Its free field left-handed solution is:

$$\psi_{CLL}{}^{+a}(x_r, x_i) = \sum_{\pm s} \int d^2p_r dp^+ d^3p_i\, N_{CLL}{}^+(p)\theta(p^+)\delta((p_i{}^3(p^+ - p^-)/\surd 2 + \mathbf{p}_{r\perp}\cdot\mathbf{p}_{i\perp})/m^2)\cdot$$
$$\cdot[b_{CLL}{}^+(p,a,s)u_{CLL}{}^a(p,a,s)e^{-i(p.x + p^*.x^*)/2} + d_{CLL}{}^{+\dagger}(p,a,s)v_{CLL}{}^{+a}(p,a,s)e^{+i(p.x + p^*.x^*)/2}] \qquad (3.196)$$

and its right-handed solution is

$$\psi_{CLR}{}^{+a}(x_r, x_i) = \sum_{\pm s} \int d^2p_r dp^+ d^3p_i\, N_{CLR}{}^+(p)\theta(p^+)\delta((p_i{}^3(p^+ - p^-)/\surd 2 + \mathbf{p}_{r\perp}\cdot\mathbf{p}_{i\perp})/m^2)\cdot$$

$$\cdot[b_{CLR}{}^+(p,a,s)u_{CLR}{}^{+a}(p,a,s)e^{-i(p.x+p^*.x^*)/2} + d_{CLR}{}^{+\dagger}(p,a,s)v_{CLR}{}^{+a}(p,a,s)c^{+i(p.x+p^*.x^*)/2}] \qquad (3.207)$$

The free left-handed Feynman propagator arranged into the form of a spectral integral is

$$iS^+{}_{CLLF}{}^{ab}(x,y) = -\delta^{ab}\int dM\, C^-R^+(\gamma^0\partial/\partial x^0 + (\nabla_r - i\nabla_i)\cdot\boldsymbol{\gamma} - m)R^-C^+\delta'(\nabla_r\cdot\nabla_i/m^2)\cdot$$
$$\cdot J_2(\mathbf{x_i} - \mathbf{y_i}, M^2)\Delta_{FT}(x - y, M) \qquad (3.205)$$

with ∇_r and ∇_i derivatives with respect to $\mathbf{x_r}$ and $\mathbf{x_i}$ and where

$$\Delta_{FT}(x - y, M) = (2\pi)^{-4}\int d^4p_r \exp[-ip^0(x^0 - y^0) + i\mathbf{p}_r\cdot(\mathbf{x_r} - \mathbf{y_r})]/(p_r{}^2 + M^2 + i\varepsilon) \qquad (3.206)$$

and

$$J_2(\mathbf{x_i}, M^2) = (2\pi)^{-3}\int d^3p_i\, \delta(M^2 - \mathbf{p_i}^2 - m^2) \exp[-i\mathbf{p_i}\cdot(\mathbf{x_i} - \mathbf{y_i})] \qquad (3.204)$$

$$= (2\pi)^{-2}|\mathbf{x_i} - \mathbf{y_i}|^{-1}\theta(M^2 - m^2)\sin((M^2 - m^2)^{1/2}|\mathbf{x_i} - \mathbf{y_i}|)$$

The free right-handed Feynman propagator arranged into the form of a spectral integral is

$$iS^{+}{}_{CLRF}{}^{ab}(x, y) = \delta^{ab}\int dM\, C^{+}R^{+}(\gamma^{0}\partial/\partial x^{0} + (\nabla_r - i\nabla_i)\cdot\gamma - m)R^{-}C^{-}\delta'(\nabla_r\cdot\nabla_i/m^2)\cdot$$
$$\cdot J_2(\mathbf{x_i} - \mathbf{y_i}, M^2)\Delta_{FT}(x - y, M) \qquad (3.213)$$

with ∇_r and ∇_i derivatives with respect to $\mathbf{x_r}$ and $\mathbf{x_i}$, and where

$$\Delta_{FT}(x - y, M) = (2\pi)^{-4}\int d^4p_r \exp[-ip^0(x^0 - y^0) + i\mathbf{p_r}\cdot(\mathbf{x_r} - \mathbf{y_r})]/(p_r^{\,2} + M^2 + i\varepsilon)$$
$$(3.214)$$

3.9 First Step towards a One Generation Standard Model

Thus we have found a set of four fermion species that corresponds to the known fermions of one fermion generation. In subsequent chapters we will derive the one generation model in detail. Then we will introduce three generations with mixing to complete the derivation of the form of the Standard Model. Then the only remaining major issue will be the values of the coupling constants and other numerical parameters.

The overall pattern that begins to emerge from the developments in this chapter divides particles and interactions into two categories (as seen in Nature):

Particles with real 4-Momenta	Complexons (Complex 3-Momenta)
Leptons	color quarks
SU(2)⊗U(1) Vector Bosons	Color SU(3) gluons
Higgs Particles	Possibly Higgs Particles

We will explore these issues in detail in the following chapters. But basically the leptons, SU(2)⊗U(1) Vector Bosons and a set of Higgs particles appear to be based on the Left-handed Extended Lorentz group. These particles have real energies and momenta although some are "normal" and some are tachyons.

The other category of particles, complexons, emerges from our study of L_C. These particles have real energies and complex 3-momenta. In perturbation theory the loop integrations of loops of these particles would consist of a 7-fold integration over energy and complex 3-momenta with corresponding 7-fold delta functions to enforce energy-momentum conservation. As pointed out earlier the complex 3-momenta of these types of fermions has an SU(3) symmetry that it is natural to generalize to local color SU(3). (The other category of fermions lacks this global SU(3) symmetry just as leptons lack color SU(3).) Thus we see the beginnings of the structure of the Standard Model in this chapter on spin ½ particles. The following chapters lead to a detailed derivation of the form of the Standard Model.

Appendix 3-A. Leptonic Tachyon Spinors

The general form of the solutions of the free tachyon Dirac equation eq. 3.18 can be written

$$\psi_T{}^r(x) = e^{-i\chi_r p\cdot x} w^r(p) \qquad (3\text{-A}.1)$$

where $\chi_r = +1$ for r = 1, 2 and $\chi_r = -1$ for r = 3, 4. Denoting the spinors $w^r(p) = w^r(0)$ for a particle is at rest in a frame (E = m) we see they can take the form

$$w^r(0) = \begin{bmatrix} \delta_{1r} \\ \delta_{2r} \\ \delta_{3r} \\ \delta_{4r} \end{bmatrix} \qquad (3\text{-A}.2)$$

where Kronecker deltas appear in the brackets. From eq. 3.15 we find

$$S_L(\Lambda_L(\omega, \mathbf{u}))w^r(0) = w_T{}^r(p) \qquad (3\text{-A}.3)$$

Using eq. 3.11 for $S_L(\Lambda_L(\omega, \mathbf{u}))$ and

$$\mathbf{p} = m\mathbf{v}\gamma_s \qquad\qquad E = m\gamma_s \qquad (3\text{-A}.4)$$

we see that eq. 3-A.3 implies the columns of the resulting $S_L(\Lambda_L(\omega, \mathbf{u}))$ matrix are

$$\begin{array}{cccc} \underline{w_T{}^3(p)} & \underline{w_T{}^4(p)} & \underline{w_T{}^1(p)} & \underline{w_T{}^2(p)} \end{array}$$

$$S_L(\Lambda_L(\omega, \mathbf{u})) = \begin{bmatrix} \cosh(\omega_L/2) & 0 & \sinh(\omega_L/2)p_z/p & \sinh(\omega_L/2)p_-/p \\ 0 & \cosh(\omega_L/2) & \sinh(\omega_L/2)p_+/p & -\sinh(\omega_L/2)p_z/p \\ \sinh(\omega_L/2)p_z/p & \sinh(\omega_L/2)p_-/p & \cosh(\omega_L/2) & 0 \\ \sinh(\omega_L/2)p_+/p & -\sinh(\omega_L/2)p_z/p & 0 & \cosh(\omega_L/2) \end{bmatrix}$$

(3-A.5)

based on the superluminal transformation of positive energy states to negative energy states (eqs. 3.15 and 3.16) with $p_\pm = p_x \pm ip_y$ and where $p = |\mathbf{p}|$. It is easy to verify

$$(i\not{p} - \chi_r m)w_T^{\,r}(p) = 0 \qquad (3\text{-A.6})$$

where $\chi_r = -1$ for r = 1, 2 and $\chi_r = +1$ for r = 3, 4.

The spinors that we defined in eq. 2.10 can be generalized in a manner similar to Dirac spinors. We will use a similar notation to the Dirac spinor notation:

$$\begin{aligned} u_T(p, s) &= w_T^{\,1}(p) \\ u_T(p, -s) &= w_T^{\,2}(p) \\ v_T(p, s) &= w_T^{\,3}(p) \\ v_T(p, -s) &= w_T^{\,4}(p) \end{aligned} \qquad (3\text{-A.7})$$

We define "double dagger" spinors:

$$\begin{aligned} u_T^{\ddagger}(p, s) &= u_T^{\dagger}(p, s)i\gamma\cdot\mathbf{p}/|\mathbf{p}| \\ u_T^{\ddagger}(p, -s) &= u_T^{\dagger}(p, -s)i\gamma\cdot\mathbf{p}/|\mathbf{p}| \\ v_T^{\ddagger}(p, s) &= v_T^{\dagger}(p, s)i\gamma\cdot\mathbf{p}/|\mathbf{p}| \\ v_T^{\ddagger}(p, -s) &= v_T^{\dagger}(p, -s)i\gamma\cdot\mathbf{p}/|\mathbf{p}| \end{aligned} \qquad (3\text{-A.8})$$

.

where † indicates Hermitian conjugate, which appear in important spinor "completeness" sums:

$$\sum_{\pm s} u_{T\alpha}(p, s)u_{T\ \beta}^{\ \ddagger}(p, s) = (2m)^{-1}(i\not{p} - m)_{\alpha\beta} \qquad (3\text{-A.9})$$

$$\sum_{\pm s} v_{T\alpha}(p, s)v_{T\ \beta}^{\ \ddagger}(p, s) = (2m)^{-1}(i\not{p} + m)_{\alpha\beta} \qquad (3\text{-A.10})$$

or

$$\sum_{\pm s} u_{T\alpha}(p, s)u_{T\ \beta}^{\ \dagger}(p, s) = -i(2m)^{-1}[(i\not{p} - m)\gamma\cdot\mathbf{p}/|\mathbf{p}|]_{\alpha\beta} \qquad (3\text{-A.11})$$

$$\sum_{\pm s} v_{T\alpha}(p, s)v_{T\ \beta}^{\ \dagger}(p, s) = -i(2m)^{-1}[(i\not{p} + m)\gamma\cdot\mathbf{p}/|\mathbf{p}|]_{\alpha\beta} \qquad (3\text{-A.12})$$

Lastly we define light-front, left-handed tachyon spinors by

$$u_{TL}^{\ \ +}(p, s) = C^- R^+ S_L(\Lambda_L(\omega, \mathbf{u}))w^1(0)$$

$$u_{TL}^{+}(p, -s) = C^{-} R^{+} S_L(\Lambda_L(\omega, \mathbf{u}))w^2(0) \qquad (3\text{-A.13})$$
$$v_{TL}^{+}(p, s) = C^{-} R^{+} S_L(\Lambda_L(\omega, \mathbf{u}))w^3(0)$$
$$v_{TL}^{+}(p, -s) = C^{-} R^{+} S_L(\Lambda_L(\omega, \mathbf{u}))w^4(0)$$

$$u_{TL}^{+\dagger}(p, s) = w^{1T}(0)\, S_L^{\dagger}(\Lambda_L(\omega, \mathbf{u}))\, R^{+}C^{-}$$
$$u_{TL}^{+\dagger}(p, -s) = w^{2T}(0)\, S_L^{\dagger}(\Lambda_L(\omega, \mathbf{u}))R^{+}C^{-} \qquad (3\text{-A.14})$$
$$v_{TL}^{+\dagger}(p, s) = w^{3T}(0)\, S_L^{\dagger}(\Lambda_L(\omega, \mathbf{u}))R^{+}C^{-}$$
$$v_{TL}^{+\dagger}(p, -s) = w^{4T}(0)\, S_L^{\dagger}(\Lambda_L(\omega, \mathbf{u}))R^{+}C^{-}$$

where the superscript "T" indicates the transpose and † indicates Hermitian conjugate.

Appendix 3-B. Proof of the Reality of the Leptonic Tachyon Action

The tachyon Lagrangian density and action are

$$\mathcal{L}_T = \psi_T{}^S(\gamma^\mu\partial/\partial x^\mu - m)\psi_T(x) \tag{3.19}$$

$$I = \int d^4x\mathcal{L}_T \tag{3.21}$$

where

$$\psi_T{}^S = \psi_T{}^\dagger\, i\gamma^0\gamma^5 \tag{3.20}$$

The complex conjugate of the tachyon Lagrangian density is

$$\mathcal{L}_T{}^* = -\psi_T{}^T\, i\gamma^0\gamma^5(\gamma^{\mu *}\partial/\partial x^\mu - m)\psi_T{}^*(x) \tag{3-B.1}$$

where the superscript ^{T} indicates the transpose. Eq. 3-B.1 can be expressed as a transpose:

$$\mathcal{L}_T{}^* = -i[\psi_T{}^\dagger(\gamma^{\mu\dagger}\overleftarrow{\partial}/\partial x^\mu - m)\gamma^5\gamma^0\psi_T(x)]^T \tag{3-B.2}$$

$$= -i[\psi_T{}^\dagger\gamma^5\gamma^0(-\gamma^\mu\overleftarrow{\partial}/\partial x^\mu - m)\psi_T(x)]^T \tag{3-B.3}$$

$$= [\psi_T{}^\dagger i\gamma^0\gamma^5(-\gamma^\mu\overleftarrow{\partial}/\partial x^\mu - m)\psi_T(x)]^T \tag{3-B.4}$$

$$= \psi_T{}^\dagger i\gamma^0\gamma^5(-\gamma^\mu\overleftarrow{\partial}/\partial x^\mu - m)\psi_T(x) \tag{3-B.5}$$

since eq. 3-B.4 is the transpose of a 1 by 1 matrix. Upon performing a partial integration in the action we find

$$I^* = \int d^4x[\psi_T{}^\dagger i\gamma^0\gamma^5(\gamma^\mu\partial/\partial x^\mu - m)\psi_T(x)] = I \tag{3-B.6}$$

4. Integer Spin Tachyons

4.0 Introduction

In this chapter we will begin by developing the quantum field theory of bosonic tachyons with real energy and 3-momenta. Then in section 4.4 we will consider complexon bosons of two types: one type with positive m^2, and the other type tachyonic ($m^2 < 0$). Both types have real energies and complex 3-momenta. They are the analogues of the complexon fermions with real energies and complex 3-momenta that were derived in the preceding chapter using L_C boosts.

4.1 Massive, Scalar Tachyons

The case of massive scalar tachyons would normally be the starting point for the discussion of tachyons. But the importance of spin ½ tachyons in the Standard Model led us to consider spin ½ tachyons first. We now turn to free, neutral, spin 0 tachyons, which we anticipate satisfies the tachyon equivalent of the Klein-Gordon equation:

$$(\Box - m^2)\phi_T(x) = 0 \tag{4.1}$$

where

$$\Box = \partial/\partial x_\mu \partial/\partial x^\mu \tag{4.2}$$

(The charged scalar tachyon case is analogous.)

Eq. 4.1 can be derived using the canonical procedure from the Lagrangian density and action

$$\mathcal{L}_T = ½[\partial\phi_T/\partial x^\mu \partial\phi_T/\partial x_\mu + m^2\phi_T{}^2] \tag{4.3}$$

$$I = \int d^4x \mathcal{L}_T$$

We can canonically second quantize this theory using light-front coordinates. The Lagrangian density then becomes

$$\mathcal{L}_T = \partial\phi_T/\partial x^+ \partial\phi_T/\partial x^- - ½\partial\phi_T/\partial x^i \partial\phi_T/\partial x^i + ½m^2\phi_T{}^2 \tag{4.4}$$

The conjugate momentum is

$$\pi_T = \partial\mathcal{L}/\partial(\partial^- \phi_T) = \partial^+ \phi_T \equiv \partial\phi_T/\partial x^- \tag{4.5}$$

and the equal x^+ commutation relations[90] are

$$[\pi_T(x), \phi_T(y)] = -i2^{-\frac{1}{2}}\delta(x^- - y^-)\delta^2(\mathbf{x}_\perp - \mathbf{y}_\perp) \tag{4.6}$$

We provisionally define the Fourier expansion of ϕ_T as

$$\phi_T(x) = \int d^2p dp^+ N_T(p)\theta(p^+)[a_T(p)e^{-ip.x} + a_T^\dagger(p)e^{+ip.x}] \tag{4.7}$$

where $N_T(p)$ is

$$N_T(p) = [(2\pi)^3 p^+]^{-\frac{1}{2}} \tag{4.8}$$

and the Fourier component operator *commutation* relations are

$$[a_T(q), a_T^\dagger(p)] = 2^{-\frac{1}{2}}\delta^2(\mathbf{q} - \mathbf{p})\delta(q^+ - p^+) \tag{4.9}$$
$$[a_T(q), a_T(p)] = [a_T(q), a_T(p)] = 0$$

We now calculate

$$[\pi_T(x), \phi_T(y)] = [\partial\phi_T(x)/\partial x^-, \phi_T(y)]$$
$$= \int d^2p dp^+ \int d^2p' dp'^+ N_T(p)N_T(p')\theta(p^+)\theta(p'^+)\cdot$$
$$\cdot\{-ip^+[a_T(p), a_T^\dagger(p')]e^{+ip'.y - ip.x} + ip^+[a_T^\dagger(p), a_T(p')]e^{-ip'.y + ip.x}\}$$
$$= -i2^{-\frac{1}{2}}\int d^2p_\perp \int_0^\infty dp^+ N_T^{+2}(p)p^+\{e^{+ip^+(y^- - x^-) - ip_\perp\cdot(y_\perp - x_\perp)} + e^{-ip^+(y^- - x^-) + ip_\perp\cdot(y_\perp - x_\perp)}\}$$

---

[90] Feinberg (G. Feinberg, Phys. Rev. **159**, 1089 (1967)) and others have suggested that scalar tachyons obey anti-commutation relations because a Lorentz transformation can change a positive energy to a negative energy (and vice versa). However in light-front coordinates a Lorentz or Superluminal boost in the z direction does not change the sign of the equivalent of energy p^-. Boosts in other directions may change the sign of p^-. However the light-front is a particular choice of variables in a specific frame. Since perturbative and other calculations lead to covariant results we can do all calculations on the light-front, and then, after expressing the results in covariant form, transform to any other reference frame. Then tachyon scattering events seen in the new coordinate system should be in agreement with the corresponding covariant calculation of the event. Therefore scalar tachyon quantization using commutators is acceptable and has the advantage of conforming to the general quantization program for scalar particles.

$$= -i2^{-1/2}\int d^2p_\perp \int_{-\infty}^{\infty} dp^+ (2\pi)^{-3}\, e^{+ip^+(y^- - x^-) - i\mathbf{p}_\perp\cdot(\mathbf{y}_\perp - \mathbf{x}_\perp)}$$

$$= -i2^{-1/2}\delta(x^- - y^-)\delta^2(\mathbf{x}_\perp - \mathbf{y}_\perp) \tag{4.10}$$

verifying the equal x^+ commutation relation.

Scalar Tachyon Feynman Propagator

The scalar tachyon Feynman propagator is defined by

$$i\Delta_{TF}(x - y) = \theta(x^+ - y^+)<0|\phi_T(x)\ \phi_T(y)|0> + \theta(y^+ - x^+)<0|\phi_T(y)\phi_T(x)|0>$$

$$= \int d^2p dp^+ \int d^2p' dp'^+\ N_T(p)N_T(p')\theta(p^+)\theta(p'^+)\cdot$$

$$\cdot\{\theta(x^+ - y^+)<0|a_T(p)a_T^\dagger(p')|0>e^{+ip'.y - ip.x} +$$

$$+ \theta(y^+ - x^+)<0|a_T(p')a_T^\dagger(p)|0>e^{-ip'.y + ip.x}\}$$

$$= \int d^2p_\perp \int_0^\infty dp^+ N_T^{+2}(p)\{\theta(x^+ - y^+)e^{+ip^-(y^+ - x^+) + ip^+(y^- - x^-) - i\mathbf{p}_\perp\cdot(\mathbf{y}_\perp - \mathbf{x}_\perp)} +$$

$$+ \theta(y^+ - x^+)e^{-ip^-(y^+ - x^+) - ip^+(y^- - x^-) + i\mathbf{p}_\perp\cdot(\mathbf{y}_\perp - \mathbf{x}_\perp)}\}$$

$$= i\int d^4p(2\pi)^{-4}e^{-ip.(x - y)}/(p^2 + m^2 + i\varepsilon) \tag{4.11}$$

with the expected tachyon pole term.

4.2 Massive Vector Tachyons

The case of massive vector tachyons is of some interest since massive vector bosons, W and Z bosons, have been found in nature. Therefore there is a possibility that, hitherto undiscovered, massive vector tachyons may exist in nature and might eventually be created by particle accelerators. In this section we will second quantize a massive tachyon vector boson in light-front coordinates.

We begin with a standard, neutral, free, massive vector boson Lagrangian with the sign of the mass term changed to make it a tachyon vector boson Lagrangian:

$$\mathcal{L}_{TVB} = -\tfrac{1}{4}\, F_T^{\mu\nu}(x)F_{T\mu\nu}(x) - \tfrac{1}{2}\, m^2 V_T^{\mu} V_{T\mu} \tag{4.12}$$

where

$$F_{T\mu\nu} = (\partial V_{T\mu}/\partial x^\nu - \partial V_{T\nu}/\partial x^\mu) \tag{4.13}$$

The equations of motion are

$$\partial F_T{}^{\mu\nu}/\partial x^\nu - m^2 V_T{}^\mu = 0 \tag{4.14}$$

Eq. 4.14 implies the subsidiary condition

$$\partial V_T{}^\mu/\partial x^\mu = 0 \tag{4.15}$$

which, in turn, implies

$$(\Box - m^2)V_T{}^\mu = 0 \tag{4.16}$$

where

$$\Box = \partial/\partial x_\mu \partial/\partial x^\mu \tag{4.2}$$

as previously.

Eq. 4.16 is immediately recognizable as a tachyon equation for each component. We now transform the Lagrangian to light-front coordinates and proceed to quantize. Using the previous definition of light-front variables we define the fields:

$$\begin{aligned}
F_T{}^{+-} &= \partial^+ V_T{}^- - \partial^- V_T{}^+ \\
F_T{}^{+j} &= \partial^+ V_T{}^j - \partial^j V_T{}^+ \\
F_T{}^{-j} &= \partial^- V_T{}^j - \partial^j V_T{}^- \\
F_T{}^{ij} &= \partial^i V_T{}^j - \partial^j V_T{}^i \\
V_T{}^- &= 2^{-1/2}(V_T{}^0 - V_T{}^3) \\
V_T{}^+ &= 2^{-1/2}(V_T{}^0 + V_T{}^3)
\end{aligned} \tag{4.17}$$

The light-front equivalent of the Lagrangian (eq. 4.12) is:

$$\mathcal{L}_{TVB} = -½(\partial V_{T\mu}/\partial x^\nu \partial V_T{}^\mu/\partial x_\nu - \partial V_T{}^\mu/\partial x_\nu \partial V_{T\nu}/\partial x^\mu) - ½m^2 V_T{}^\mu V_{T\mu}$$

After using the constraint eq. 4.15 and discarding a total divergence, we see

$$\mathcal{L}_{TVB} \equiv -½\partial V_{T\mu}/\partial x^\nu \partial V_T{}^\mu/\partial x_\nu - ½m^2 V_T{}^\mu V_{T\mu} \tag{4.18}$$

which becomes

$$\begin{aligned}
\mathcal{L}_{TVB} \equiv & -\partial^+ V_T{}^- \partial^- V_T{}^+ - \partial^+ V_T{}^+ \partial^- V_T{}^- + \partial^+ V_T{}^i \partial^- V_T{}^i + \partial^i V_T{}^+ \partial^i V_T{}^- + \\
& + ½\partial^i V_T{}^j \partial^i V_T{}^j - ½m^2(2V_T{}^+ V_T{}^- - V_T{}^i V_T{}^i)
\end{aligned} \tag{4.19}$$

using light-front coordinates with implied sums over i and j. The resulting equations of motion are

$$(\Box - m^2)V_T^{-} = 0 \qquad (4.20)$$
$$(\Box - m^2)V_T^{+} = 0$$
$$(\Box - m^2)V_T^{i} = 0$$

for i = 1, 2.

The conjugate spacelike-surface momenta are

$$\pi^{\mu} = \partial\mathcal{L}/\partial(\partial^0 V_T^{\mu}) = -\,\partial V_T^{\mu}/\partial x^0 \qquad (4.21)$$

and the conjugate light-front momenta are

$$\pi^{+} = \partial\mathcal{L}/\partial(\partial^{-} V_T^{+}) = -\,\partial^{+} V_T^{-} \equiv -\,\partial V_T^{-}/\partial x^{-} \qquad (4.22)$$
$$\pi^{-} = \partial\mathcal{L}/\partial(\partial^{-} V_T^{-}) = -\,\partial^{+} V_T^{+} \equiv -\,\partial V_T^{+}/\partial x^{-} \qquad (4.23)$$
$$\pi^{i} = \partial\mathcal{L}/\partial(\partial^{-} V_T^{i}) = \partial^{+} V_T^{i} \equiv \partial V_T^{i}/\partial x^{-} \qquad (4.24)$$

The equal x^+ commutation relations are

$$[\pi_T^{\,a}(x), V_T^{\,b}(y)] = -i2^{-\frac{1}{2}}\delta^{3ab}(x - y) \qquad (4.25)$$

where

$$\delta^{3ab}(x - y) = \int d^2k\,dk^{+}\,e^{i[k^{+}(x^{-} - y^{-}) - \mathbf{k}\cdot(\mathbf{x} - \mathbf{y})]}[g^{ab} + k^a k^b/m^2]/(2\pi)^3 \qquad (4.26)$$

where $\mathbf{k} = (k^1, k^2)$, and $g^{-+} = g^{+-} = 1 = -g^{11} = = -g^{22}$ with all other $g^{ab} = 0$. The equal x^+ commutation relations satisfy the constraint:

$$\partial_a[\pi_T^{\,a}(x), V_T^{\,b}(y)] = \partial_b[\pi_T^{\,a}(x), V_T^{\,b}(y)] = 0 \qquad (4.27)$$

implied by eq. 4.15.

Next we define the Fourier expansion of V_T^{μ} as

$$V_T^{\,\mu}(x) = \sum_s \int d^2k\,dk^{+} N_{TV}(k)\theta(k^{+})\varepsilon^{\mu}(k, s)[a_T(k, s)e^{-ik.x} + a_T^{\dagger}(k, s)e^{+ik.x}] \quad (4.28)$$

where $k^2 = 2k^+k^- - k^{i\,2} = -m^2$, and where $N_{TV}(k)$ is

$$N_{TV}(k) = [(2\pi)^3 k^+]^{-½} \qquad (4.29)$$

There are three spin orientations: two transverse orientations and a longitudinal orientation, s = ±1, 0. The spin polarization vector satisfies

$$k_\mu \varepsilon^\mu(k, s) = 0 \qquad (4.30)$$

It also satisfies the normalization condition

$$\sum_s \varepsilon^\mu(k, s)\varepsilon^\upsilon(k, s) = -(g^{\mu\nu} + k^\mu k^\nu/m^2) \qquad (4.31)$$

The Fourier component operator commutation relations are

$$[a_T(q, s), a_T^\dagger(p, s')] = 2^{-½}\delta_{ss'}\delta^2(\mathbf{q} - \mathbf{p})\delta(q^+ - p^+) \qquad (4.32)$$

$$[a_T(q, s), a_T(p, s')] = [a_T(q, s), a_T(p, s')] = 0$$

Eqs. 4.28 – 4.32 imply the commutation relations eqs. 4.25.

Tachyon Vector Boson Feynman Propagator

The tachyon vector boson Feynman propagator is defined by

$$i\Delta_{TF}(x - y)^{\mu\nu} = \theta(x^+ - y^+)<0|V_T^\mu(x)V_T^\nu(y)|0> +$$
$$+ \theta(y^+ - x^+)<0|V_T^\nu(y)V_T^\mu(x)|0> \qquad (4.33)$$

and is equal to

$$= -i\int \frac{d^4k\, e^{-ik.(x-y)}\, (g^{\mu\nu} + k^\mu k^\nu/m^2)}{(2\pi)^4\,(k^2 + m^2 + i\varepsilon)}$$

The propagator displays the tachyon poles as expected.

4.3 Massive Spin 2 Tachyons – Massive Tachyon Gravitons

Gravitons – the quanta of the gravitation – are massless as far as we know. Massive gravitons have been a subject of a number of theoretical investigations. While there is no evidence for massive gravitons there is evidence that the universe in the large has additional forces at play that affect the rotation of galaxies and seem to be producing an accelerating expansion of the universe. Therefore it is sensible to consider the possibility that massive spin 2 tachyons may exist that could play a role in the understanding of unusual features of the universe in the large. Since the effect of new forces seems to be seen only at distances comparable to the size of galaxies or greater, it is possible that massive spin 2 tachyons may have a small mass of the order of [1/L] where L is the galactic radius of galaxies such as our galaxy.

4.4 L_C Spin 0 Bosons with Complex 3-Momenta

In this section we will develop the quantum field theory of complexon bosons with real energies and complex 3-momenta that are the bosonic analogues of complexon fermions with real energies and complex 3-momenta that we discovered using L_C boosts in chapter 3.

Complexon Scalar Bosons with Complex 3-Momenta

The energy and momentum of these bosons satisfy the mass shell condition

$$p^2 = p^{0\,2} - (\mathbf{p_r} + i\mathbf{p_i})\cdot(\mathbf{p_r} + i\mathbf{p_i}) = p^{0\,2} - \mathbf{p_r}\cdot\mathbf{p_r} + \mathbf{p_i}\cdot\mathbf{p_i} = m^2 \qquad (3.98)$$

where the complex 3-momentum is $\mathbf{p} = \mathbf{p_r} + i\mathbf{p_i}$ and

$$\mathbf{p_r}\cdot\mathbf{p_i} = 0 \qquad (4.34)$$

The wave equation for these bosons is

$$(\partial^2/\partial t^2 - (\nabla_r + i\nabla_i)\cdot(\nabla_r + i\nabla_i) + m^2)\phi_C(t, \mathbf{x_r}, \mathbf{x_i}) = 0 \qquad (4.35)$$

with the subsidiary condition (corresponding to eq. 3.102a)

$$\nabla_r\cdot\nabla_i\phi_C(t, \mathbf{x_r}, \mathbf{x_i}) = 0 \qquad (4.36)$$

The Lagrangian with equation of motion eq. 4.35 is

$$\mathcal{L}_{\phi C} = \tfrac{1}{2}[D^\mu\phi_C D_\mu\phi_C - m^2\phi_C{}^2] \qquad (4.37)$$

where

$$D_0 = \partial/\partial x^0$$
$$D_k = \partial/\partial x_r{}^k + i\,\partial/\partial x_i{}^k \qquad (4.38)$$

and its action[91] is

$$I_{\phi C} = \int d^7x\,\mathcal{L}_{\phi C} \qquad (4.39)$$

[91] The action can be shown to be real by performing partial integrations on the imaginary terms and then using the subsidiary condition. The Lagrangian density resulting from these manipulations is real.

The conjugate momentum is

$$\pi_C = \partial\mathcal{L}/\partial(\partial_0\phi_C) = \partial\phi_C/\partial t \equiv \partial_0\phi_C \tag{4.40}$$

and the equal time commutation relations are

$$[\pi_C(x), \phi_C(y)] = ½\, i\delta'(\nabla_r\cdot\nabla_i/m^2)\delta^3(\mathbf{x}_r - \mathbf{y}_r)\delta^3(\mathbf{x}_i - \mathbf{y}_i) \tag{4.41}$$

where the mass squared in the delta-function could be omitted in the case of a zero mass particle. The Fourier expansion of ϕ_C is

$$\phi_C(x) = \int d^3p_r d^3p_i N_C(p)\delta(\mathbf{p}_r\cdot\mathbf{p}_i/m^2)[a_C(p)e^{-i(p.x^* + p^*.x)/2} + a_C^\dagger(p)e^{+i(p.x^* + p^*.x)/2}] \tag{4.42}$$

based on the analogous fermion case. Eq. 4.42 can be expressed as

$$\phi_C(x) = \int d^3p_r d^3p_i N_C(p)\delta(\mathbf{p}_r\cdot\mathbf{p}_i/m^2)[a_C(p)e^{-i(p_0x_0 - \mathbf{p}_r\cdot\mathbf{x}_r + \mathbf{p}_i\cdot\mathbf{x}_i)} + a_C^\dagger(p)e^{+i(p_0x_0 - \mathbf{p}_r\cdot\mathbf{x}_r + \mathbf{p}_i\cdot\mathbf{x}_i)}] \tag{4.43}$$

where

$$N_C(p) = [1/((2\pi)^6p^0)]^{-½} \tag{4.44}$$

The Fourier component operator *commutation* relations are

$$\begin{aligned}[a_C(q), a_C^\dagger(p)] &= \delta^3(\mathbf{q}_r - \mathbf{p}_r)\delta^3(\mathbf{q}_i - \mathbf{p}_i)\\ [a_C(q), a_C(p)] &= [a_C(q), a_C(p)] = 0\end{aligned} \tag{4.45}$$

We now calculate the equal-time commutator

$$\begin{aligned}[\pi_C(x), \phi_C(y)] &= [\partial\phi_C(x)/\partial x^0, \phi_C(y)]\\ &= \int d^3p_r d^3p_i \int d^3p'_r d^3p'_i N_C(p)N_C(p')\delta(\mathbf{p}_r\cdot\mathbf{p}_i/m^2)\, \delta(\mathbf{p}'_r\cdot\mathbf{p}'_i/m^2)\cdot\\ &\quad\cdot\{-ip^0[a_C(p), a_C^\dagger(p')]e^{+i(\mathbf{p}_r\cdot\mathbf{x}_r - \mathbf{p}_i\cdot\mathbf{x}_i) - i(\mathbf{p}'_r\cdot\mathbf{y}_r - \mathbf{p}'_i\cdot\mathbf{y}_i)} +\\ &\quad + ip^0[a_C^\dagger(p), a_C(p')]e^{-i(\mathbf{p}_r\cdot\mathbf{x}_r - \mathbf{p}_i\cdot\mathbf{x}_i) + i(\mathbf{p}'_r\cdot\mathbf{y}_r - \mathbf{p}'_i\cdot\mathbf{y}_i)}\}\end{aligned}$$

$$= ½ i\int d^3p_r d^3p_i N_C^{\,2}(p)\delta'(\mathbf{p}_r\cdot\mathbf{p}_i/m^2)p^0\{e^{-i(\mathbf{p}_r\cdot(\mathbf{y}_r - \mathbf{x}_r) - \mathbf{p}_i\cdot(\mathbf{y}_i - \mathbf{x}_i))} + e^{+i(\mathbf{p}_r\cdot(\mathbf{y}_r - \mathbf{x}_r) - \mathbf{p}_i\cdot(\mathbf{y}_i - \mathbf{x}_i))}\}$$

$$= \tfrac{1}{2}\, i\delta'(\nabla_r\cdot\nabla_i/m^2)\delta^3(\mathbf{x_r} - \mathbf{y_r})\delta^3(\mathbf{x_i} - \mathbf{y_i})$$

verifying the equal time commutation relation eq. 4.41.

Complexon Scalar boson Feynman Propagator

The complexon scalar Feynman propagator is defined by

$$i\Delta_{CF}(x - y) = \theta(x^0 - y^0)\langle 0|\phi_C(x)\phi_C(y)|0\rangle + \theta(y^0 - x^0)\langle 0|\phi_C(y)\phi_C(x)|0\rangle$$

$$= \int d^3p_r d^3p_i \int d^3p'_r d^3p'_i N_C(p)N_C(p')\delta(\mathbf{p_r}\cdot\mathbf{p_i}/m^2)\,\delta(\mathbf{p'_r}\cdot\mathbf{p'_i}/m^2)\cdot$$
$$\cdot\{\theta(x^0 - y^0)\langle 0|a_C(p)a_C^\dagger(p')|0\rangle e^{-i(p\cdot x^* + p^*\cdot x)/2 + i(p'\cdot y^* + p'^*\cdot y)/2} +$$
$$+ \theta(y^0 - x^0)\langle 0|a_C(p')a_C^\dagger(p)|0\rangle e^{+i(p\cdot x^* + p^*\cdot x)/2 - i(p'\cdot y^* + p'^*\cdot y)/2}\}$$

$$= \int d^3p_r d^3p_i N_C^{\,2}(p)[\delta(\mathbf{p_r}\cdot\mathbf{p_i}/m^2)]^2\{\theta(x^0 - y^0)e^{-i(p\cdot(x^* - y^*) + p^*\cdot(x - y))/2} +$$
$$+ \theta(y^0 - x^0)e^{+i(p\cdot(x^* - y^*) - p^*\cdot(x - y))/2}\}$$

$$= -\tfrac{1}{2}\int d^3p_r d^3p_i N_C^{\,2}(p)\delta'(\mathbf{p_r}\cdot\mathbf{p_i}/m^2)\{\theta(x^0 - y^0)e^{-ip_0(x_0 - y_0) + i(\mathbf{p_r}\cdot(\mathbf{y_r} - \mathbf{x_r}) - \mathbf{p_i}\cdot(\mathbf{y_i} - \mathbf{x_i}))} +$$
$$+ \theta(y^0 - x^0)e^{+ip_0(x_0 - y_0) - i(\mathbf{p_r}\cdot(\mathbf{y_r} - \mathbf{x_r}) - \mathbf{p_i}\cdot(\mathbf{y_i} - \mathbf{x_i}))}\}$$

$$= -\tfrac{1}{2}\int d^3p_r d^3p_i\, (2\pi)^{-6}\delta'(\mathbf{p_r}\cdot\mathbf{p_i}/m^2)\{\theta(x^0 - y^0)e^{-ip_0(x_0 - y_0) + i(\mathbf{p_r}\cdot(\mathbf{y_r} - \mathbf{x_r}) - \mathbf{p_i}\cdot(\mathbf{y_i} - \mathbf{x_i}))} +$$
$$+ \theta(y^0 - x^0)e^{+ip_0(x_0 - y_0) - i(\mathbf{p_r}\cdot(\mathbf{y_r} - \mathbf{x_r}) - \mathbf{p_i}\cdot(\mathbf{y_i} - \mathbf{x_i}))}\}/p^0$$

$$= i\int d^4p_r d^3p_i\, (2\pi)^{-7}\delta'(\mathbf{p_r}\cdot\mathbf{p_i}/m^2)e^{-ip_0(x_0 - y_0) + i(\mathbf{p_r}\cdot(\mathbf{y_r} - \mathbf{x_r}) - \mathbf{p_i}\cdot(\mathbf{y_i} - \mathbf{x_i}))}/(p^2 - m^2 + i\varepsilon) \tag{4.46}$$

Eq. 4.46 can be put in the form of a spectral integral similar to eq. 3.149 but without the γ-matrix denominator.

Tachyonic Complexon Scalar Bosons

The energy and momentum of these bosons satisfy the mass shell condition

$$p^2 = p^{0\,2} - (\mathbf{p_r} + i\mathbf{p_i})\cdot(\mathbf{p_r} + i\mathbf{p_i}) = p^{0\,2} - \mathbf{p_r}\cdot\mathbf{p_r} + \mathbf{p_i}\cdot\mathbf{p_i} = -m^2 \tag{4.47}$$

where the complex 3-momentum is $\mathbf{p} = \mathbf{p_r} + i\mathbf{p_i}$ and satisfies eq. 4.34.

The wave equation for these bosons is

$$(\partial^2/\partial t^2 - (\nabla_r + i\nabla_i)\cdot(\nabla_r + i\nabla_i) - m^2)\phi_{CT}(t, \mathbf{x_r}, \mathbf{x_i}) = 0 \tag{4.48}$$

with the subsidiary condition (corresponding to eq. 3.102a)

$$\nabla_{\mathbf{r}}\cdot\nabla_{\mathbf{i}}\phi_{CT}(t, \mathbf{x_r}, \mathbf{x_i}) = 0 \tag{4.49}$$

The Lagrangian with equation of motion eq. 4.35 is

$$\mathcal{L}_{\phi CT} = ½[D^{\mu}\phi_{CT}D_{\mu}\phi_{CT} + m^2\phi_{CT}^{\ 2}] \tag{4.50}$$

where

$$D_0 = \partial/\partial x^0$$
$$D_k = \partial/\partial x_r^{\ k} + i\ \partial/\partial x_i^{\ k} \tag{4.51}$$

and its action[92] is

$$I_{\phi CT} = \int d^7x\mathcal{L}_{\phi CT} \tag{4.52}$$

As in previous cases we must second quantize the theory using light-front coordinates to avoid the problems of earlier tachyon theories. Therefore we use the light-front variables of eqs. 3.184 – 3.186:

$$x^{\pm} = (x^0 \pm x_r^{\ 3})/\surd 2 \tag{3.184}$$
$$\partial/\partial x^{\pm} \equiv \partial^{\mp} \equiv (\partial/\partial x^0 \pm \partial/\partial x_r^{\ 3})/\surd 2$$

with the "transverse" real coordinate variables, $x_r^{\ 1}$ and $x_r^{\ 2}$, and imaginary coordinate variables $x_i^{\ 1}$, $x_i^{\ 2}$, and $x_i^{\ 3}$.

The inner product of two 4-vectors has the form

$$x\cdot y = x^+y^- + y^+x^- + i[y_i^{\ 3}(x^+ - x^-) + x_i^{\ 3}(y^+ - y^-)]/\surd 2 + x_i^{\ 3}y_i^{\ 3} - (\mathbf{x}_{\mathbf{r}\perp} - i\mathbf{x}_{\mathbf{i}\perp})\cdot(\mathbf{y}_{\mathbf{r}\perp} - i\mathbf{y}_{\mathbf{i}\perp}) \tag{3.185}$$

with

$$\mathbf{x}_{\mathbf{r}\perp} = (x_r^{\ 1}, x_r^{\ 2}) \qquad \mathbf{x}_{\mathbf{i}\perp} = (x_i^{\ 1}, x_i^{\ 2}) \tag{3.186}$$
$$\mathbf{y}_{\mathbf{r}\perp} = (y_r^{\ 1}, y_r^{\ 2}) \qquad \mathbf{y}_{\mathbf{i}\perp} = (y_i^{\ 1}, y_i^{\ 2})$$

where $x = (x^0, \mathbf{x} = \mathbf{x_r} - i\mathbf{x_i})$ and $y = (y^0, \mathbf{y} = \mathbf{y_r} - i\mathbf{y_i})$.

In light-front variables the Lagrangian (eq. 4.50) becomes

$$\mathcal{L}_{\phi CT} = \partial\phi_{CT}/\partial x^+\partial\phi_{CT}/\partial x^- - ½\nabla_{\mathbf{r}\perp}\phi_{CT}\cdot\nabla_{\mathbf{r}\perp}\phi_{CT} + ½\nabla_{\mathbf{i}}\phi_{CT}\cdot\nabla_{\mathbf{i}}\phi_{CT} + ½m^2\phi_{CT}^{\ 2} \tag{4.53}$$

---

[92] The action can be shown to be real by performing partial integrations on the imaginary terms and then using the subsidiary condition. The Lagrangian density resulting from these manipulations is real.

with subsidiary condition eq. 4.49. The conjugate momentum is

$$\pi_{CT} = \partial\mathcal{L}/\partial(\partial^{-}\phi_{CT}) = \partial^{+}\phi_{CT} \equiv \partial\phi_{CT}/\partial x^{-} \tag{4.54}$$

and the equal x^+ commutation relations are

$$[\pi_{CT}(x), \phi_{CT}(y)] = i(\tfrac{1}{2})^{3/2}\delta'(\nabla_r\cdot\nabla_i/m^2)\delta(x^- - y^-)\delta^2(\mathbf{x}_\perp - \mathbf{y}_\perp)\delta^3(\mathbf{x}_i - \mathbf{y}_i) \tag{4.55}$$

The Fourier expansion of ϕ_{CT} is

$$\phi_{CT}(x) = \int d^2p dp^+ \int d^3p_i N_{CT}(p)\delta(\mathbf{p_r}\cdot\mathbf{p_i}/m^2)\theta(p^+)[a_{CT}(p)e^{-i(p.x^* + p^*.x)/2} + \\ + a_{CT}^\dagger(p)e^{+i(p.x^* + p^*.x)/2}] \tag{4.56}$$

where p^2 satisfies eq. 4.47 and where $N_{CT}(p)$ is

$$N_{CT}(p) = [(2\pi)^3p^+]^{-1/2} \tag{4.57}$$

and the Fourier component operator *commutation* relations are

$$[a_{CT}(q), a_{CT}^\dagger(p)] = 2^{-1/2}\delta^2(\mathbf{q} - \mathbf{p})\delta(q^+ \quad p^+) \tag{4.58}$$
$$[a_{CT}(q), a_{CT}(p)] = [a_{CT}(q), a_{CT}(p)] = 0$$

We now calculate

$$[\pi_{CT}(x), \phi_{CT}(y)] = [\partial\phi_{CT}(x)/\partial x^-, \phi_{CT}(y)]$$
$$=\int d^2p dp^+ \int d^2p' dp'^+ \int d^3p_i \int d^3p'_i N_{CT}(p)N_{CT}(p')\theta(p^+)\theta(p'^+)\delta(\mathbf{p_r}\cdot\mathbf{p_i}/m^2)\delta(\mathbf{p'_r}\cdot\mathbf{p'_i}/m^2)\cdot \\ \cdot\{-ip^+[a_{CT}(p), a_{CT}^\dagger(p')]e^{-i(p.x^* + p^*.x)/2 + i(p'.y^* + p'^*.y)/2} + \\ + ip^+[a_{CT}^\dagger(p), a_{CT}(p')]e^{+i(p.x^* + p^*.x)/2 - i(p'.y^* + p'^*.y)/2}\}$$

$$= -i2^{-1/2}\int d^2p_\perp \int_0^\infty dp^+ \int d^3p_i N_{CT}^{+2}(p)[\delta(\mathbf{p_r}\cdot\mathbf{p_i}/m^2)]^2 p^+ \{e^{+ip^+(y^- - x^-) - i\mathbf{p}_\perp\cdot(\mathbf{y}_\perp - \mathbf{x}_\perp) + i\mathbf{p}_i\cdot(\mathbf{y}_i - \mathbf{x}_i)} + \\ + e^{-ip^+(y^- - x^-) + i\mathbf{p}_\perp\cdot(\mathbf{y}_\perp - \mathbf{x}_\perp) - i\mathbf{p}_i\cdot(\mathbf{y}_i - \mathbf{x}_i)}\}$$

$$= -i2^{-1/2}\int d^2p_\perp \int_{-\infty}^{\infty} dp^+ \int d^3p_i (2\pi)^{-6}\, [\delta(\mathbf{p_r}\cdot\mathbf{p_i}/m^2)]^2 e^{+ip^+(y^- - x^-) - i\mathbf{p}_\perp\cdot(\mathbf{y}_\perp - \mathbf{x}_\perp) + i\mathbf{p}_i\cdot(\mathbf{y}_i - \mathbf{x}_i)}$$

$$= (\tfrac{1}{2})^{3/2}\, i\delta'(\nabla_r\cdot\nabla_i/m^2)\delta(x^- - y^-)\delta^2(\mathbf{x}_\perp - \mathbf{y}_\perp)\delta^3(\mathbf{x}_i - \mathbf{y}_i) \qquad (4.59)$$

verifying the equal x^+ commutation relation.

Scalar Tachyonic Complexon Feynman Propagator

The scalar tachyonic complexon Feynman propagator is defined by

$$i\Delta_{CTF}(x - y) = \theta(x^+ - y^+)<0|\phi_{CT}(x)\,\phi_{CT}(y)|0> + \theta(y^+ - x^+)<0|\phi_{CT}(y)\phi_{CT}(x)|0>$$
$$= i\int d^4p_r d^3p_i (2\pi)^{-7}\delta'(\mathbf{p}_r\cdot\mathbf{p}_i/m^2)e^{-ip^+(x^- - y^-) - ip^-(x^+ - y^+) + i\mathbf{p}_\perp\cdot(\mathbf{x}_\perp - \mathbf{y}_\perp) - i\mathbf{p}_i\cdot(\mathbf{x}_i - \mathbf{y}_i)}/(p^2 + m^2 + i\varepsilon) \qquad (4.60)$$

Eq. 4.60 can be placed in the form of a spectral integral.

4.5 Complexon Vector Bosons with Complex 3-Momenta

In this section we will second quantize massive complexon vector bosons (with complex 3-momenta) with positive m^2, and negative m^2 (tachyonic complexons).

Complexon Vector Bosons

We begin with a standard, neutral, free, massive vector boson Lagrangian with one change the spatial derivatives are complex:

$$\mathcal{L}_{CVB} = -\tfrac{1}{4} F_C^{\mu\nu}(x)F_{C\mu\nu}(x) + \tfrac{1}{2} m^2 V_C^{\mu} V_{C\mu} \qquad (4.61)$$

where

$$F_{C\mu\nu} = D_\nu V_{C\mu} - D_\mu V_{C\nu} \qquad (4.62)$$

where D_μ is defined by eq. 4.51. The equations of motion are

$$D_\nu F_C^{\mu\nu} + m^2 V_C^{\mu} = 0 \qquad (4.63)$$

Eq. 4.63 implies the subsidiary condition

$$D_\mu V_C^{\mu} = 0 \qquad (4.64)$$

which, in turn, implies

$$(\Box + m^2)V_C^{\mu} = 0 \qquad (4.64)$$

where

$$\Box = D_\mu D^\mu \qquad (4.65)$$

In addition we have the subsidiary condition on the complex 3-momenta

$$\nabla_r \cdot \nabla_i V_C^\mu = 0 \qquad (4.66)$$

The conjugate momenta are

$$\pi^\mu = \partial\mathcal{L}/\partial(\partial^0 V_C^\mu) = -\partial V_C^\mu/\partial x^0 \qquad (4.67)$$

The equal time commutation relations are

$$[\pi_C^{\mu}(x), V_C^{\nu}(y)] = \tfrac{1}{2} i\delta'(\nabla_r \cdot \nabla_i/m^2)\delta^{\mu\nu}(x - y) \qquad (4.68)$$

where

$$\delta^{\mu\nu}(x - y) = \int d^3k_r d^3k_i e^{+i\mathbf{k}\cdot(\mathbf{x}-\mathbf{y})}[g^{\mu\nu} - k^\mu k^\nu/m^2]/(2\pi)^6 \qquad (4.69)$$

where $k = (k^0, \mathbf{k} = \mathbf{k}_r + i\mathbf{k}_i)$, $x = (x^0, \mathbf{x} = \mathbf{x}_r - i\mathbf{x}_i)$ and $y = (y^0, \mathbf{y} = \mathbf{y}_r - i\mathbf{y}_i)$.

The equal time commutation relations satisfy the constraint:

$$\partial_\mu[\pi_C^{\mu}(x), V_C^{\nu}(y)] = \partial_\nu[\pi_C^{\mu}(x), V_C^{\nu}(y)] = 0 \qquad (4.70)$$

implied by eq. 4.64.

The Fourier expansion of V_C^μ is

$$V_C^{\mu}(x) = \sum_s \int d^3k_r d^3k_i N_{CV}(k)\varepsilon^\mu(k, s)\delta(\mathbf{k}_r \cdot \mathbf{k}_i/m^2)[a_C(k, s)e^{-i(k.x^* + k^*.x)/2} + \\ + a_C^{\dagger}(k, s)e^{+i(k.x^* + k^*.x)/2}] \qquad (4.71)$$

where $N_{CV}(k)$ is

$$N_{CV}(k) = [(2\pi)^3 k^0]^{-1/2} \qquad (4.72)$$

There are three spin orientations: two transverse orientations and a longitudinal orientation, $s = \pm 1, 0$. The spin polarization vector satisfies

$$k_\mu \varepsilon^\mu(k, s) = 0 \qquad (4.73)$$

It also satisfies the normalization condition

$$\sum_s \varepsilon^\mu(k, s)\varepsilon^\upsilon(k, s) = -(g^{\mu\nu} - k^\mu k^\nu/m^2) \qquad (4.74)$$

The Fourier component operator commutation relations are

$$[a_C(q, s), a_C^{\dagger}(p, s')] = \delta_{ss'}\delta^3(\mathbf{q}_r - \mathbf{p}_r)\delta^3(\mathbf{q}_i - \mathbf{p}_i) \qquad (4.75)$$

$$[a_C(q, s), a_C(p, s')] = [a_C(q, s), a_C(p, s')] = 0$$

Eqs. 4.71 – 4.75 imply the commutation relations eqs. 4.68.

Complexon Vector Boson Feynman Propagator

The complexon vector boson Feynman propagator is defined by

$$i\Delta_{CF}(x - y)^{\mu\nu} = \theta(x^0 - y^0)\langle 0|V_C^{\mu}(x)V_C^{\nu}(y)|0\rangle + \\ + \theta(y^0 - x^0)\langle 0|V_C^{\nu}(y)V_C^{\mu}(x)|0\rangle \qquad (4.76)$$

and is equal to

$$= \tfrac{1}{2}\, i \int \frac{d^4k_r d^3k_i\, \delta'(\mathbf{k_r}\cdot\mathbf{k_i}/m^2) e^{-ik.(x - y)}\, (g^{\mu\nu} - k^{\mu}k^{\nu}/m^2)}{(2\pi)^7\, (k^2 - m^2 + i\varepsilon)}$$

Tachyonic Complexon Vector Bosons

This case is similar to the previous case except for the change in the sign of the mass squared, which necessitates light-front quantization. The tachyonic complexon vector boson Lagrangian is:

$$\mathcal{L}_{CTVB} = -\tfrac{1}{4}\, F_{CT}^{\mu\nu}(x)F_{CT\mu\nu}(x) - \tfrac{1}{2}\, m^2 V_{CT}^{\mu}V_{CT\mu} \qquad (4.77)$$

where

$$F_{CT\mu\nu} = D_{\nu}V_{CT\mu} - D_{\mu}V_{CT\nu} \qquad (4.78)$$

where D_{μ} is defined by eq. 4.51. The equations of motion are

$$D_{\nu}F_{CT}^{\mu\nu} - m^2V_{CT}^{\mu} = 0 \qquad (4.79)$$

Eq. 4.79 implies the subsidiary condition

$$D_{\mu}V_{CT}^{\mu} = 0 \qquad (4.80)$$

which, in turn, implies

$$(\Box - m^2)V_{CT}^{\mu} = 0 \tag{4.81}$$

where $\Box = D_\mu D^\mu$ by eq. 4.65. In addition we have the subsidiary condition on the complex 3-momenta:

$$\nabla_r \cdot \nabla_i V_{CT}^{\mu} = 0 \tag{4.82}$$

Eq. 4.81 is immediately recognizable as a tachyon equation for each component. We now transform the Lagrangian to light-front coordinates and proceed to quantize. Based on the previous definition of light-front variables we define the fields:

$$\begin{aligned}
F_{CT}^{+-} &= \partial^+ V_{CT}^- - \partial^- V_{CT}^+ \\
F_{CT}^{+j} &= \partial^+ V_{CT}^j - \partial^j V_{CT}^+ \\
F_{CT}^{-j} &= \partial^- V_{CT}^j - \partial^j V_{CT}^- \\
F_{CT}^{ij} &= \partial^i V_{CT}^j - \partial^j V_{CT}^i \\
V_{CT}^- &= 2^{-1/2}(V_{CT}^0 - V_{CT}^3) \\
V_{CT}^+ &= 2^{-1/2}(V_{CT}^0 + V_{CT}^3)
\end{aligned} \tag{4.83}$$

The light-front equivalent of the Lagrangian (eq. 4.77) is:

$$\mathcal{L}_{CTVB} = -\tfrac{1}{2}(D^\nu V_{CT\mu} D_\nu V_{CT}^{\mu} - D_\nu V_{CT}^{\mu} D_\mu V_{CT}^{\nu} - \tfrac{1}{2}m^2 V_{CT}^{\mu} V_{CT\mu} \tag{4.84}$$

After using the constraint eq. 4.80 and discarding a total divergence, we see

$$\mathcal{L}_{CTVB} \equiv -\tfrac{1}{2}D^\nu V_{CT\mu} D_\nu V_{CT}^{\mu} - \tfrac{1}{2}m^2 V_{CT}^{\mu} V_{CT\mu} \tag{4.85}$$

which becomes

$$\begin{aligned}
\mathcal{L}_{CTVB} \equiv & -\partial^+ V_{CT}^- \partial^- V_{CT}^+ - \partial^+ V_{CT}^+ \partial^- V_{CT}^- + \partial^+ V_{CT}^k \partial^- V_{CT}^k + E^i V_{CT}^+ E^i V_{CT}^- + \\
& + \tfrac{1}{2}E^i V_{CT}^k E^i V_{CT}^k - \tfrac{1}{2}m^2(2V_{CT}^+ V_{CT}^- - V_{CT}^k V_{CT}^k)
\end{aligned} \tag{4.86}$$

using light-front coordinates with implied sums over i = 1, 2, 3 and k = 1, 2; where

$$E^1 = \partial/\partial x_r^1 + i\partial/\partial x_i^1 \qquad E^2 = \partial/\partial x_r^2 + i\partial/\partial x_i^2 \qquad E^3 = i\partial/\partial x_i^3$$

The resulting equations of motion are

$$(\Box - m^2)V_{CT}^{\;-} = 0 \tag{4.87}$$
$$(\Box - m^2)V_{CT}^{\;+} = 0$$
$$(\Box - m^2)V_{CT}^{\;i} = 0$$

for i = 1, 2 in agreement with eq. 4.81.

The conjugate spacelike-surface momenta are

$$\pi^\mu = \partial\mathcal{L}/\partial(\partial^0 V_{CT}{}^\mu) = -\,\partial V_{CT}{}^\mu/\partial x^0 \tag{4.88}$$

and the conjugate light-front momenta are

$$\pi^{+} = \partial\mathcal{L}/\partial(\partial^{-} V_{CT}^{\;+}) = -\,\partial^+ V_{CT}^{\;-} \equiv -\,\partial V_{CT}^{\;-}/\partial x^- \tag{4.89}$$
$$\pi^{-} = \partial\mathcal{L}/\partial(\partial^{-} V_{CT}^{\;-}) = -\,\partial^+ V_{CT}^{\;+} \equiv -\,\partial V_{CT}^{\;+}/\partial x^- \tag{4.90}$$
$$\pi^{i} = \partial\mathcal{L}/\partial(\partial^{-} V_{CT}^{\;i}) = \partial^+ V_{CT}^{\;i} \equiv \partial V_{CT}^{\;i}/\partial x^- \tag{4.91}$$

The equal x^+ commutation relations are

$$[\pi_{CT}^{\;\mu}(x), V_{CT}^{\;\nu}(y)] = +i2^{-3/2}\delta^{6\mu\nu}(x - y) \tag{4.92}$$

where

$$\delta^{6\mu\nu}(x - y) = \int d^2k_r dk_r^{\;+} d^3k_i \delta'(\mathbf{k_r}\cdot\mathbf{k_i}/m^2) e^{i[k_r^{\;+}(x^- - y^-) - \mathbf{k}_{r\perp}\cdot(\mathbf{x}_{r\perp} - \mathbf{y}_{r\perp}) + \mathbf{k}_i\cdot(\mathbf{x}_i - \mathbf{y}_i)]}\cdot [g^{\mu\nu} + k^\mu k^\nu/m^2]/(2\pi)^6 \tag{4.93}$$

where $\mathbf{k}_{r\perp} = (k_r^{\;1}, k_r^{\;2})$, and $g^{-+} = g^{+-} = 1 = -g^{11} = = -g^{22}$ with all other $g^{ab} = 0$ with variables defined as in eqs. 3.184 – 3.186. (In normal coordinates the 3-vector $\mathbf{k} = \mathbf{k_r} + i\mathbf{k_i}$.) The equal x^+ commutation relations satisfy the constraint:

$$\partial_\mu[\pi_{CT}^{\;\mu}(x), V_{CT}^{\;\nu}(y)] = \partial_\nu[\pi_{CT}^{\;\mu}(x), V_{CT}^{\;\nu}(y)] = 0 \tag{4.94}$$

as implied by eq. 4.80.

Next we define the Fourier expansion of $V_{CT}{}^\mu$ as

$$V_{CT}^{\;\mu}(x) = \sum_s \int d^2k dk^+ d^3k_i N_{CTV}(k)\delta(\mathbf{k_r}\cdot\mathbf{k_i}/m^2)\theta(k^+)\varepsilon^\mu(k, s)\cdot$$

$$\cdot[a_{CT}(k, s)e^{-i(k.x^* + k^*.x)/2} + a_{CT}^{\dagger}(k, s)e^{+i(k.x^* + k^*.x)/2}] \quad (4.95)$$

where $x = (x^0, \mathbf{x} = \mathbf{x_r} - \mathbf{ix_r})$, $k = (k^0, \mathbf{k} = \mathbf{k_r} + \mathbf{ik_i})$, and where $k^2 = -m^2$, and where $N_{CTV}(k)$ is

$$N_{CTV}(k) = [(2\pi)^3 k^+]^{-1/2} \quad (4.96)$$

There are three spin orientations: two transverse orientations and a longitudinal orientation, s = ±1, 0. The spin polarization vector satisfies

$$k_\mu \varepsilon^\mu(k, s) = 0 \quad (4.97)$$

It also satisfies the normalization condition

$$\sum_s \varepsilon^\mu(k, s)\varepsilon^\upsilon(k, s) = -(g^{\mu\nu} + k^\mu k^\nu/m^2) \quad (4.98)$$

The Fourier component operator commutation relations are

$$[a_{CT}(q, s), a_{CT}^{\dagger}(p, s')] = 2^{-1/2}\delta_{ss'}\delta^2(\mathbf{q} - \mathbf{p'})\delta(q^+ - p^+) \quad (4.99)$$
$$[a_{CT}(q, s), a_{CT}(p, s')] = [a_{CT}(q, s), a_{CT}(p, s')] = 0$$

Eqs. 4.95 – 4.99 imply the commutation relations eqs. 4.92.

Tachyonic Complexon Vector Boson Feynman Propagator

The tachyonic complexon vector boson Feynman propagator is defined by

$$i\Delta_{CTF}(x - y)^{\mu\nu} = \theta(x^+ - y^+)<0|V_{CT}^{\mu}(x)V_{CT}^{\nu}(y)|0> +$$
$$+ \theta(y^+ - x^+)<0|V_{CT}^{\nu}(y)V_{CT}^{\mu}(x)|0> \quad (4.100)$$

and is equal to

$$= \tfrac{1}{2} i \int \frac{d^4k_r d^3k_i\, \delta'(\mathbf{k_r}\cdot\mathbf{k_i}/m^2)e^{-ik.(x-y)}\,(g^{\mu\nu} + k^\mu k^\nu/m^2)}{(2\pi)^7\,(k^2 + m^2 + i\varepsilon)}$$

with $k = (k^0, \mathbf{k} = \mathbf{k_r} + \mathbf{ik_i})$.

5. Free Tachyon Discrete Symmetries: C, P, T, and CPT

5.0 Introduction

In the first part of this chapter we will consider the behavior of quantum tachyon theories (developed from Extended Lorentz group (and L_C) boosts) under the discrete transformations P, C and T. In sections 5.8 and 5.9 we will consider the behavior of quantum field theories of complexons under the discrete transformations.

5.1 Tachyons and the Discrete Symmetries: C, P, and T

The discrete (improper) transformations, parity, time reversal and charge conjugation, are of major importance in analyzing the structure of ElectroWeak Theory and the Standard Model. In this chapter we will examine these transformations with respect to tachyon particles.

First, bosonic tachyons (spins 0, 1, and 2) have discrete transformation properties similar to ordinary bosons and so will not be considered further. The interested reader should read standard texts on this topic and notice the sign of the squared mass does not introduce any distinctive differences between tachyon and normal bosons.

In the case of fermions (odd half integer spin particles) there is a difference between tachyon fermions and conventional fermions. We will consider the case of spin ½ tachyons. Fundamental tachyon fermions of higher spin (should any exist) would also have distinctively different P, C, and T transformation properties.

We will use the manifestly covariant Lagrangian

$$\mathcal{L} = \psi_T^\dagger i\gamma^0\gamma^5(\gamma^\mu\partial/\partial x^\mu - m)\psi_T(x) \tag{5.1}$$

with equations of motion

$$(\gamma^\mu\partial/\partial x^\mu - m)\psi_T(x) = 0 \tag{5.2}$$

and anti-commutator

$$\{\psi_{T\,a}^{\dagger}(x), \psi_{Tb}(x')\} = -[\gamma^5]_{ab}\,\delta^3(x - x') \tag{5.3}$$

as the starting points of our discussion.

5.2 Parity

In defining the parity transformation for spin ½ tachyons we try to retain as much similarity as possible to the Dirac spin ½ fermion parity transformation. By definition the parity transformation changes $\mathbf{x} \rightarrow -\mathbf{x}$. In the case of a Dirac field if the transformation is defined as:

$$\mathcal{P}\psi(\mathbf{x}, t)\mathcal{P}^{-1} = \gamma^0\psi(-\mathbf{x}, t) \tag{5.4}$$

then the free Dirac field Lagrangian, field equation and anti-commutators are invariant under this transformation.

If we now consider a spin ½ tachyon field and assume the same general form for the transformation:

$$\mathcal{P}\psi_T(\mathbf{x}, t)\mathcal{P}^{-1} = \gamma^0\psi_T(-\mathbf{x}, t) \tag{5.5}$$

then we find

$$\mathcal{P}\mathcal{L}(\mathbf{x}, t)\mathcal{P}^{-1} = -\mathcal{L}(-\mathbf{x}, t) \tag{5.6}$$

$$(\gamma^\mu\partial/\partial x'^\mu - m)\psi_T(-\mathbf{x}, t) = 0 \tag{5.7}$$

$$\{\psi_{T\,a}^{\dagger}(x'), \psi_{Tb}(y')\} = [\gamma^5]_{ab}\,\delta^3(x' - y') \tag{5.8}$$

where x' = $(-\mathbf{x}, t)$ and y' = $(-\mathbf{y}, t)$. The Lagrangian and the anti-commutation relations change sign under the parity transformation. Therefore the physics of tachyons is not invariant under parity. This fact is directly evidenced by eq. 3.45 where the expression of the Lagrangian in terms of left-handed and right-handed fields shows the Lagrangian changes sign under the interchange of left and right handed fields (an effect of the parity transformation). Thus spin ½ tachyon theory, like nature, violates parity.

Note that parity violation is intrinsic to tachyons – even free tachyons. The discussion of the Standard Model in the following chapters will associate tachyon parity violation with Standard Model parity violation.

5.3 Charge Conjugation

The charge conjugation transformation is connected to the interchange of particle and antiparticle. If we assume that a spin ½ tachyon has charge and is coupled to the electromagnetic field then (assuming the usual gauge coupling) the tachyon Lagrangian becomes

$$\mathcal{L} = \psi_T^{\dagger}i\gamma^0\gamma^5[\gamma^\mu(\partial/\partial x^\mu + ieA_\mu) - m]\psi_T(x) \tag{5.9}$$

If the theory were charge conjugation invariant then a unitary operator $\mathcal{C}$ would exist that would change the sign of the electromagnetic current $j^\mu(x)$, and the electromagnetic field A(x, t), while leaving the Lagrangian invariant:

$$\mathcal{C} j^\mu(x) \mathcal{C}^{-1} = -\, j^\mu(x) \qquad ???? \qquad (5.10)$$

$$\mathcal{C} \mathbf{A}(\mathbf{x}, t) \mathcal{C}^{-1} = -\mathbf{A}(\mathbf{x}, t) \qquad ???? \qquad (5.11)$$

$$\mathcal{C} \mathcal{L}(\mathbf{x}, t) \mathcal{C}^{-1} = \mathcal{L}(\mathbf{x}, t) \qquad ???? \qquad (5.12)$$

The tachyon electromagnetic current implied by the Lagrangian eq. 5.9 is

$$j_T{}^\mu(x) = \tfrac{1}{2} e[\psi_T{}^\dagger \gamma^0 \gamma^5, \gamma^\mu \psi_T] \qquad (5.13)$$

where we antisymmetrize as in the case of Dirac fermions.

We extend the standard charge conjugation transformation[93] *with one modification* from that of a conventional Dirac spin ½ field to the case of spin ½ tachyons:

$$\psi_{TC}(\mathbf{x}, t) = \mathcal{C}_T \psi_T(\mathbf{x}, t) \mathcal{C}_T{}^{-1} = C_T \bar{\psi}_T{}^T(\mathbf{x}, t) = C_T \gamma^0 \psi_T{}^*(\mathbf{x}, t) \qquad (5.14)$$

where $\psi_{TC}(\mathbf{x}, t)$ is the antitachyon field of opposite charge, where the superscript T indicating the transpose, and where the tachyon charge conjugation matrix (which differs from the Dirac field analogue) is

$$C_T = i\gamma^2\gamma^5\gamma^0 \qquad \text{and} \qquad C_T{}^{-1} = i\gamma^0\gamma^5\gamma^2 = -C_T = C_T{}^\dagger \qquad (5.15)$$

The Hermitian conjugate of the anti-tachyon field is

$$\psi_{TC}{}^\dagger(\mathbf{x}, t) = \mathcal{C}_T \psi_T{}^\dagger(\mathbf{x}, t) \mathcal{C}_T{}^{-1} = \bar{\psi}_T{}^*(\mathbf{x}, t) C_T{}^{-1} = \psi_T{}^T(\mathbf{x}, t) \gamma^0 C_T{}^{-1} \qquad (5.16)$$

Under this transformation we find the tachyon Lagrangian, field equation, and anti-commutator are invariant under charge conjugation:

$$[\gamma^\mu(\partial/\partial x^\mu + ieA_\mu) - m]\psi_{TC}(x) = 0 \qquad (5.17)$$

using $\mathcal{C}_T A_\mu \mathcal{C}_T{}^{-1} = -A_\mu$, and for the Lagrangian in eq. 5.9

$$\mathcal{C}_T \mathcal{L}(\mathbf{x}, t) \mathcal{C}_T{}^{-1} = \mathcal{L}(\mathbf{x}, t) \qquad (5.18)$$

The charge conjugate anti-commutator is

---

[93] In the Dirac representation. See for example Bjorken (1965) p. 115, or Kaku (1993) p. 117.

$$\{\psi_{TC}{}^{\dagger}{}_{a}(x), \psi_{TCb}(y)\} = -[\gamma^5]_{ab}\, \delta^3(x - y) \tag{5.19}$$

The charge conjugated current

$$j_{TC}{}^{\mu}(x) = \tfrac{1}{2} e[\psi_{TC}{}^{\dagger}\gamma^0\gamma^5, \gamma^{\mu}\psi_{TC}] = -\, j_T{}^{\mu}(x) \tag{5.20}$$

so that

$$\mathcal{C}_T j_T{}^{\mu}(x) A_{\mu} \mathcal{C}_T{}^{-1} = j_T{}^{\mu}(x) A_{\mu} \tag{5.21}$$

resulting in the tachyon charge conjugation invariant Lagrangian eq. 5.9. Thus tachyons and antitachyons can be expected to have the same charge and mass.

5.4 CP Transformation

The CP transformation has been of major theoretical and experimental interest for some time. Experimentally[94] CP violation has been found in certain sectors: K meson and B meson decays.

Since we have seen that tachyons violate parity and do not violate charge conjugation we can see that tachyons inherently violate CP invariance.

5.5 Time Reversal

Time reversal invariance is also a significant theoretical and experimental issue. The standard Dirac fermion time reversal transformation is:

$$\mathfrak{T}\psi(\mathbf{x}, t)\mathfrak{T}^{-1} = T\psi(\mathbf{x}, -t) \tag{5.22}$$

where $\mathfrak{T} = \mathcal{U}K$ where $\mathcal{U}$ is a unitary operator and K is the operator that takes the complex conjugate of all c-numbers, and where

$$T = i\gamma^1\gamma^3 \tag{5.23}$$

Due to the form of $\mathcal{L}$ (eq. 5.9), which assumes an electromagnetic interaction for the purpose of illustration, we find that the *tachyon* time reversal transformation is

$$\mathfrak{T}_T\psi_T(\mathbf{x}, t)\mathfrak{T}_T{}^{-1} = T_T\psi_T(\mathbf{x}, -t) \tag{5.24}$$

where $\mathfrak{T}_T = \mathcal{U}_T K$ where $\mathcal{U}_T$ is a unitary operator and K is the operator that takes the complex conjugate of all c-numbers, and where

$$T_T = i\gamma^5\gamma^1\gamma^3 \tag{5.25}$$

---

[94] B. Aubert et al, BaBar-PUB-07/001, arXiv:hp-ex/0702046 (2007) and references therein.

The matrix T_T satisfies:

$$T_T^{-1} = -i\gamma^3\gamma^1\gamma^5 = T_T \tag{5.26}$$

$$T_T\gamma^\mu T_T^{-1} = -\gamma_\mu \tag{5.27}$$

Under time reversal the current satisfies

$$\mathfrak{T}_T j_{T\mu}(\mathbf{x}, t)\mathfrak{T}_T^{-1} = j_T{}^\mu(\mathbf{x}, -t) \tag{5.28}$$

If we assume the electromagnetic field satisfies

$$\mathfrak{T}\mathbf{A}(\mathbf{x}, t)\mathfrak{T}^{-1} = -\mathbf{A}(\mathbf{x}, -t)$$

under time reversal, then the tachyon Lagrangian

$$\mathfrak{L} = \psi_T{}^\dagger i\gamma^0\gamma^5[\gamma^\mu(\partial/\partial x^\mu + ieA_\mu) - m]\psi_T(x) \tag{5.29}$$

satisfies

$$\mathfrak{T}_T\mathfrak{L}(\mathbf{x}, t)\mathfrak{T}_T^{-1} = \mathfrak{L}(\mathbf{x}, -t) \tag{5.30}$$

under time reversal. Although the action changes by a translation in time, Poincaré translation invariance implies the action is invariant. The equation of motion derived from the Lagrangian eq. 5.9 is also invariant under the tachyon time reversal transformation.

Thus we find the dynamics of the tachyon Lagrangian theory (eq. 5.9) to be invariant under the tachyon time reversal transformation.

5.6 Tachyon CPT Non-Invariance

The question of CPT invariance has long been of theoretical and experimental interest. For conventional particle theories the CPT Theorem implies CPT invariance under very general conditions. We will examine the case of CPT invariance of a model tachyon theory with Lagrangian eq. 5.9. For *Dirac* fermions

$$\mathfrak{C}\mathfrak{P}\mathfrak{T}\psi_a(\mathbf{x}, t)\mathfrak{T}^{-1}\mathfrak{P}^{-1}\mathfrak{C}^{-1} = i[\psi^\dagger(-\mathbf{x}, -t)\gamma^5]_a = i[\gamma^5\psi^*(-\mathbf{x}, -t)]_a \tag{5.31}$$

For spin ½ *tachyons*

$$\mathfrak{C}_T\mathfrak{P}\mathfrak{T}_T\psi_{Ta}(\mathbf{x}, t)\mathfrak{T}_T^{-1}\mathfrak{P}^{-1}\mathfrak{C}_T^{-1} = -i[\psi^\dagger(-\mathbf{x}, -t)\gamma^5]_a \tag{5.32}$$

where a is a spinor index. More succinctly,

$$\mathcal{C}_T\mathcal{P}\mathcal{T}_T\psi_T(\mathbf{x}, t)\mathcal{T}_T^{-1}\mathcal{P}^{-1}\mathcal{C}_T^{-1} = -i\gamma^5\psi^*(-\mathbf{x}, -t) \tag{5.33}$$

Eq. 5.33 differs only by a phase from eq. 5.31.

Therefore one might think that bilinear combinations of ψ and $\psi^\dagger$ which of necessity must be factors in a Lagrangian will result in the cancellation of the –1 factors upon CPT transformation.

However the free field Lagrangian

$$\mathcal{L} = \psi_T^\dagger(x)i\gamma^0\gamma^5(\gamma^\mu\partial/\partial x^\mu - m)\psi_T(x) \tag{5.34}$$

changes to

$$\mathcal{L}_{CPT} = \psi_T^\dagger(-x)i\gamma^0\gamma^5(\gamma^\mu\partial/\partial x^\mu + m)\psi_T(-x) \tag{5.35}$$

The mass term violates CPT invariance. Thus massive spin ½ tachyons inherently violate CPT invariance. If all spin ½ particles start out massless and acquire masses through spontaneous symmetry breaking then the breaking of CPT invariance by massive, spin ½ tachyons is another consequence of spontaneous symmetry breaking.

5.7 Microcausality and Tachyons

Since the CPT Theorem does not hold for spin ½ tachyons it is of interest to consider Jost's Theorem: CPT invariance is equivalent to weak local commutativity, which is a weak form of microcausality. In the case of spin ½ tachyons *weak local commutativity* (or weak microcausality) is defined as

$$<0|\{\psi_T^\dagger(x), \psi_T(y)\}|0> = 0 \quad \text{for} \quad (x-y)^2 < 0$$

(i.e. space-like $(x - y)^2$).

The absence of CPT invariance leads us to inquire if microcausality, and/or weak microcausality, still hold?

To answer this question we evaluate the left-handed field commutator $\{\psi_{TL}^{+\dagger}(x), \psi_{TL}^{+}(y)\}$ to see if the normal microcausality condition holds:

$$\{\psi_{TL}^{+\dagger}(x), \psi_{TL}^{+}(y)\} = 0 \quad \text{for} \quad (x-y)^2 < 0 \tag{5.36}$$

We insert the Fourier expansions:

$$\{\psi_{TL}{}^{+}{}_{a}(x), \psi_{TL}{}^{+\dagger}{}_{b}(y)\} = \sum_{\pm s,s'} \int d^2p dp^+ \int d^2p' dp'^+ N_{TL}^+(p)N_{TL}^+(p')\theta(p^+)\theta(p'^+)\cdot$$
$$\cdot[\{b_{TL}^{+\dagger}(p',s'), b_{TL}^{+}(p,s)\}u_{TL}{}^{+}{}_{a}(p,s)u_{TL}{}^{+\dagger}{}_{b}(p',s')e^{+ip'\cdot y - ip\cdot x} +$$

$$+ \{d_{TL}^{\;+}(p',s'), d_{TL}^{\;+\dagger}(p,s)\} v_{TL\;a}^{\;+}(p,s) v_{TL\;b}^{\;+\dagger}(p',s') e^{-ip'.y + ip.x}]$$

$$= \sum_{\pm s} \int d^2p dp^+ N_{TL}^{\;+2}(p)\theta(p^+)[u_{TL\;a}^{\;+}(p,s) u_{TL\;b}^{\;+\;\dagger}(p,s) e^{+ip.(y-x)} + $$
$$+ v_{TL\;a}^{\;+}(p,s) v_{TL\;b}^{\;+\dagger}(p,s) e^{-ip.(y-x)}]$$

$$= i\int d^2p dp^+\theta(p^+) N_{TL}^{\;+2}(p)(2m|\mathbf{p}|)^{-1}\{[C^-R^+(-i\not{p} + m)\boldsymbol{\gamma}\cdot\mathbf{p}R^+C^-]_{ab} e^{+ip.(y-x)} +$$
$$+ [C^-R^+(-i\not{p} - m)\boldsymbol{\gamma}\cdot\mathbf{p}R^+C^-]_{ab} e^{-ip.(y-x)}\}$$

$$= ½[C^-R^+]_{ab} \int d^2p_\perp \int_0^\infty dp^+(2\pi)^{-3}\,(e^{+ip.(y-x)} + e^{-ip.(y-x)})$$

where $p\cdot(y - x) = p^-(y^+ - x^+) + p^+(y^- - x^-) - \mathbf{p}_\perp\cdot(\mathbf{y}_\perp - \mathbf{x}_\perp)$. Since $p^2 = -m^2$, the integral can be rewritten, after letting $p^\mu = -p^\mu$, as

$$= ½[C^-R^+]_{ab} \int d^2p_\perp \int_{-\infty}^\infty dp^+(2\pi)^{-3}\, e^{-ip.(y-x)}$$

where $p^- = (p_\perp^{\;2} + m^2)/(2p^+)$. For spacelike $(x - y)^2 < 0$ we can always choose a coordinate system where $y^+ - x^+ = 0$ with the result

$$\{\psi_{TL}^{\;+\dagger}(x), \psi_{TL}^{\;+}(y)\} = 2^{-1}C^-R^+\delta(y^- - x^-)\delta^2(\mathbf{y} - \mathbf{x}) \quad \underline{\text{if}} \quad y^+ - x^+ = 0 \qquad (5.37)$$

Therefore

$$\{\psi_{TL}^{\;+\dagger}(x), \psi_{TL}^{\;+}(y)\} = 0 \quad \text{for} \quad (x - y)^2 < 0 \qquad (5.38)$$

Consequently, free left-handed (or right-handed) tachyons with light-front quantization satisfy the microcausality condition.

5.8 C, P and T for Spin ½ Complexons ($m^2 > 0$)

In this section we will examine the behavior under C, P, and T of the spin ½ complexon quantum fields[95] that we found in chapter 3. The Lagrangian that we defined for this type of fermion was

$$\mathcal{L} = \bar{\psi}_C(i\gamma^\mu D_\mu - m)\psi_C(x) \qquad (3.127)$$

[95] We will not consider the case of integer spin quantum fields with complex 3-momenta since they have straightforward transformation properties under C, P and T.

where $\bar{\psi}_C = \psi_C^{\dagger}\gamma^0$ and

$$\psi_C^{\dagger} = [\psi_C(\mathbf{x}_r, \mathbf{x}_i)]^{\dagger} \Big|_{\mathbf{x}_i = -\mathbf{x}_i} \tag{3.128}$$

$$D_0 = \partial/\partial x^0$$
$$D_k = \partial/\partial x^k + i\,\partial/\partial x_i^k \tag{3.129}$$

with $x^k = x_r^k$ for k = 1,2, 3. The invariant action (under real Lorentz transformations) is

$$I = \int d^7x \mathscr{L} \tag{3.130}$$

Parity

Under a parity transformation

$$\mathscr{P}\psi_C(\mathbf{x}, t)\mathscr{P}^{-1} = \gamma^0\psi_C(-\mathbf{x}, t) \tag{5.39}$$

we find

$$\mathscr{P}\mathscr{L}(\mathbf{x}, t)\mathscr{P}^{-1} = \mathscr{L}(-\mathbf{x}, t) \tag{5.40}$$

$$(i\gamma^{\mu}D'_{\mu} - m)\psi_C(-\mathbf{x}, t) = 0 \tag{5.41}$$

where $x = x_r - ix_i$. Parity is conserved.

Charge Conjugation

To examine the fermion Lagrangian under charge conjugation we introduce an electromagnetic field of the kind

$$\mathscr{L} = \psi_C^{\dagger}\gamma^0[i\gamma^{\mu}(D_{\mu} - ieA_{\mu}) - m]\psi_C(x) \tag{5.42}$$

$$j_C^{\mu}(x) = ½\,[\psi_C^{\dagger}\gamma^0, \gamma^{\mu}\psi_C] \tag{5.42}$$

$$\psi_{C\mathscr{C}}(\mathbf{x}, t) = \mathscr{C}\psi_C(\mathbf{x}, t)\mathscr{C}^{-1} = C\bar{\psi}_C^{\,T}(\mathbf{x}, t) = C\gamma^0\psi_C^{\,*}(\mathbf{x}, t) \tag{5.43}$$

where $\psi_{C\mathscr{C}}(\mathbf{x}, t)$ is the antiparticle field of opposite charge, where the <u>super</u>script T indicating the transpose, and where the charge conjugation matrix (which is the same as the Dirac field analogue) is

$$C = i\gamma^2\gamma^0 = -C^T \qquad \text{and} \qquad C^{-1} = i\gamma^0\gamma^2 = -C = C^{\dagger} \tag{5.44}$$

The Hermitian conjugate of the antiparticle field is

$$\psi_{C\mathcal{C}}{}^{\dagger}(\mathbf{x}, t) = \mathcal{C}\psi_{C}{}^{\dagger}(\mathbf{x}, t)\big|_{\mathbf{x_i}=-\mathbf{x_i}}\mathcal{C}^{-1} = \bar{\psi}_{C}{}^{*}(\mathbf{x}, t)\big|_{\mathbf{x_i}=-\mathbf{x_i}}C^{-1} = \psi_{C}{}^{T}(\mathbf{x}, t)\big|_{\mathbf{x_i}=-\mathbf{x_i}}\gamma^{0}C^{-1} \tag{5.45}$$

Applying the charge conjugation transformation we find

$$\mathcal{C}j^{\mu}(x)\mathcal{C}^{-1} = -\,j^{\mu}(x) \tag{5.46}$$

and if $\mathcal{C}\mathbf{A}(\mathbf{x}, t)\mathcal{C}^{-1} = -\mathbf{A}(\mathbf{x}, t)$ then the action

$$I = \int d^7x\mathcal{L}(\mathbf{x}, t) = \int d^7x\ \mathcal{C}\mathcal{L}(\mathbf{x}, t)\mathcal{C}^{-1} \tag{5.47}$$

is invariant under charge conjugation. Note we change the sign of the imaginary coordinates $\mathbf{x_i} \rightarrow -\mathbf{x_i}$ in the action integration to obtain eq. 5.47.

Time Reversal

We use a modification of the standard Dirac fermion time reversal transformation:

$$\mathfrak{T}\psi_C(\mathbf{x}, t)\mathfrak{T}^{-1} = T\psi_C(\mathbf{x}^*, -t) \tag{5.48}$$

where $\mathfrak{T} = \mathcal{U}K$ where $\mathcal{U}$ is a unitary operator and K is the operator that takes the complex conjugate of all c-numbers, where $\mathbf{x} = \mathbf{x_r} - i\mathbf{x_i}$, and where

$$T = i\gamma^1\gamma^3 \tag{5.49}$$

Assuming the Lagrangian $\mathcal{L}$ (eq. 5.42) we find the time reversal transformation gives

$$I = \int d^7x\mathfrak{T}_T\mathcal{L}(\mathbf{x}, t)\mathfrak{T}_T{}^{-1} = \int d^7x\mathcal{L}(\mathbf{x}, -t) \tag{5.50}$$

under time reversal. Although the action changes by a translation in time, Poincaré time translation invariance implies the action is invariant. Note we change the sign of the imaginary coordinates $\mathbf{x_i} \rightarrow -\mathbf{x_i}$ in the action integration to obtain eq. 5.50.

CP Invariance

Since we have seen that particles with complex 3-momenta do not violate parity and do not violate charge conjugation we can see that CP is not violated.

CPT Invariance

The combined impact of the above subsections shows CPT is not violated by particles with complex 3-momenta.

5.9 C, P and T for Complexon Tachyons

In this section we will examine the behavior under C, P, and T of the spin ½ tachyonic complexon quantum fields. The Lagrangian that we defined for this type of fermion was

$$\mathcal{L} = \bar{\psi}_{CT}(\gamma^\mu D_\mu - m)\psi_{CT}(x) \tag{5.51}$$

where $\bar{\psi}_{CT} = \psi_{CT}{}^\dagger i\gamma^0\gamma^5$ and

$$\psi_{CT}{}^\dagger = [\psi_{CT}(\mathbf{x_r}, \mathbf{x_i})]^\dagger \big|_{\mathbf{x_i} = -\mathbf{x_i}} \tag{3.128}$$

$$\begin{aligned} D_0 &= \partial/\partial x^0 \\ D_k &= \partial/\partial x^k + i\,\partial/\partial x_i{}^k \end{aligned} \tag{3.129}$$

with $x^k = x_r{}^k$ for k = 1,2, 3. The invariant action (under real Lorentz transformations) is

$$I = \int d^7x\mathcal{L} \tag{3.130}$$

Parity

Under a parity transformation

$$\mathcal{P}\psi_{CT}(\mathbf{x}, t)\mathcal{P}^{-1} = \gamma^0\psi_{CT}(-\mathbf{x}, t) \tag{5.52}$$

we find

$$\mathcal{P}\mathcal{L}(\mathbf{x}, t)\mathcal{P}^{-1} = -\mathcal{L}(-\mathbf{x}, t) \tag{5.53}$$

Parity is *not* conserved.

Charge Conjugation

To examine the fermion Lagrangian under charge conjugation we introduce an electromagnetic field of the kind

$$\mathcal{L} = \psi_{CT}{}^\dagger\, i\gamma^0\gamma^5[\gamma^\mu(D_\mu - ieA_\mu) - m]\psi_{CT}(x) \tag{5.54}$$

$$j_{CT}{}^\mu(x) = \tfrac{1}{2}[\psi_{CT}{}^\dagger\gamma^0\gamma^5, \gamma^\mu\psi_{CT}] \tag{5.55}$$

$$\psi_{CT\mathcal{C}}(\mathbf{x}, t) = \mathcal{C}_T\psi_{CT}(\mathbf{x}, t)\mathcal{C}_T^{-1} = C_T\bar{\psi}_{CT}{}^T(\mathbf{x}, t) = C_T\gamma^0\psi_{CT}{}^*(\mathbf{x}, t) \tag{5.56}$$

where $\psi_{CT\mathcal{C}}(\mathbf{x}, t)$ is the antiparticle field of opposite charge, where the superscript T indicating the transpose, and where the charge conjugation matrix (which is the same as the Dirac field analogue) is

$$C_T = i\gamma^2\gamma^5\gamma^0 \qquad \text{and} \qquad C_T^{-1} = i\gamma^0\gamma^5\gamma^2 = -C_T = C_T^\dagger \tag{5.15}$$

The Hermitian conjugate of the antiparticle field is

$$\begin{aligned}\psi_{CT\mathcal{C}}{}^\dagger(\mathbf{x}, t) &= \mathcal{C}_T\psi_{CT}{}^\dagger(\mathbf{x}, t)\big|_{\mathbf{x_i}=-\mathbf{x_i}}\mathcal{C}_T^{-1} \\ &= \bar{\psi}_{CT}{}^*(\mathbf{x}, t)\big|_{\mathbf{x_i}=-\mathbf{x_i}}C_T^{-1} = \psi_{CT}{}^T(\mathbf{x}, t)\big|_{\mathbf{x_i}=-\mathbf{x_i}}\gamma^0 C_T^{-1}\end{aligned} \tag{5.57}$$

Applying the charge conjugation transformation we find

$$\mathcal{C}_T j^\mu(x)\mathcal{C}_T^{-1} = -j^\mu(x) \tag{5.58}$$

and if $\mathcal{C}\mathbf{A}(\mathbf{x}, t)\mathcal{C}^{-1} = -\mathbf{A}(\mathbf{x}, t)$ then the action

$$I = \int d^7x\mathcal{L}(\mathbf{x}, t) = \int d^7x\mathcal{C}_T\mathcal{L}(\mathbf{x}, t)\mathcal{C}_T^{-1} \tag{5.59}$$

is *invariant* under charge conjugation. Note we change the sign of the imaginary coordinates $\mathbf{x_i} \rightarrow -\mathbf{x_i}$ in the action integration to obtain eq. 5.59.

Time Reversal

We use a modification of the standard Dirac fermion time reversal transformation:

$$\mathcal{T}_T\psi_{CT}(\mathbf{x}, t)\mathcal{T}_T^{-1} = T_T\psi_{CT}(\mathbf{x}^*, -t) \tag{5.60}$$

where $\mathcal{T}_T = \mathcal{U}_T K$ where $\mathcal{U}_T$ is a unitary operator and K is the operator that takes the complex conjugate of all c-numbers, where $\mathbf{x} = \mathbf{x_r} - i\mathbf{x_i}$, and where

$$T_T = i\gamma^5\gamma^1\gamma^3 \tag{5.25}$$

The matrix T_T satisfies:

$$T_T^{-1} = -i\gamma^3\gamma^1\gamma^5 = T_T \tag{5.26}$$

$$T_T\gamma^\mu T_T^{-1} = -\gamma_\mu \tag{5.27}$$

Assuming the Lagrangian $\mathcal{L}$ (eq. 5.51) we find the time reversal transformation gives

$$I = \int d^7x \mathfrak{T}_T \mathcal{L}(\mathbf{x}, t) \mathfrak{T}_T^{-1} = \int d^7x \mathcal{L}(\mathbf{x}, -t) \qquad (5.61)$$

under time reversal. Although the action changes by a translation in time, Poincaré time translation invariance implies the action is *invariant*. Note we change the sign of the imaginary coordinates $\mathbf{x}_i \rightarrow -\mathbf{x}_i$ in the action integration to obtain eq. 5.50.

CP Invariance

Since we have seen that tachyons with complex 3-momenta violate parity and do not violate charge conjugation we can see that tachyons inherently violate CP invariance.

CPT Invariance

The combined impact of the above subsections shows CPT is violated by tachyons with complex 3-momenta.

5.10 Summary of C, P, T, CP and CPT for Spin ½ Particles

In this section we summarize the results of the preceding sections. The letter I indicates invariant under the transformation; The letter N indicates not invariant under the transformation.

	P	C	T	CP	CPT
$M^2 > 0$	I	I	I	I	I
Complexon ($M^2 > 0$)	I	I	I	I	I
Tachyons	N	I	I	N	N
Tachyonic Complexons	N	I	I	N	N

Table 5.1. Summary of the C, P and T properties of the four types of spin ½ fermions.

Appendix 5-A. Tachyons and the Problem of the Uniformity of the Universe

A major cosmological problem has been the large-scale uniformity of the universe. The standard argument begins with the observation that if we trace back the various regions of the universe to very early times, parts of the universe that are now quite similar could not be in interaction with each other in the distant past because interactions are limited by the speed of light. Thus it is difficult to understand how these regions could be so similar today without any interaction between them that would have established uniformity. One proposed resolution of this difficulty is the inflationary scenario of Guth and others.

However, another solution to the uniformity problem presents itself. If we consider the universe at extremely early times when quarks were not confined, then tachyonic quarks and neutrinos could be the mechanism to resolve the uniformity problem since tachyons can exceed the speed of light. A detailed model of this possibility remains to be created.

6. Dirac Equation Generalizations to Left-handed Extended Lorentz and Lc Group Covariance

6.0 Introduction

In this chapter we begin by generalizing the Dirac equation to an equation that has Extended Lorentz Group covariance. This equation describes a doublet consisting of a normal Dirac particle wave function and a tachyon wave function. We identify these doublets with ElectroWeak lepton doublets.

Then starting in section 6.5 we consider a generalized equation (covariant under the Lc group) for doublets of particles with complex 3-momenta consisting of a particle with positive mass squared and a tachyon. Because of an inherent SU(3) symmetry we identify these doublets with quark ElectroWeak doublets. The SU(3) symmetry leads us to identify the each quark in a doublet as a color SU(3) triplet, and leads thence to SU(3) color quark confinement (described in a subsequent chapter).

6.1 Transformations of Dirac and Tachyon Equations

A Left-handed Extended Lorentz boost of the Dirac equation transforms the Dirac equation into the spin ½ tachyon equation, and vice versa:

$$S_L(\Lambda_L(\omega, \mathbf{u}))\psi(x) \rightarrow \psi_T'(x') \qquad (6.1a)$$
$$S_L(\Lambda_L(\omega, \mathbf{u}))\psi_T(x) \rightarrow \psi'(x')$$

and

$$S_L(\Lambda_L(\omega, \mathbf{u}))(\gamma^\mu\partial/\partial x^\mu - m)S_L^{-1}(\Lambda_L(\omega, \mathbf{u})) = (i\gamma^\mu\partial/\partial x'^\mu - m) \qquad (6.1b)$$
$$S_L(\Lambda_L(\omega, \mathbf{u}))\gamma^5(i\gamma^\mu\partial/\partial x^\mu - m)\gamma^5 S_L^{-1}(\Lambda_L(\omega, \mathbf{u})) = (\gamma^\mu\partial/\partial x'^\mu - m)$$

where

$$x'^\mu = i\Lambda_L{}^\mu{}_\nu(\omega, \mathbf{u})x^\nu \qquad (6.1c)$$
$$\partial/\partial x'^\mu = -i\Lambda_L{}^\nu{}_\mu(\omega, \mathbf{u})\partial/\partial x^\nu$$

by eq. 2.10b which in matrix form is

$$X' = E(\mathbf{v})X = i\Lambda_L(\mathbf{v})X \qquad (2.10b)$$

Eqs. 6.1a – 6.1c imply

$$S_L(\Lambda_L(\omega, \mathbf{u}))(\gamma^\mu\partial/\partial x^\mu - m)\psi_T(x) = (i\gamma^\mu\partial/\partial x'^\mu - m)S_L(\Lambda_L(\omega, \mathbf{u}))\psi_T(x)$$
$$= (i\gamma^\mu\partial/\partial x'^\mu - m)\psi'(x') \qquad (6.1d)$$

and

$$S_L(\Lambda_L(\omega, \mathbf{u}))\gamma^5(i\gamma^\mu\partial/\partial x^\mu - m)\psi(x) = (\gamma^\mu\partial/\partial x'^\mu - m)S_L(\Lambda_L(\omega, \mathbf{u}))\gamma^5\psi(x)$$
$$= (\gamma^\mu\partial/\partial x'^\mu - m)\psi_T'(x') \qquad (6.1e)$$

where

$$\psi'(x') = S_L(\Lambda_L(\omega, \mathbf{u}))\psi_T(x) \qquad (6.1f)$$

and

$$\psi_T'(x') = S_L(\Lambda_L(\omega, \mathbf{u}))\gamma^5\psi(x) \qquad (6.1g)$$

Thus the Dirac equation is also not Left-handed Extended Lorentz covariant.

6.2 Doublet Extended Dirac Equations

We will now consider the issue of generalizing the Dirac equation so that the extended equation is covariant under both Lorentz transformations and Left-handed Extended Lorentz transformations – the set of extended Lorentz transformations.

The only obvious method to obtain an extended Dirac equation that is covariant under extended Lorentz transformations is to define an 8 × 8 matrix generalization. Let

$$đ(x) = \begin{bmatrix} (\gamma^\mu\partial/\partial x^\mu - m) & 0 \\ 0 & (i\gamma^\mu\partial/\partial x^\mu - m) \end{bmatrix} \qquad (6.2)$$

be an 8 × 8 matrix operator with the 4 × 4 matrix elements shown, and let

$$\Psi(x) = \begin{bmatrix} \psi_T(x) \\ \psi(x) \end{bmatrix} \qquad (6.3)$$

be an 8 component column vector composed of a Dirac field and a tachyon field. Then the extended Dirac equation is

$$đ(x)\Psi(x) = 0 \qquad (6.4)$$

We now define the 8 × 8 Left-handed Extended Lorentz transformation

$$S_{L8}(\Lambda_L(v)) = \begin{bmatrix} 0 & S_L(\Lambda_L(v))\gamma^5 \\ S_L(\Lambda_L(v)) & 0 \end{bmatrix} \qquad (6.5)$$

with inverse transformation

$$S_{L8}^{-1}(\Lambda_L(v)) = \begin{bmatrix} 0 & S_L^{-1}(\Lambda_L(v)) \\ \gamma^5 S_L^{-1}(\Lambda_L(v)) & 0 \end{bmatrix} \tag{6.6}$$

Note: we use the notations $S_L(\Lambda_L(v))$ and $S_L(\Lambda_L(\omega, \mathbf{u}))$ interchangeably. Applying S_{L8} to eq. 6.4 yields

$$0 = S_{L8}(\Lambda_L(v))đ(x)\Psi'(x) = đ(x')\Psi''(x') \tag{6.7}$$

where

$$\Psi'(x') = \begin{bmatrix} S_L\gamma^5\psi(x) \\ S_L\psi_T(x) \end{bmatrix} = \begin{bmatrix} \psi_T'(x') \\ \psi'(x') \end{bmatrix} \tag{6.8}$$

Thus the extended Dirac equation is covariant under generalized Left-handed Extended Lorentz transformations such as eqs. 6.5 - 6.6. Covariance requires the tachyon and the Dirac particle must have the same absolute value for the mass.

It is easy to show that the extended Dirac equation eq. 6.4 is also covariant under conventional Lorentz transformations in the 8×8 representation:

$$S_8(\Lambda(v)) = \begin{bmatrix} S(\Lambda(v)) & 0 \\ 0 & S(\Lambda(v)) \end{bmatrix} \tag{6.9}$$

with inverse

$$S_8^{-1}(\Lambda(v)) = \begin{bmatrix} S^{-1}(\Lambda(v)) & 0 \\ 0 & S^{-1}(\Lambda(v)) \end{bmatrix} \tag{6.10}$$

and

$$S_{8A}(\Lambda(v)) = \begin{bmatrix} 0 & S(\Lambda(v)) \\ S(\Lambda(v)) & 0 \end{bmatrix} \tag{6.11}$$

with inverse transformation

$$S_{8A}{}^{-1}(\Lambda(v)) = \begin{bmatrix} 0 & S^{-1}(\Lambda(v)) \\ S^{-1}(\Lambda(v)) & 0 \end{bmatrix} \tag{6.12}$$

Under a conventional Lorentz transformation we find

$$0 = S_8(\Lambda(v))\text{đ}(x)\Psi(x) = \text{đ}(x')\Psi'(x') \tag{6.13}$$

$$0 = S_{8A}(\Lambda(v))\text{đ}(x)\Psi(x) = \text{đ}(x')\Psi'(x')$$

The Lagrangian density that corresponds to our 8-dimensional construction is

$$\mathcal{L}_8 = \overline{\Psi}(x)\text{đ}(x)\Psi(x) \tag{6.14}$$

where

$$\overline{\Psi}(x) = \Psi^\dagger\Gamma^0 \tag{6.15}$$

and

$$\Gamma^0 = \begin{bmatrix} i\gamma^0\gamma^5 & 0 \\ 0 & \gamma^0 \end{bmatrix} \tag{6.16}$$

The action is

$$I = \int d^4x\mathcal{L}_8 \tag{6.17}$$

is invariant under Lorentz transformations S_8 and S_{8A}.

The Hamiltonian density for the 8-dimensional theory is

$$\mathcal{H}_8(x) = \begin{bmatrix} i\psi_T{}^\dagger\gamma^5(\boldsymbol{\alpha}\cdot\nabla + \beta m)\psi_T & 0 \\ 0 & \psi^\dagger(-i\boldsymbol{\alpha}\cdot\nabla + \beta m)\psi \end{bmatrix} \tag{6.18}$$

6.3 Non-Invariance of the Extended Free Action under a Left-handed Extended Lorentz Transformation

The action 6.17 is not invariant under a Left-handed Extended Lorentz transformation. The fundamental cause of this non-invariance is the three dimensional nature of space. In the case of Dirac particles one can define a Lorentz invariant action because time is one- dimensional. Thus one can use $\psi^\dagger\gamma^0 = \bar{\psi}$ to form the Dirac field Lagrangian and action. A key factor in the Lorentz invariance is the relation between the inverse and Hermitian conjugate of the spinor boost operator

$$\gamma^0 S^{-1}\gamma^0 = S^\dagger \qquad (6.19)$$

In the case of the tachyon Lagrangian and action, a Left-handed Extended Lorentz invariance is not possible because the tachyonic equivalent to eq. 6.19 is[96]

$$S_L^{-1}(\Lambda(\mathbf{v}))\boldsymbol{\gamma}\cdot\mathbf{p}/|\mathbf{p}| = i\gamma^0 S_L^\dagger(\Lambda(\mathbf{v})) \qquad (6.20)$$

where $\mathbf{p} = m\gamma_s\mathbf{v}$. The appearance of $\boldsymbol{\gamma}\cdot\mathbf{p}/|\mathbf{p}|$ in eq. 6.20 precludes the invariance of the free tachyon action.

We will now show the effect of a Left-handed Extended Lorentz transformation (eqs. 6.5 and 6.6) on the Lagrangian density eq. 6.14. The two non-zero parts of the Lagrangian density $\mathcal{L}_8$ (eq. 6.14) are

$$\mathcal{L}_1 = \psi_T^\dagger i\gamma^0\gamma^5(\gamma^\mu\partial/\partial x^\mu - m)\psi_T(x) \qquad (6.21)$$

and

$$\mathcal{L}_2 = \psi^\dagger\gamma^0(i\gamma^\mu\partial/\partial x^\mu - m)\psi(x) \qquad (6.22)$$

The effect of the transformation eqs. 6.5 and 6.6 on these terms is

$$\begin{aligned}\mathcal{L}_1' &= \psi_T^\dagger i\gamma^0\gamma^5 S_L^{-1}S_L(\gamma^\mu\partial/\partial x^\mu - m)\, S_L^{-1}S_L\psi_T(x)\\ &= \psi_T^\dagger i\gamma^0\gamma^5 S_L^{-1}(i\gamma^\mu\partial/\partial x'^\mu - m)S_L\psi_T(x)\\ &= -\psi_T^\dagger S_L^\dagger\gamma^5(\boldsymbol{\gamma}\cdot\mathbf{p}/|\mathbf{p}|)(i\gamma^\mu\partial/\partial x'^\mu - m)S_L\psi_T(x)\\ &= \psi'^\dagger(x')(\boldsymbol{\gamma}\cdot\mathbf{p}/|\mathbf{p}|)\gamma^5(i\gamma^\mu\partial/\partial x'^\mu - m)\psi'(x') \end{aligned} \qquad (6.23)$$

using eqs. 6.20, 6.1f and 6.1g; and

$$\begin{aligned}\mathcal{L}_2' &= \psi^\dagger\gamma^0\gamma^5 S_L^{-1}S_L\gamma^5(i\gamma^\mu\partial/\partial x^\mu - m)\gamma^5 S_L^{-1}S_L\gamma^5\psi(x)\\ &= \psi^\dagger\gamma^0\gamma^5 S_L^{-1}(\gamma^\mu\partial/\partial x'^\mu - m)S_L\gamma^5\psi(x)\\ &= i\psi^\dagger\gamma^5 S_L^\dagger(\boldsymbol{\gamma}\cdot\mathbf{p}/|\mathbf{p}|)(\gamma^\mu\partial/\partial x'^\mu - m)S_L\gamma^5\psi(x)\end{aligned}$$

---

[96] This relation is derivable from eqs. 3.11 and 3.12.

$$= i\psi_T'^{\dagger}(x')(\boldsymbol{\gamma}\cdot\mathbf{p}/|\mathbf{p}|)(\gamma^\mu\partial/\partial x'^\mu - m)\psi_T'(x') \qquad (6.24)$$

where $\psi'(x')$ is a solution of the Dirac equation obtained by a Left-handed Extended Lorentz boost (by $\mathbf{v} = \mathbf{p}/(\gamma m)$) of a tachyon field, and where $\psi_T'(x')$ is a solution of the tachyon equation obtained by a Left-handed Extended Lorentz boost (by $\mathbf{v} = \mathbf{p}/(\gamma m)$) of a Dirac field. Eqs. 6.23 and 6.24 clearly show that $\mathcal{L}_8$ is *not* invariant under Left-handed Extended Lorentz transformations.

Consequently the action of eq. 6.17 is only invariant under inhomogeneous Lorentz transformations. *This state of affairs is actually an advantage when we derive features of the Standard Model because it will be seen to prevent any interplay between unbroken ElectroWeak SU(2) rotations and Left-handed Extended Lorentz transformations.*

6.4 The Diracian Dilemma – To what do Left-handed Extended Lorentz Boost Particles Correspond? Leptons

The development of this 8-dimensional formalism, and in particular, the "bi-spinor" wave function consisting of a Dirac spinor and a tachyon spinor, raises the question, "Is there a particle interpretation for the "bi-spinor" wave function?" Dirac faced a similar issue in 1928-1930 with the negative energy states of the Dirac equation. He developed "hole theory" which eventually led to the interpretation of holes in the sea of filled negative energy states as *positrons*. We now face the same problem: With what pairs of particles do we identify the doublets consisting of a Dirac particle and a tachyon?

The obvious natural interpretation of these 8-spinors is ElectroWeak isodoublets such as:

$$\Psi_\ell(x) = \begin{bmatrix} \psi_{\ell T} \\ \\ \psi_\ell \end{bmatrix} \sim \begin{bmatrix} \nu \\ \\ e \end{bmatrix} \qquad (6.25)$$

for leptons where "e" represents a charged lepton and ν represents a neutrino. With this interpretation we can introduce SU(2)⊗U(1) gauge interactions and develop one-generation, leptonic ElectroWeak theory naturally.

6.5 To what do Complexons Correspond? Quarks

We have identified two of the four types of spin ½ fermions as leptons. The remaining two types of spin ½ fermions – complexons – ψ_C (eq. 3.101) and ψ_{CT} (eqs. 3.163, 3.196 and 3.207) seem to naturally correspond with quarks since their equations of motion and wave functions have a natural SU(3) symmetry (eqs. 3.113 – 3.126). We therefore associate a color SU(3) symmetry with these two types of spin ½ complexons. The Electroweak doublet of quarks then is

$$\Psi_q^{\ a}(x) = \begin{bmatrix} \psi_C^{\ a} \\ \\ \psi_{CT}^{\ a} \end{bmatrix} \sim \begin{bmatrix} u^a \\ \\ d^a \end{bmatrix} \tag{6.26}$$

where u is an "up" type quark and d is a "down" type quark.[97]

The rationale for constructing quark doublets is the same as the leptonic case: We wish to define a generalization of the "Dirac-like" equations of motion that is covariant under L_C boosts.

6.6 Quark Doublets

We assume that quark doublets consist of a complexon and a tachyonic complexon[98] and to this extent they mirror lepton doublets. In this section we will develop a generalized free complexon equation and describe its features.

Summary of L_C Boosts to Generate Spin ½ Equations

We begin by recapitulating L_C boost features for coordinates and spinors:

$$\Lambda_C(\mathbf{v_c}) = \exp[i\omega\hat{\mathbf{w}}\cdot\mathbf{K}] \tag{2.61}$$
$$\omega = (\omega_r^{\ 2} - \omega_i^{\ 2} + 2i\omega_r\omega_i\ \hat{\mathbf{u}}_r\cdot\hat{\mathbf{u}}_i)^{1/2} \tag{2.62}$$
$$\hat{\mathbf{w}} = (\omega_r\hat{\mathbf{u}}_r + i\omega_i\hat{\mathbf{u}}_i)/\omega \tag{2.63}$$
$$\hat{\mathbf{w}}\cdot\hat{\mathbf{w}} = \hat{\mathbf{u}}_r\cdot\hat{\mathbf{u}}_r = \hat{\mathbf{u}}_i\cdot\hat{\mathbf{u}}_i = 1 \tag{2.64a}$$
$$\mathbf{v_c} = \hat{\mathbf{w}}\tanh(\omega) \tag{2.64b}$$

The corresponding L_C spinor boost for $m^2 > 0$ particles with complex 3-momenta is

---

[97] While the lepton situation is clear in the sense that charged leptons cannot be tachyons since their masses are known (Thus only tachyonic neutrinos are the only currently allowed possibility.), the quark situation is somewhat unclear. We have provisionally chosen the "down" type of quark (d, s, and b) as tachyonic. The association of bound states of these quarks such as the K^0 and B^0 systems which are known to have CP violation, and the CP violation engendered by tachyons, encourages this interpretation.

In addition, $W^{\pm}$ charge asymmetry in $p\bar{p}$ collisions indicate the d sea in a proton is greater than the u sea (K. Abe et al, PRL **74**, 850 (1995)) as does the asymmetry of Drell-Yan production in deep inelastic scattering on p and n targets (A. Baldit et al, Phys. Lett. **B332**, 244 (1994)). These results are to be expected since there is no mass gap for a d tachyon sea while there is a mass gap for a u Dirac particle sea. Complexon quarks may explicate the discrepancies between theory and experiment in the spin structure functions of the parton model for nucleons.

[98] The global SU(3) symmetry of complexons makes their identification with quarks reasonable. However, the complexon theory that we develop and use for quark dynamics in the Standard Model is not required. Our Standard Model could use Dirac fermion dynamics for the up-type quarks and tachyon dynamics for down-type quarks. Then the (broken) Left-handed Extended Lorentz group would be the basic space-time group rather than L_C. We choose to use complexon dynamics for quarks because they have an internal SU(3)-like structure suggestive of color SU(3). More importantly, their spin dynamics is different and thus may resolve the differences between theory and experiment for the deep inelastic parton spin-dependent structure functions. Nevertheless, quarks could be similar to leptons in this regard and form a doublet of a Dirac fermion and an ordinary tachyon. Whether quarks are complexons or not is an experimental question!

$$S_C(\omega, \mathbf{v_c}) = \exp(-i\omega\sigma_{0k}\hat{w}_k/2) = \exp(-\omega\gamma^0\boldsymbol{\gamma}\cdot\hat{\mathbf{w}}/2)$$
$$= \cosh(\omega/2)I + \sinh(\omega/2)\gamma^0\boldsymbol{\gamma}\cdot\hat{\mathbf{w}} \tag{3.78}$$

with inverse transformation

$$S_C^{-1}(\omega, \mathbf{v_c}) = \gamma^2\gamma^0 K^{-1} S_C^\dagger K\gamma^0\gamma^2 = \gamma^2\gamma^0 S_C^{\ T}\gamma^0\gamma^2 = \exp(\omega\gamma^0\boldsymbol{\gamma}\cdot\hat{\mathbf{w}}/2)$$
$$= \cosh(\omega/2)I - \sinh(\omega/2)\gamma^0\boldsymbol{\gamma}\cdot\hat{\mathbf{w}} \tag{3.79}$$

The Dirac-like complexon equation resulting from the boost is

$$[i\gamma^0\partial/\partial t + i\boldsymbol{\gamma}\cdot(\nabla_\mathbf{r} + i\nabla_\mathbf{i}) - m]\psi_C(t, \mathbf{x_r}, \mathbf{x_i}) = 0 \tag{3.101}$$

where $\mathbf{x} = \mathbf{x_r} - i\mathbf{x_i}$. The subsidiary condition is

$$\nabla_\mathbf{r}\cdot\nabla_\mathbf{i}\ \psi_C(t, \mathbf{x_r}, \mathbf{x_i}) = 0 \tag{3.102a}$$

The L_C coordinate boost that leads to $m^2 < 0$ tachyonic complexons with complex 3-momenta is

$$\Lambda_{CL}(\mathbf{v_c}) \equiv \Lambda_{CL}(\omega, \hat{\mathbf{w}}) = \exp[i(\omega + i\pi/2)\hat{\mathbf{w}}\cdot\mathbf{K}] \tag{3.151}$$

where

$$\omega = (\omega_\mathbf{r}^{\ 2} - \omega_\mathbf{i}^{\ 2})^{1/2} \tag{2.62}$$
$$\hat{\mathbf{w}} = (\omega_\mathbf{r}\hat{\mathbf{u}}_\mathbf{r} + i\omega_\mathbf{i}\hat{\mathbf{u}}_\mathbf{i})/\omega \tag{2.63}$$
$$\hat{\mathbf{w}}\cdot\hat{\mathbf{w}} = \hat{\mathbf{u}}_\mathbf{r}\cdot\hat{\mathbf{u}}_\mathbf{r} = \hat{\mathbf{u}}_\mathbf{i}\cdot\hat{\mathbf{u}}_\mathbf{i} = 1 \tag{2.64a}$$
$$\mathbf{v_c} = \hat{\mathbf{w}}\tanh(\omega + i\pi/2) = \hat{\mathbf{w}}\ \text{cotanh}(\omega) \tag{3.152}$$
$$\omega_L = \omega + i\pi/2$$

The L_C spinor boost for tachyonic complexons is

$$S_{CL}(\Lambda_{CL}(\omega, \hat{\mathbf{w}})) = \exp(-i\omega_L\sigma_{0i}\hat{w}_i/2) = \exp(-\omega_L\gamma^0\boldsymbol{\gamma}\cdot\hat{\mathbf{w}}/2)$$
$$= \cosh(\omega_L/2)I + \sinh(\omega_L/2)\gamma^0\boldsymbol{\gamma}\cdot\hat{\mathbf{w}} \tag{3.154}$$

The resulting Dirac-like tachyonic complexon equation is

$$[\gamma^0\partial/\partial t + \boldsymbol{\gamma}\cdot(\nabla_\mathbf{r} + i\nabla_\mathbf{i}) - m]\psi_{CL}(t, \mathbf{x_r}, \mathbf{x_i}) = 0 \tag{3.163a}$$

with the subsidiary condition

$$\nabla_{\mathbf{r}}\cdot\nabla_{\mathbf{i}}\, \psi_{CL}(t, \mathbf{x_r}, \mathbf{x_i}) = 0 \tag{3.164}$$

L_C Boosts between Complexons and Tachyonic Complexons

An L_C spinor boost of a complexon can change it into a tachyonic complexon and vice versa:

$$\begin{aligned} &S_{CL}(\Lambda_{CL}(\omega, \hat{\mathbf{w}}))\psi_C(x) \to \psi_{CT}'(x') \\ &S_{CL}(\Lambda_{CL}(\omega, \hat{\mathbf{w}}))\psi_{CT}(x) \to \psi_C'(x') \end{aligned} \tag{6.27}$$

Similarly the differential operator used in the equations of motion can also be transformed.

$$\begin{aligned} &S_{CL}(\Lambda_{CL}(\omega, \hat{\mathbf{w}}))(\gamma^\mu D_\mu - m)S_{CL}^{-1}(\Lambda_{CL}(\omega, \hat{\mathbf{w}})) = (i\gamma^\mu D'_\mu - m) \\ &S_{CL}(\Lambda_{CL}(\omega, \hat{\mathbf{w}}))\gamma^5(i\gamma^\mu D_\mu - m)\gamma^5 S_{CL}^{-1}(\Lambda_{CL}(\omega, \hat{\mathbf{w}})) = (\gamma^\mu D'_\mu - m) \end{aligned} \tag{6.28}$$

where

$$\begin{aligned} x'^\mu &= i\Lambda_{CL}{}^\mu{}_\nu(\omega, \mathbf{u})x^\nu \\ D'_\mu &= -i\Lambda_{CL}{}^\nu{}_\mu(\omega, \mathbf{u})D_\nu \end{aligned} \tag{6.29}$$

or in matrix form

$$X' = E_{CL}(\omega, \hat{\mathbf{w}})X \equiv i\Lambda_{CL}(\omega, \hat{\mathbf{w}})X \tag{6.30}$$

Eqs. 6.27 – 6.29 imply

$$\begin{aligned} S_{CL}(\Lambda_{CL}(\omega, \hat{\mathbf{w}}))(\gamma^\mu D_\mu - m)\psi_{CT}(x) &= (i\gamma^\mu D'_\mu - m)S_{CL}(\Lambda_{CL}(\omega, \hat{\mathbf{w}}))\psi_{CT}(x) \\ &= (i\gamma^\mu D'_\mu - m)\psi_C'(x') \end{aligned} \tag{6.31}$$

and

$$\begin{aligned} S_{CL}(\Lambda_{CL}(\omega,\hat{\mathbf{w}}))\gamma^5(i\gamma^\mu D_\mu - m)\psi_C(x) &= (\gamma^\mu D'_\mu - m)S_{CL}(\Lambda_{CL}(\omega,\hat{\mathbf{w}}))\gamma^5\psi_C(x) \\ &= (\gamma^\mu D'_\mu - m)\psi_{CT}'(x') \end{aligned} \tag{6.32}$$

where

$$\psi_C'(x') = S_{CL}(\Lambda_{CL}(\omega, \hat{\mathbf{w}}))\psi_{CT}(x) \tag{6.33}$$

and

$$\psi_{CT}'(x') = S_{CL}(\Lambda_{CL}(\omega, \hat{\mathbf{w}}))\gamma^5\psi_C(x) \tag{6.34}$$

Thus neither complexon dynamical equation is L_C covariant.

Doublet Dynamical Equation for Complexons

We will now consider the issue of generalizing the complexon dynamical equations so that the generalized equation is covariant under both Lorentz transformations and L_C boosts.

The only obvious method to obtain a generalized equation that is covariant under L_C boosts is to define an 8×8 matrix generalization. Let

$$đ_C(x) = \begin{bmatrix} (i\gamma^\mu D_\mu - m) & 0 \\ 0 & (\gamma^\mu D_\mu - m) \end{bmatrix} \tag{6.35}$$

be an 8×8 matrix operator with the 4×4 matrix elements shown, and let

$$\Psi_C(x) = \begin{bmatrix} \psi_C(x) \\ \psi_{CT}(x) \end{bmatrix} \tag{6.36}$$

be an 8 component column vector composed of a complexon field and a tachyonic complexon field. Then the generalized complexon equation is

$$đ_C(x)\Psi_C(x) = 0 \tag{6.37}$$

We now define the 8×8 Left-handed L_C boost transformation

$$S_{CL8} \equiv S_{CL8}(\Lambda_{CL}(\omega, \hat{\mathbf{w}})) = \begin{bmatrix} 0 & S_{CL}(\Lambda_{CL}(\omega, \hat{\mathbf{w}})) \\ S_{CL}(\Lambda_{CL}(\omega, \hat{\mathbf{w}}))\gamma^5 & 0 \end{bmatrix} \tag{6.38}$$

with inverse transformation

$$S_{CL8}{}^{-1} \equiv S_{CL8}{}^{-1}(\Lambda_{CL}(\omega, \hat{\mathbf{w}})) = \begin{bmatrix} 0 & \gamma^5 S_{CL}{}^{-1}(\Lambda_{CL}(\omega, \hat{\mathbf{w}})) \\ S_{CL}{}^{-1}(\Lambda_{CL}(\omega, \hat{\mathbf{w}})) & 0 \end{bmatrix} \tag{6.39}$$

Applying S_{CL8} to eq. 6.37 yields

$$0 = S_{CL8}đ_C(x)\Psi_C(x) = đ_C(x')\Psi_C'(x') \tag{6.40}$$

where

$$\Psi_C'(x') = \begin{bmatrix} S_{CL8}\psi_{CT}(x) \\ \\ S_{CL8}\gamma^5\psi_C(x) \end{bmatrix} = \begin{bmatrix} \psi_C'(x') \\ \\ \psi_{CT}'(x') \end{bmatrix} \tag{6.41}$$

Thus the generalized complexon equation is covariant under L_C boosts. Covariance requires the complexon, and tachyonic complexon, must have the same absolute value for the mass.

It is easy to show that the generalized complexon equation is also covariant under conventional Lorentz transformations represented as 4 × 4 diagonal blocks in an 8 × 8 matrix representation. (The demonstration is analogous to eqs. 6.9 – 6.13.)

The Lagrangian density that corresponds to our 8-dimensional construction is

$$\mathcal{L}_{C8} = \overline{\Psi}_C(x)\not{d}_C(x)\Psi_C(x) \tag{6.42}$$

where

$$\overline{\Psi}_C(x) = \Psi_C^{\dagger}\big|_{\mathbf{x_i} = -\mathbf{x_i}}\Gamma_C^{\,0} \tag{6.43}$$

and

$$\Gamma_C^{\,0} = \begin{bmatrix} \gamma^0 & 0 \\ \\ 0 & i\gamma^0\gamma^5 \end{bmatrix} \tag{6.44}$$

The action is

$$I = \int d^4x\,\mathcal{L}_{C8} \tag{6.45}$$

is invariant under Lorentz transformations S_8 and S_{8A} (eqs. 6.9 – 6.12).

The Hamiltonian density for the 8-dimensional theory is

$$\mathcal{H}_{C8}(x) = \begin{bmatrix} \psi_C^{\dagger}(-i\boldsymbol{\alpha}\cdot\nabla_C + \beta m)\psi_C & 0 \\ \\ 0 & i\psi_{CT}^{\dagger}\gamma^5(\boldsymbol{\alpha}\cdot\nabla_C + \beta m)\psi_{CT} \end{bmatrix} \tag{6.46}$$

where

$$\nabla_C = \mathbf{D} \tag{6.47}$$

is the spatial vector part of D^{μ}.

Non-Invariance of the Generalized Free Complexon Action under an L_C Boost

The action 6.45 is not invariant under L_C boosts. The reason is similar to that of section 6.3 for the "leptonic" type of particle: there is no simple relation between the Hermitian conjugate of an L_C spinor boost and its inverse (similar to eq. 6.19 for the Dirac boost case).

Consequently the action of eq. 6.45 is only invariant under inhomogeneous Lorentz transformations. *This state of affairs is again an advantage when we derive features of the Standard Model because it prevents any interplay between unbroken ElectroWeak SU(2) rotations and* L_C *transformations in the complexon (quark) sector.*

7. Derivation of One Generation Leptonic Standard Model Features

7.1 The Standard Model Leptonic Sector as a Consequence of the L_c Group

The development of the Standard Model in the 1960's and 1970's was a major step forward in our understanding of the major forces of nature. However the strange, and, in many physicists' opinions, unattractive form of the theory led particle theorists to conclude that it was a provisional theory that would eventually be replaced by a more elegant fundamental theory. Work in these directions has focused upon 1) embedding the Standard Model within a larger ("elegant") symmetry group; 2) embedding the Standard Model within a space-time with extra dimensions that generate the Standard Model through some mechanism such as the Kaluza-Klein mechanism; and 3) viewing the Standard Model as somehow a "low energy" phenomenology that emerges from a Superstring Theory.

In this chapter we will take an alternate view. We will derive the leptonic sector of the Standard Model based on L_c covariance of the equations of motion. Thus, unlike previous efforts, we view the Standard Model as in a natural form dictated by L_c group covariance and certain other fundamental physical requirements. Beauty being in the eye of the beholder we shall endeavor to show the attractiveness of the derivation of the Standard Model from covariance under an extension of the Lorentz group to boosts (and rotations) beyond the speed of light.

The naturalness of the derivation, and its close connection to the left-handed L_c *group, strongly suggest the Standard Model, which was grown by theorists from experiment, has an undeniable genuine quality that will likely survive the passage of time. The basis of the derivation in a more fundamental theoretical framework raises the hope that we have found a newer, deeper level of understanding of elementary particle dynamics.*

7.2 Assumptions for the Leptonic Sector of the Standard Model

We will make certain assumptions that provide a basis for the derivation of most aspects of the leptonic sector the Standard Model (The quark sector is derived in a subsequent chapter.):

1. One generation of leptons is assumed in this chapter. (The three generation case is treated later.)
2. The form of the equations of motion of the unbroken leptonic sector of the Standard Model are determined by covariance under the L_c group, and local gauge symmetries. Electron number is not gauged.

3. Leptonic matter is composed of spin ½ Dirac particles (electrons) with charge -1 and tachyons (neutrinos) of charge zero as well as their anti-particles.[99]
4. A neutrino is a tachyon with a non-zero bare mass that may be changed by symmetry breaking effects.
5. Gauge fields are massless "before" spontaneous symmetry breaking and are thus conventional gauge fields without a tachyon equivalent in the theory.
6. L_c group covariance, and gauge symmetries of the dynamical equations of motion, are spontaneously broken through the appearance of additional fermion and vector boson mass terms generated by a mechanism such as the Higgs mechanism.

7.3 Derivation of the Leptonic Sector of the Standard Model

The steps of the derivation are:

A. L_c group covariance of the dynamical equations of motion requires that spin ½ particles be described by a generalization of the Dirac equation to an 8 × 8 matrix form in eqs 6.2 – 6.18 based on a doublet consisting of a Dirac particle and a tachyon.

B. We identify the Dirac particle with a charged lepton and the tachyon with a neutrino. The bare masses of these particles have the same numeric value (before symmetry breaking).

C. The leptonic sector free field Lagrangian (without gauge fields introduced yet) is explicitly

$$\mathscr{L}_{\text{freelep}} = \Psi^\dagger(x)\begin{bmatrix} \gamma^0\gamma^5 i(\gamma^\mu\partial/\partial x^\mu - m) & 0 \\ 0 & \gamma^0(i\gamma^\mu\partial/\partial x^\mu - m)\end{bmatrix}\Psi(x) \qquad (7.1)$$

Focussing on the derivative term we see that it can be put in the form

$$\Psi^\dagger(x)\gamma^0\begin{bmatrix}\gamma^5 & 0 \\ 0 & I_4\end{bmatrix} i\gamma^\mu\partial/\partial x^\mu\,\Psi(x) = \Psi^\dagger(x)\gamma^0[C^+I - C^-\sigma_3]i\gamma^\mu\partial/\partial x^\mu\Psi(x) \qquad (7.2)$$

where I_4 is a 4×4 identity matrix, where I and σ_3 are 2 × 2 matrices, and where C^+ and C^- are defined in eq. 3.34. The Pauli matrices σ_i are

[99] If neutrinos are Majorana particles then the derivation must be modified.

$$\sigma_1 = \begin{bmatrix} 0 & 1 \\ 1 & 0 \end{bmatrix} \quad \sigma_2 = \begin{bmatrix} 0 & -i \\ i & 0 \end{bmatrix} \quad \sigma_3 = \begin{bmatrix} 1 & 0 \\ 0 & -1 \end{bmatrix} \tag{7.3}$$

Expression 7.2 can be re-expressed in terms of left-handed and right-handed fields as

$$\Psi_L^{\dagger}(x)\gamma^0 i\gamma^{\mu}\partial/\partial x^{\mu}\Psi_L(x) - \Psi_R^{\dagger}(x)\gamma^0 i\gamma^{\mu}\partial/\partial x^{\mu}\sigma_3\Psi_R(x) \tag{7.4}$$

D. At this point we are in a position to introduce couplings to gauge fields. In view of the doublet nature of the fields $\Psi_L(x)$ and $\Psi_R(x)$ it would appear, at first glance, that the symmetry group of the gauge fields would be $SU(2)_L \otimes SU(2)_R$. However the right-handed tachyon field in expression 7.4 has the wrong sign in the Lagrangian, as has been noted in the previous discussion of the free tachyon Lagrangian and anti-commutator. Consequently the right-handed tachyon field *cannot* have trilinear or higher order couplings. If it did have such interactions then it would rapidly degrade to lower and lower energy by the emission of particles since right-handed leptonic tachyons can exist in principle as free particles (modulo possible Higgs terms). (In this regard the situation is similar to that of time-like photons, except that the set of tachyon physical states cannot be defined in a manner analogous to Gupta-Bleuler electrodynamics where the time-like and longitudinal photons "cancel" each other so that only transverse photons have physical effects.) *Thus there must be no right-handed leptonic tachyon interactions*[100] *(thus apparently excluding the Higgs mechanism for neutrino masses but not mass renormalization.)*

The doublet nature of the left-handed sector implies at least local SU(2) symmetries implemented with a covariant derivative.

The restricted nature of the right-handed leptonic sector indicates that *only* the Dirac particle in the "right-handed doublet" can have an interaction. Also the appearance of σ_3 in the right-handed term in expression 7.4 breaks SU(2) invariance <u>if</u> the left-handed covariant derivative (eq. 7.5 below) were substituted for $\partial/\partial x^{\mu}$ in the right-handed term. Thus a U(1) local gauge field interaction, restricted to the Dirac field member of the right-handed doublet and coupling to both members of the left-handed doublet, is the only allowed possibility. Without the U(1) interaction, the "right-handed doublet" would have no trilinear or higher order elementary particle interactions and would be physically irrelevant (except gravitationally) in the unbroken gauge theory before spontaneous breakdown.

Putting these symmetries together we obtain a left-handed covariant derivative implementing local SU(2)⊗U(1) invariance:

---

[100] Right-handed neutrinos must interact with gravitons due to their mass-energy and the universality of the gravitational interaction. The extreme weakness of the gravitational interaction mitigates this effect.

$$D_{L\mu} = \partial/\partial x^{\mu} + \tfrac{1}{2} i g_2 \boldsymbol{\sigma}\cdot\mathbf{W}_{\mu} + \tfrac{1}{2} i g' B_{\mu} \qquad (7.5)$$

and a right-handed covariant derivative[101]

$$D_{R\mu} = \partial/\partial x^{\mu}\sigma_3 + \tfrac{1}{2} i g' B_{\mu}|Q|\sigma_3$$

$$= \partial/\partial x^{\mu}\sigma_3 + \tfrac{1}{2} i g' B_{\mu}|Q| \qquad (7.6)$$

where Q is the charge operator using the relation $|Q|\sigma_3 = |Q|$ for leptons. We use the absolute value of Q in order to achieve consistency in form with the right-handed quark sector described in the next chapter. As a result expression 7.4 becomes

$$\Psi_L^{\dagger}(x)\gamma^0 i\gamma^{\mu} D_{L\mu}\Psi_L(x) - \Psi_R^{\dagger}(x)\gamma^0 i\gamma^{\mu} D_{R\mu}\Psi_R(x) \qquad (7.7)$$

Thus the leptonic sector of the Lagrangian[102] (modulo mass/Higgs terms) is

$$\begin{aligned}\mathcal{L}_{\text{lep1}} &= \Psi_L^{\dagger}\gamma^0 i\gamma^{\mu} D_{L\mu}\Psi_L - \Psi_R^{\dagger}\gamma^0 i\gamma^{\mu} D_{R\mu}\Psi_R \qquad (7.8)\\ &= \Psi_L^{\dagger}\gamma^0 i\gamma^{\mu} D_{L\mu}\Psi_L + \overline{\psi}_{eR}\gamma^0 i\gamma^{\mu} D_{R\mu}\psi_{eR} - \overline{\psi}_{\upsilon R}\gamma^0 i\gamma^{\mu}\partial/\partial x^{\mu}\psi_{\upsilon R}\\ &= \Psi_L^{\dagger}\gamma^0 i\gamma^{\mu} D_{L\mu}\Psi_L + \overline{\psi}_{eR}\gamma^0 i\gamma^{\mu}(\partial/\partial x^{\mu} + \tfrac{1}{2} i g' B_{\mu})\psi_{eR} - \overline{\psi}_{\upsilon R}\gamma^0 i\gamma^{\mu}\partial/\partial x^{\mu}\psi_{\upsilon R}\end{aligned}$$

where we identify the tachyon as a neutrino and the Dirac particle as a charged lepton such as the electron. Our leptonic sector Lagrangian is now the usual leptonic sector ElectroWeak Lagrangian with a tachyonic neutrino.

E. Local gauge invariance prior to symmetry breaking: The gauge field sector has the usual Yang-Mills Lagrangian terms and the B field has Lagrangian terms similar to that of the QED Lagrangian.

F. Spontaneous symmetry breaking of gauge symmetry, and of L_c group covariance, via the Higgs mechanism (or an alternative mechanism) can be implemented in such a way as to give the electron its known mass as well as the massive vector bosons. Since spontaneous symmetry breaking breaks L_c covariance to Lorentz covariance it is a moot point whether the Higgs sector exhibits a similar covariance.

That concludes the derivation of leptonic sector of the Standard Model. We have shown that the form of the leptonic sector of the Standard Model is fundamental in nature and

---

[101] The coupling constants are defined by $e = -g'\cos\theta_W = g_2\sin\theta_W$.

[102] Note that the gauge fields do not have a tachyon equivalent since they are initially massless prior to spontaneous symmetry breaking.

based on L_c group covariance of the equations of motion (or L_C covariance). Thus it is not correct to view the Standard Model as the result of the breakdown of a larger internal symmetry group to SU(2)⊗U(1). On the contrary, the SU(2)⊗U(1) interaction form is a consequence of the L_c covariance of the free dynamical equations.

8. Can Neutrinos Be Tachyons?

Recently the three species of neutrinos have been found to have masses in neutrino oscillation experiments although the masses are very small. If we denote the three neutrinos as m_1, m_2, and m_3; and let $\Delta m_{ij}^2 = m_j^2 - m_i^2$, then[103]

$$\Delta m_{12}^2 \cong 8 \times 10^{-5}\ \text{eV}^2$$
$$\Delta m_{23}^2 \cong 2.8 \times 10^{-3}\ \text{eV}^2$$

It is claimed that the sign of each Δm^2 is unambiguously determined.[104] However the mass difference values so obtained can be interpreted as differences in tachyon masses as well as differences in "normal" particle masses. The dependence of neutrino oscillations on masses squared results from the time evolution of mixed neutrino states. We consider mixtures of two neutrino states in the vacuum (for the sake of simplicity):

$$|\nu_a\rangle = \cos\theta\, |\nu_e\rangle - \sin\theta\, |\nu_\mu\rangle$$

$$|\nu_b\rangle = \sin\theta\, |\nu_e\rangle + \cos\theta\, |\nu_\mu\rangle$$

with differing masses m_a and m_b.[105] The phase factors determining the time dependence of $|\nu_a\rangle$ and $|\nu_b\rangle$ generate the neutrino oscillations from which the mass relations were obtained. The phase of the kth state is

$$|\nu_k(t)\rangle \sim e^{-im_k^2 t/(2p)}$$

for normal neutrinos. In the case of tachyons the factor in the exponential changes sign:

Tachyonic Neutrinos

$$E = (\mathbf{p}^2 - m^2)^{1/2} \cong p - m^2/(2p)$$

"Normal" Neutrinos

$$E = (\mathbf{p}^2 + m^2)^{1/2} \cong p + m^2/(2p)$$

Thus tachyonic neutrinos would exhibit the time dependence

---

[103] S. M. Bilenky, "Neutrino Masses, Mixing and Oscillations", arXiv:hep-ph/050175 (October 13, 2005).

[104] A. B. McDonald, "Evidence for Neutrino Oscillations I", p. 8 arXiv:nucl-ex/0412005 (December, 2005).

[105] L. Wolfenstein, Phys. Rev. **D17**, 2369 (1978).

$$|\nu_k(t)> \sim e^{+im_{Tk}^{2}t/(2p)}$$

where we use the mass subscript "T" to indicate a tachyon.

As a result, if neutrinos are "normal" the experimental results above would suggest the neutrino masses satisfy

$$m_3^2 > m_2^2 > m_1^2$$

On the other hand, if neutrinos are tachyons the experimental results above would suggest the value of the tachyon neutrino masses satisfies

$$m_1^2 > m_2^2 > m_3^2$$

which is an inverted spectrum compared to hypothetical "normal" neutrinos. When one considers the fact that tachyon masses squared are negative we see that an ordering of the tachyon neutrino mass spectrum, consistent with experimental neutrino results, is:

A Negative Neutrino m^2 Spectrum

$m^2 = 0$ ---------

m_1^2---------

m_2^2---------

m_3^2----------

9. Derivation of One Generation Quark Sector Standard Model Features

9.1 Quark Sector of the Standard Model

The derivation of the form of the quark sector of the Standard Model is very similar to the preceding derivation of the form of the leptonic sector – but with some important points of difference. The primary difference is that we identify quarks with complexons. (This is a plausible assumption in view of the appearance of global SU(3) symmetry within complexon fields. Since complexons have very different spin characteristics compared to Dirac fields, complexon quarks may resolve difficulties in parton spin studies.)

9.2 Quark Sector Assumptions

We will make assumptions that will provide a basis for a derivation of most aspects of single generation, quark sector Standard Model:

1. One generation of quarks is assumed in this chapter. A three generation Standard Model is developed in a subsequent chapter.
2. The equations of motion of the unbroken form of the Standard Model are determined by covariance under the complex group L_C, and local gauge symmetries. Baryon number is not gauged.
3. Quarks are composed of spin ½ complexons and tachyonic complexons.
4. Interacting spin ½ baryons must exist.
5. The strong interaction gauge fields are massless non-Abelian vector complexons and are thus without a tachyon equivalent.
6. L_C covariance and gauge symmetries of the dynamical equations are spontaneously broken through the appearance of mass terms generated by a mechanism such as the Higgs mechanism.

9.3 Derivation of the Form of the Standard Model Quark Sector

The derivation:

A. L_C covariance of the dynamical equations of motion requires that spin ½ particles be described by a complexon equation[106] generalized to an 8×8 matrix form similar to

[106] The global SU(3) symmetry of complexons makes their identification with quarks reasonable. However, the complexon theory that we develop and use for quark dynamics in the Standard Model is not required. Our Standard Model could use Dirac fermion dynamics for the up-type quarks and tachyon dynamics for down-type quarks. Then the (broken) Left-handed Extended Lorentz group would be the basic space-time group rather than L_C. We choose to use complexon dynamics for quarks because they have an internal SU(3)-like structure suggestive of color SU(3). More importantly, their spin dynamics is different and thus may resolve the differences between theory and

eqs 6.35 – 6.37 based on a doublet consisting of a complexon and a tachyonic complexon. However in the case of quarks the complexon is the top component in the doublet.

B. Thus the 8 × 8 complexon quark matrix formalism is:

$$đ_q(x)\Psi_q(x) = 0 \tag{9.1}$$

where

$$đ_q(x) = \begin{bmatrix} (i\gamma^\mu D_\mu - m_0) & 0 \\ 0 & (\gamma^\mu D_\mu - m_0) \end{bmatrix} \tag{9.2}$$

and

$$\Psi_q(x) = \begin{bmatrix} \psi_{Cu}(x) \\ \psi_{CTd}(x) \end{bmatrix} \tag{9.3}$$

The upper 4-component field is a u-type complexon field and the lower 4-component Field is a tachyonic complexon field. The generalized complexon equation eq. 9.1 is covariant under 8 × 8 doublet L_C transformations such as eqs. 6.38 – 6.41.

The generalized fermion equation eq. 9.2 is also covariant under conventional Lorentz transformations in the 8 × 8 representation.

The free quark sector Lagrangian density that corresponds to the 8-dimensional fermion equation eq. 9.1 is

$$\mathscr{L}_{\text{freeQuark}} = \overline{\Psi}_q(x)đ_q(x)\Psi_q(x) \tag{9.4}$$

where

$$\overline{\Psi}_q(x) = \Psi_q^{\dagger}\Gamma_q^{0} \tag{9.5a}$$

with

$$\Psi_q^{\dagger} = (\Psi_q^{\dagger})|_{\mathbf{x_i} = -\mathbf{x_i}} \tag{9.5b}$$

---

experiment for the deep inelastic parton spin-dependent structure functions. Nevertheless, quarks could be similar to leptons in this regard and form a doublet of a Dirac fermion and an ordinary tachyon. Whether quarks are complexons or not is an experimental question!

Where the † in the parentheses on the right side of eq. 9.5b indicates Hermitian conjugation, and

$$\Gamma_q{}^0 = \begin{bmatrix} \gamma^0 & 0 \\ 0 & i\gamma^0\gamma^5 \end{bmatrix} \tag{9.6}$$

C. Having seen that L_C covariance requires doublets composed of a complexon and a tachyonic complexon with the same absolute value for the bare mass m_0 we now **identify the complexon with a "u-type" quark and the tachyonic complexon with a "d-type" quark (an arbitrary choice[107])**. If we assume there is one quark generation (temporarily in this chapter), and do not introduce gauge fields as yet, then we have the free field Lagrangian explicitly as

$$\mathcal{L}_{\text{freeQuark}} = \Psi_q{}^{\dagger} \begin{bmatrix} \gamma^0(i\gamma^\mu D_\mu - m_0) & 0 \\ 0 & \gamma^0\gamma^5 i(\gamma^\mu D_\mu - m_0) \end{bmatrix} \Psi_q \tag{9.7}$$

with

$$\Psi_q(x) = \begin{bmatrix} \psi_{Cu} \\ \psi_{CTd} \end{bmatrix} \tag{9.8}$$

Focusing on the derivative term we see that it can be put in the form

$$\Psi_q{}^{\dagger}\gamma^0 \begin{bmatrix} I_4 & 0 \\ 0 & \gamma^5 \end{bmatrix} i\gamma^\mu D_\mu \Psi_q = \Psi_q{}^{\dagger}\gamma^0[C^+ I + C^- \sigma_3] i\gamma^\mu D_\mu \Psi_q \tag{9.9}$$

where I_4 is a 4×4 identity matrix, where I, and the Pauli matrix σ_3, are 2 × 2 matrices, and C^+ and C^- are defined in eq. 3.34. The Pauli matrices σ_i are defined by eq. 7.3.

Expression 9.13 can be expressed in terms of left-handed and right-handed

---

[107] Since CP violation seems to be more associated with d, s, and b bound states such as the K^0 and B^0 (and their antiparticle) systems, and since tachyon dynamics violates CP invariance there is some justification for this choice although a detailed analysis remains to be done to confirm this choice.

fields as

$$\Psi_{qL}{}^{\dagger}\gamma^0 i\gamma^\mu D_\mu \Psi_{qL} + \Psi_{qR}{}^{\dagger}\gamma^0 i\gamma^\mu D_\mu \sigma_3 \Psi_{qR} \tag{9.10}$$

D. At this point we are in a position to introduce couplings to gauge fields. In view of the doublet nature of the fields $\Psi_{qL}(x)$ and $\Psi_{qR}(x)$ it would again appear, at first glance, that the symmetry group of the gauge fields would be $SU(2)_L \otimes SU(2)_R$. However the right-handed tachyonic complexon field in expression 9.10 has the wrong sign in the Lagrangian, as has been noted in the previous discussion of the free tachyon Lagrangian and anti-commutator (in the preceding chapter). Consequently, **if quarks were *not* confined,** the right-handed quark tachyon field could not have trilinear or higher order couplings. **If free tachyon quarks existed, and had interactions, then they would rapidly degrade to lower and lower energy by the emission of particles.**

However, because of quark confinement in bound states with discrete energy levels, a bound tachyon quark can only emit particles if a lower energy bound state exists. As a result right-handed tachyon quarks can have interactions, such as the electromagnetic interaction, because quark confinement "tames" their propensity to emit particles due to the "wrong sign" in the Lagrangian.

Again there is an analogy to Gupta-Bleuler QED quantization. In Gupta-Bleuler quantization physical states are required to have equal numbers of time-like and longitudinal photons thus canceling their physical effects. Similarly, right-handed tachyon quarks are required to be bound to other quarks by quark confinement to avoid continuous emission of particles.[108] **Since interactions are allowed for right-handed tachyon quarks the Higgs mechanism can be used to change their mass.**

The doublet nature of the left-handed sector implies at least local SU(2) symmetries. The appearance of σ_3 in the right-handed term in expression 9.10 breaks SU(2) invariance if the left-handed covariant derivative (eq. 9.11 below) were substituted for D_μ in the right-handed term. Thus the right-handed fields can only have a U(1) local gauge field interaction, and are SU(2) singlets. Together we thus obtain a left-handed covariant derivative implementing local SU(2)⊗U(1) covariance:

$$D_{qL\mu} = D_\mu + \tfrac{1}{2} i g_2 \boldsymbol{\sigma}\cdot\mathbf{W}_\mu + i g_1 B_\mu/6 \tag{9.11}$$

[108] Therefore quark confinement is required in order to have a properly formulated quark sector. Another interaction – the strong interaction – is required for quark confinement. Presently there is only one accepted mechanism for quark confinement – through a non-Abelian gauge coupling. (Higher derivative theories with quark confinement are in disfavor.) An additional non-Abelian symmetry must be introduced for quarks. As discussed later, SU(3) appears to be a natural choice.

and a right-handed covariant derivative[109]

$$D_{qR\mu} = D_\mu \sigma_3 + ig_1 B_\mu |Q| \tag{9.12}$$

where |Q| is the absolute value of the charge operator (with u eigenvalue 2/3 and d eigenvalue 1/3). The absolute value is used in order to compensate for the minus sign in front of the right-handed tachyon (d quark) term. As a result expression 9.10 becomes

$$\Psi_{qL}{}^\dagger(x)\gamma^0 i\gamma^\mu D_{qL\mu}\Psi_{qL}(x) + \Psi_{qR}{}^\dagger(x)\gamma^0 i\gamma^\mu D_{qR\mu}\Psi_{qR}(x) \tag{9.13}$$

Thus the quark sector of the Lagrangian[110] (modulo mass/Higgs, and strong interaction, terms) is

$$\mathscr{L}_{\text{quark1}} = \Psi_{qL}{}^{\dagger}\gamma^0 i\gamma^\mu D_{qL\mu}\Psi_{qL} + \Psi_{qR}{}^{\dagger}\gamma^0 i\gamma^\mu D_{qR\mu}\Psi_{qR} \tag{9.14}$$

$$= \Psi_{qL}{}^{\dagger}\gamma^0 i\gamma^\mu D_{qL\mu}\Psi_{qL} + \overline{\psi}_{uR} i\gamma^\mu D_{qR\mu}\psi_{uR} - \overline{\psi}_{dR} i\gamma^\mu D_{qR\mu}\psi_{dR}$$

$$= \Psi_{qL}{}^{\dagger}\gamma^0 i\gamma^\mu D_{qL\mu}\Psi_{qL} + \overline{\psi}_{uR} i\gamma^\mu (D_\mu + \tfrac{2}{3} ig_1 B_\mu)\psi_{uR} -$$
$$- \overline{\psi}_{dR} i\gamma^\mu (D_\mu + \tfrac{1}{3} ig_1 B_\mu)\psi_{dR}$$

where we *provisionally* identify the tachyonic complexon as a d-type quark and the complexon as a u-type quark. Our quark sector Lagrangian is now the usual Standard Model quark sector Lagrangian (modulo mass/Higgs, and strong interaction, terms) except that d quarks are tachyonic complexons.

E. The SU(2) gauge field sector has the usual Yang-Mills Lagrangian terms, and the B field is a U(1) Abelian gauge field.

F. Spontaneous symmetry breaking of gauge symmetry, and of L_C covariance, via the Higgs mechanism can be implemented in such a way as to give the quarks, and massive vector bosons, their "known" masses. Since spontaneous symmetry breaking breaks L_C covariance of the dynamical equations of motion to Lorentz covariance it is a moot point whether the Higgs sector is manifestly L_C covariant or not.

---

[109] The quark SU(2) coupling constant is, by gauge invariance, required to have the same value as the leptonic SU(2) coupling constant. The U(1) coupling constants are not required to be the same in both sectors and, in fact, are different. The coupling constants here are defined by $e = g_1\cos\theta_W = g_2\sin\theta_W$.

[110] Note that the gauge fields do not have a tachyon equivalent since they are initially massless prior to spontaneous symmetry breaking.

Thus L_C covariance of the quark equations of motion generate most of the "unusual" features of the quark sector of the Standard Model.

9.4 Color SU(3)

Our derivation yields the general form of the ElectroWeak quark sector of the Standard Model with a dynamics modified by the presence of complex spatial 3-momenta. This difference is a positive for the theory. In the discussion of eqs. 3.113 – 3.126 we showed that complexons have an SU(3) symmetry. The same discussion applies to tachyonic complexons.

Thus, although the discussion in this chapter has hitherto dealt with color scalar quarks, we need only introduce a color index on the quarks to obtain quark fields in the fundamental SU(3) representation $\underline{3}$. Eqs. 3.125, 3.196 and 3.207 contain free color quark field expansions of the $\underline{3}$ representation.

The remaining step is to introduce local SU(3) Yang-Mills gauge fields that interact with the quarks through gauge covariant derivatives. Thus our quark sector has the ElectroWeak and Strong interactions of the Standard Model flowing naturally from L_C covariance of the initially free, quark dynamical equations.

It has long been known that the strong interaction must have a non-Abelian symmetry group in order to have quark confinement – a feature required by our derivation and in apparent complete agreement with experimental data.[111] If we require that spin ½ bound quark states exist (assumption 4 earlier in this chapter) as they do, then strict quark confinement would rule out an SU(2) color strong interaction. Therefore the appearance of SU(3) symmetry in complexon fields, and the requirement of the "most minimal" non-Abelian symmetry group for the color interaction, provide a strong basis for SU(3) as the color symmetry group.

As shown earlier quarks have a complex spatial 3-momentum – they are complexons and tachyonic complexons. If we think of quark gluon interactions, and in particular, perturbative diagrams with quark-gluon loops such as the simplest quark self-energy diagram, then imaginary momentum should "flow" around a loop as well as real momentum. Thus we postulate that gluons are massless spin 1 complexons (although this is not required by our derivation). Therefore we assume an SU(3) non-Abelian, massless, version of vector complexons (chapter 4). The gluon Lagrangian term is

$$\mathcal{L}_{CG} = -\tfrac{1}{4}\, F_C^{a\mu\nu}(x) F_{C\ \mu\nu}^{\ a}(x) \tag{9.15}$$

where

$$F_{C\ \mu\nu}^{\ a} = D_\nu A_{C\ \mu}^{\ a} - D_\mu A_{C\ \nu}^{\ a} + g f^{abc} A_{C\ \mu}^{\ b} A_{C\ \nu}^{\ c} \tag{9.16}$$

---

[111] Evidence is scanty at the present time for five quark states and other exotics that would seem to differ from strict confinement in three quark or quark-antiquark bound states.

where D_μ is defined by eq. 4.51, g is the coupling constant, and the f^{abc} are the SU(3) structure constants.

The theory of the strong color interaction should be developed within a path integral framework that takes account of Faddeev-Popov gauge fixing. Before doing that we must consider the effect of the complex spatial 3-vector on the non-Abelian gauge formalism.

9.5 Pure Complexon Gauge Groups

Consider a local Lie group G with the generators of its algebra satisfying the commutation relations:

$$[t^a, t^b] = if^{abc}t^c \tag{9.17}$$

If a complexon field ψ_{Ca} transforms as some representation of G then a transformed field has the form:

$$\psi_{Ca}'(x) = \Theta_{ab}(x)\psi_{Cb}(x) \tag{9.18}$$

where $x = (x^0, \mathbf{x} = \mathbf{x_r} + i\mathbf{x_i})$ and $\Theta_{ab}(x)$ is an element of G. Since G is a local group we can represent $\Theta_{ab}(x)$ in the form:

$$\Theta_{ab}(x) = [\exp(-i\varphi_c(x)t^c)]_{ab} \tag{9.19}$$

where $\varphi_c(x)$ is a local parameter dependent on x, and t^c is the matrix in the representation of G under consideration.

The gauge covariant derivative in the case of complexon fermions is

$$\mathcal{D}_\nu = D_\nu - igA_{C\nu} \tag{9.20}$$

where

$$A_{C\nu} = A_C{}^a{}_\nu t^a \tag{9.21}$$

Applying a gauge transformation to the gauge covariant derivative of a complexon fermion field $\mathcal{D}_\nu\psi_C(x)$:

$$\Theta\mathcal{D}_\nu\psi_C(x) = \Theta D_\nu\psi_C(x) - ig\Theta A_{C\nu}\Theta^{-1}\Theta\psi_C(x) \tag{9.22}$$

$$= D_\nu\psi_C'(x) - igA_C'{}_\nu\psi_C'(x) = (D_\nu\psi_C(x))'$$

where

$$\psi_C'(x) = \Theta(x)\psi_C(x) \tag{9.23}$$

we find

$$A_C'{}_\nu = (-i/g)(D_\nu\Theta(x))\Theta^{-1}(x) + \Theta(x)A_{C\nu}(x)\Theta^{-1}(x) \tag{9.24}$$

The reader will note that the form of eqs. 9.15 – 9.24 is identical to those of a conventional non-Abelian gauge field with the replacement:

$$\partial/\partial x^{\nu} \rightarrow D_{\nu} \tag{9.25}$$

with D_ν given by eq. 3.129. Thus we see that $\varphi_c(x)$, the local parameter in eq. 9.19 is dependent in general, on time, and the real and imaginary parts of the complex spatial 3-vector. Thus the formalism differs only in small ways from the conventional non-Abelian gauge theory. However, perturbation theory and non-perturbative phenomena such as instantons exhibit significant differences.

The commutator of covariant derivatives

$$F_{C\mu\nu} = F_{C}{}^{a}{}_{\mu\nu}t^{a} = (i/g)[\mathcal{D}_{\mu}, \mathcal{D}_{\nu}] \tag{9.26}$$

is itself covariant under gauge transformations:

$$F_{C}{}'_{\mu\nu} = \Theta F_{C\mu\nu}\Theta^{-1} \tag{9.27}$$

The strong interaction part of the action has terms of the form

$$I = \int d^7x[\mathcal{L}_{CG} + \bar{\Psi}_q i\gamma^{\mu}\mathcal{D}_{\mu}\Psi_q + \ldots] \tag{9.28}$$

where the "extra" three integrations are over the imaginary spatial coordinates, and Ψ_q represents a generic quark (complexon) field. The action must be supplemented with a constraint similar to those seen earlier for complexons, which ensures the reality of the Lagrangian term $\mathcal{L}_{CG}$ (and thus the corresponding Hamiltonian terms and ultimately the unitarity of the theory.) The constraint simply specifies the imaginary part of $\mathcal{L}_{CG}$ is zero:

$$\begin{aligned}[\partial A_C{}^{a\mu}/\partial x_{ik}]\mathrm{Re}\, F_C{}^{a}{}_{\mu k} &= [\partial A_C{}^{a\mu}/\partial x_{ik}](\partial A_C{}^{a}{}_{\mu}/\partial x_r{}^{k} - \partial A_C{}^{a}{}_{k}/\partial x_r{}^{\mu} + gf^{abc}A_C{}^{b}{}_{\mu}A_C{}^{c}{}_{k}) \\ &= 0 \end{aligned} \tag{9.29}$$

where r indicates the real part of the coordinate $\mathbf{x} = \mathbf{x}_r + i\mathbf{x}_i$, and $\partial/\partial x_i{}^{k}$ is the derivative of the kth component of the imaginary coordinate 3-vector $\mathbf{x}_i$. It is not a choice of gauge but rather a restriction on the dependence of the $A_{C\nu}$ field on the real and imaginary spatial 3-vectors.

If we assume that we can integrate by parts in the imaginary coordinates then we can re-express the constraint as

$$A_C^{a\mu}\ \mathrm{Re}\ \partial F_{C\ \mu k}^{\ a}/\partial x_i^{\ k} = 0 \qquad (9.30)$$

which is explicitly

$$A_C^{a\mu}[\partial^2 A_{C\ \mu}^{\ a}/\partial x_r^{\ k}\partial x_i^{\ k} - \partial^2 A_{C\ k}^{\ a}/\partial x_r^{\ \mu}\partial x_i^{\ k} + g f^{abc}\partial(A_{C\ \mu}^{\ b} A_{C\ k}^{\ c})/\partial x_i^{\ k}] = 0 \qquad (9.31)$$

by eq. 9.16 where $x_r^{\ 0} = x^0$. This restriction can be implemented within the framework of the path integral formalism by using the Faddeev-Popov Method with the introduction of ghosts in a manner similar to the use of the Faddeev-Popov Method to implement a gauge condition. The form of the restriction stated in eqs. 9.30 and 9.31 leads to second order derivative Faddeev-Popov ghost terms in the path integral while eq. 9.29 would lead to a third order derivative ghost terms in the path integral with attendant unitarity (and possibly other) issues.

9.6 Pure Gauge Complexon Path Integral Formulation and Faddeev-Popov Method

The path integral formalism for complexon non-Abelian, pure, Yang-Mills fields differs significantly from the conventional gauge field path integral formalism. The path integral for a complexon gauge field can be written symbolically as:

$$Z(J^\mu) = N\int DA_C \Delta_{FP}(A_C)\delta(F(A_C))\Delta_C(A_C)\delta(F_C(A_C))\exp\{i\int d^7y[\mathscr{L} + \\ + J^\mu(y)A_{C\mu}(y)]\} \qquad (9.32)$$

where $\delta(F(A_C))$ specifies the gauge, $\Delta_{FP}(A_C)$ is its Faddeev-Popov determinant; and $\delta(F_C(A_C))$ specifies the complexon condition (eq. 9.31) with $\Delta_C(A_C)$ the Faddeev-Popov determinant for the complexon condition. In both cases the Faddeev-Popov determinant can be calculated in the standard way.[112]

First we consider the gauge fixing delta function. Note that it can be written as a delta function in the gauge times a determinant:

$$\delta(F(A_C^{\ \omega})) = \delta(\omega - \omega_0)\left|\det \delta F(A_{C\mu}^{\ \ \omega}(x))/\delta\omega(x)\right|^{-1}\Big|_{F(A_C)=0} \qquad (9.33)$$

where ω_0 is a reference gauge, where

$$A_{C\ \mu}^{\ a\ \omega}(x) = A_{C\ \mu}^{\ a}(x) - g^{-1}D_\mu\omega^a + f^{abc}\omega^b(x)A_{C\ \mu}^{\ c}(x) \\ = A_{C\ \mu}^{\ a}(x) + \delta A_{C\ \mu}^{\ a\ \omega}(x) \qquad (9.34)$$

and

$$\mathrm{Re}\ F_{C\ \mu k}^{\ a\ \omega} = \mathrm{Re}\ F_{C\ \mu k}^{\ a} + f^{abc}\omega^b(x)F_{C\ \mu k}^{\ c} \\ = \mathrm{Re}\ F_{C\ \mu k}^{\ a} + \delta(\mathrm{Re}\ F_{C\ \mu k}^{\ a\ \omega}) \qquad (9.35)$$

---

[112] See for example Huang (1992).

under an infinitesimal gauge transformation, and where

$$\Delta_{FP}(A_C) = \left|\det \delta F(A_{C\mu}{}^{\omega}(x))/\delta\omega(x)\right|\Big|_{F(A_C) = 0,\ \omega = 0} \quad (9.36)$$

We will choose the complexon Lorentz gauge to evaluate the Faddeev-Popov determinant:

$$F^a(A_C) = D_\mu A_C{}^{a\mu}(x) = 0 \quad (9.37)$$

We find

$$F^a(A_{C\mu}{}^{\omega}(x)) = D^\mu(A_C{}^a{}_\mu(x) - g^{-1}D_\mu\omega^a(x) + f^{abc}\,\omega^b(x)A_C{}^c{}_\mu(x))$$

$$= -g^{-1}D^\mu D_\mu\omega^a(x) + f^{abc}A_C{}^c{}_\mu(x)D^\mu\omega^b(x) \quad (9.38)$$

Thus

$$\delta F^a(A_{C\mu}{}^{\omega}(x))/\delta\omega^b(x) = -g^{-1}\,\delta^{ab}D^\mu D_\mu + f^{abc}A_C{}^{c\mu}(x)D_\mu \quad (9.39)$$

and

$$\Delta_{FP}(A_C) = \left|\det\,(g^{-1}\delta^{ab}D^\mu D_\mu - f^{abc}A_C{}^{c\mu}(x)D_\mu)\right| \quad (9.40)$$

where | … | represent absolute value.

We can rewrite the Faddeev-Popov determinant as a path integral over anti-commuting c-number fields with a ghost Lagrangian:

$$\Delta_{FP}(A_C) = \int D\chi^* D\chi\, \exp[\, i\int d^7x\, \mathscr{L}^{ghost}(x)] \quad (9.41)$$

where

$$\mathscr{L}^{ghost}(x) = \chi^{a*}(x)[\delta^{ab}D^\mu D_\mu - gf^{abc}A_C{}^{c\mu}(x)D_\mu]\chi^b(x) \quad (9.42)$$

Faddeev-Popov Application to Complexon Condition

The complexon condition can be also implemented within the path integral formalism using the Faddeev-Popov Mechanism. Using the identity

$$1 = \int DA_C\Delta_C(A_C)\delta(F_C(A_C)) \quad (9.43)$$

we see that an infinitesimal gauge transformation yields eqs. 9.34 and 9.35. This enables us to relate $\Delta_C(A)$ to the determinant

$$\delta(F_C(A_C{}^{\omega})) = \left|\det \delta F_C(A_{C\mu}{}^{\omega}(x))/\delta\omega(x)\right|^{-1}\Big|_{F_C(A_C)=0,\ \omega = 0} \quad (9.44)$$

and

$$\Delta_C(A_C) = \left|\det \delta F_C(A_{C\mu}{}^{\omega}(x))/\delta\omega(x)\right|\Big|_{F_C(A_C) = 0,\ \omega = 0} \tag{9.45}$$

From eq. 9.31 we see

$$F_C(A_{C\mu}(x)) - A_C{}^{a\mu}[\partial^2 A_C{}^a{}_\mu/\partial x_r{}^k\partial x_i{}^k - \partial^2 A_C{}^a{}_k/\partial x_r{}^\mu\partial x_i{}^k + gf^{abc}\partial(A_C{}^b{}_\mu A_C{}^c{}_k)/\partial x_i{}^k] \tag{9.46}$$

with

$$F_C{}^a(A_{C\mu}) = 0 \tag{9.47}$$

Inserting eq. 9.34 and 9.35 we find

$$\begin{aligned}[\delta F_C(A_{C\mu}{}^{\omega}(x))/\delta\omega^a(x)]\big|_{F_C(A_C) = 0,\ \omega = 0} &= \delta[\delta A_C{}^{b\mu\omega}\,\mathrm{Re}\ \partial F_C{}^b{}_{\mu k}/\partial x_{ik} + \\ &\quad + A_C{}^{b\mu}\,\partial\delta(\mathrm{Re}\ F_C{}^b{}_{\mu k}{}^{\omega})/\partial x_{ik}\ ]/\delta\omega^a(x)\big|_{\omega = 0} \\ &= -g^{-1}(\mathrm{Re}\ \partial F_C{}^a{}_{\mu k}/\partial x_{ik})D^\mu - f^{abc}A_C{}^{b\mu}(\mathrm{Re}\ F_C{}^c{}_{\mu k})\partial/\partial x_{ik}\end{aligned} \tag{9.48}$$

Thus

$$\Delta_C(A_C) = \left|\det\,(g^{-1}(\mathrm{Re}\ \partial F_C{}^a{}_{\mu k}/\partial x_{ik})D^\mu + f^{abc}A_C{}^{b\mu}(\mathrm{Re}\ F_C{}^c{}_{\mu k})\partial/\partial x_{ik}\right| \tag{9.49}$$

where | … | represent absolute value.

We can rewrite this Faddeev-Popov determinant as a path integral over anti-commuting c-number fields with a ghost Lagrangian:

$$\Delta_C(A_C) = \lim_{W\to\infty} \int D\chi_C{}^* D\chi_C \exp[ir^{-2}\int d^7x\, \mathscr{L}_C{}^{ghost}(x)] \tag{9.50}$$

where r is a constant that is taken to the limit ∞, and where

$$\begin{aligned}\mathscr{L}_C{}^{ghost}(x) = \chi_C{}^*(x)\{D^\mu D_\mu + r^2 t^a[(\mathrm{Re}\ \partial F_C{}^a{}_{\mu k}/\partial x_{ik})D^\mu + \\ + gf^{abc}A_C{}^{b\mu}(\mathrm{Re}\ F_C{}^c{}_{\mu k})\partial/\partial x_{ik}]\}\chi_C(x)\end{aligned} \tag{9.51}$$

where t^a is a 3 × 3 matrix of the 3 representation of color SU(3) and $\chi_C(x)$ is a three row field in the 3 representation. *The introduction of $D^\mu D_\mu$ is based on consistency with the complexon formalism. It is needed to establish a perturbative expansion of the path integral. Its effect vanishes in the limit r → ∞ reducing ghost loops of this type to point interactions.* The reader will note that second order and third order derivative terms appear in the interaction in $\mathscr{L}_C{}^{ghost}(x^\mu)$ and raise the issue of non-renormalizable divergences. If one uses the Two-Tier approach to quantum field theory developed by Blaha (2005a) then all potential divergences disappear. *The Two-Tier formulation of the pure, complexon, Yang-Mills theory that we are discussing is finite.*

The complete pure complexon, Yang-Mills path integral is

$$Z(J^\mu) = N \int DA_C D\chi^* D\chi D\chi_C{}^* D\chi_C \Delta_{FP}(A_C)\delta(F(A_C))\Delta_C(A_C)\delta(F_C(A_C))\cdot \cdot\exp\{i\int d^7y\, [\mathscr{L} + J^\mu A_{C\mu}]\} \quad (9.52)$$

where

$$\mathscr{L} = \mathscr{L}_{CG} + \mathscr{L}^{\text{ghost}} + \mathscr{L}_C{}^{\text{ghost}} \quad (9.53)$$

with the Lagrangian terms specified by eqs. 9.15, 9.42, and 9.51.

9.7 Complexon Quark-Gluon Perturbation Theory

In this section we will give the flavor of complexon strong interaction theory by considering a simple diagram. Fig. 9.1 is a self-energy diagram for a quark in which a gluon is emitted and absorbed. If one considers the momentum then complexon particles have complex spatial momentum. From the form of the propagators and wave functions previously considered it is clear that the real part of the 3-momentum and the imaginary part of the 3-momentum will be separately conserved at each vertex. The general form of the self-energy integral corresponding to Fig. 9.1 is

$$I_{ab}(p) = \int d^7q\, P_{ab}(q, p)\exp(\text{Gaussian})/Q(q, p) \quad (9.54)$$

where a and b are SU(3) color indices, $P_{ab}(q, p)$ is a polynomial in q and p together with gamma matrix factors and color SU(3) matrix factors, and Q(q, p) is the product of a quark and a gluon propagator denominator factor. The Gaussian exponential factor appears if we use a Two-Tier formulation[113] of the complexon quark and gluon theory. This exponential factor guarantees the convergence of the integral resulting in a finite result.

Thus while perturbation theory is only partly useful in calculating strong interaction phenomena due to the "largeness" of the coupling constant in general, we see complexon perturbation theory is well defined, finite, and sensible if one uses Two-Tier quantum field theory.

[113] See Blaha (2005a) – part of this volume.

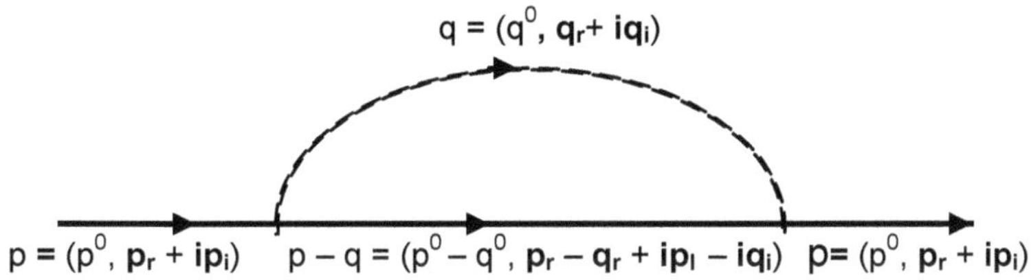

Figure 9.1. Quark self-energy diagram in which a quark emits and subsequently absorbs a gluon. The quark momentum is $p = (p^0, \mathbf{p_r} + i\mathbf{p_i})$ and the gluon momentum is $q = (q^0, \mathbf{q_r} + i\mathbf{q_i})$ with conservation of energy, and separate conservation of real spatial momentum and imaginary spatial momentum at each vertex.

The integral corresponding to Fig. 9.1 has the form:

$$I_{1ab}(p) = \int dq^0 d^3q_r d^3q_i\, P_{ab}(q, p)\exp(\text{Gaussian})/Q(q, p) \qquad (9.55)$$

and includes integrations over both real and imaginary 3-momenta.

9.8 Complexon Quark ElectroWeak Interactions

In our theory quarks are complexons with complex spatial 3-momenta. On the other hand, the ElectroWeak bosons: $W^{\pm}{}_{\mu}$, Z_{μ} and A_{μ} are not complexons and so have totally real momenta. In this section we address the issue of quark ElectroWeak perurbation theory. *The general perturbation theory rule in this situation is that real and imaginary momenta are separately conserved at each vertex.*

Fig. 9.2 shows a quark self-energy diagram in which a quark emits a photon and subsequently reabsorbs it. Since ElectroWeak bosons have real energies and momenta and since real and imaginary momenta are separately conserved at each vertex by the above stated rule, the photon momentum is real while the quark spatial 3-momentum is complex.

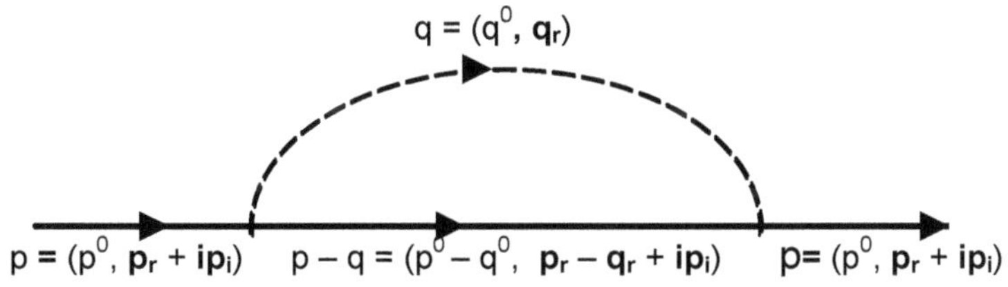

Figure 9.2. Quark self-energy diagram in which a quark emits and subsequently absorbs a photon. The quark momentum is $p = (p^0, \mathbf{p_r} + i\mathbf{p_i})$ and the photon momentum is $q = (q^0, \mathbf{q_r})$ with conservation of energy, and separate conservation of real spatial momentum and imaginary spatial momentum at each vertex.

The integral corresponding to Fig. 9.2 has the form:

$$I_{2ab}(p) = \int dq^0 d^3q_r d^3q_i \delta^3(\mathbf{q}_i) P_{ab}(q, p)\exp(\text{Gaussian})/Q(q, p) \qquad (9.56)$$
$$= \int dq^0 d^3q_r P_{ab}(q, p)\exp(\text{Gaussian})/Q(q, p)$$

However, in cases where there are complexon quark loops the imaginary quark 3-momenta affects the result. For example, Fig. 9.3 shows a complexon quark loop contribution to the photon self-energy in which the imaginary quark 3-momenta integration contributes.

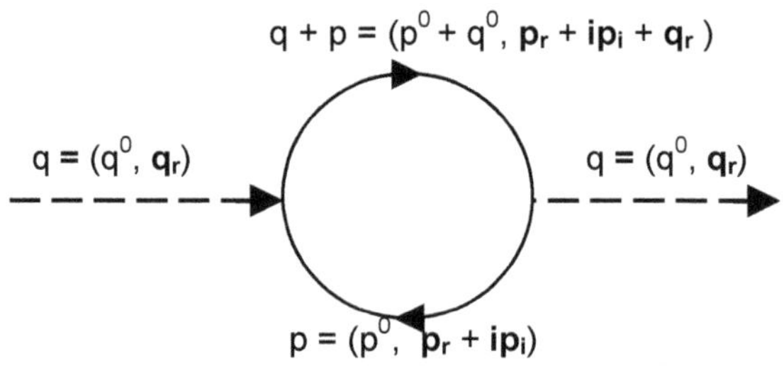

Figure 9.3. Photon self-energy diagram with a quark loop. The photon momentum is $q = (q^0, \mathbf{q}_r)$ and quark loop momentum is $p = (p^0, \mathbf{p}_r + i\mathbf{p}_i)$.

9.9 Complexon Quarks with Color SU(3)

In this chapter we have seen that the complex Lorentz group denoted L_C leads to the form of quark ElectroWeak theory and also leads naturally to color SU(3) with complexon gluons.

10. One Generation Standard Model

10.1 One Generation Standard Model

In the preceding chapters we have developed the form of the fermion sectors of a one generation Standard Model based on the L_C covariance of the equations of motion. We can thus say that physics beyond the speed of light (the light barrier) and the physics of complex spatial 3-momenta play an important role in determining the nature of the fundamental physics of the phenomena we see in our universe. Thus L_C covariance brings us to a deeper level of understanding of the nature of the Standard Model. L_C covariance makes the superficial "unusual" form of the Standard Model understandable.

While some may feel that the regime of L_C physics is an unwarrented (unphysical) extension of our sublight, real space-time, experience of nature, it is certainly a more conservative theoretical extension than positing many additional space-time dimensions, and internal symmetries, strings, and superstrings, for which no significant evidence yet exists.

It is clear from the discussion in chapter 0 that Black Holes must contain tachyons – faster-than-light particles – and there is much evidence for the existence of Black Holes. Thus we are confronted with the reality of faster-than-light physics. Contrast this reality with the hypothetical nature of the many proposed more fundamental theories for which the Standard Model is supposed to be a "low energy" approximation.

We conclude this chapter with the derived form of the leptonic and quark[114] sectors of the one generation Standard Model:

$$\mathcal{L}_{SM1} = \Psi_L^{\dagger}\gamma^0 i\gamma^{\mu} D_{L\mu}\Psi_L - \Psi_R^{\dagger}\gamma^0 i\gamma^{\mu} D_{R\mu}\Psi_R +$$

$$+ \Psi_{qL}^{\dagger}\gamma^0 i\gamma^{\mu}\mathcal{D}_{qL\mu}\Psi_{qL} + \Psi_{qR}^{\dagger}\gamma^0 i\gamma^{\mu}\mathcal{D}_{qR\mu}\Psi_{qR} + \mathcal{L}_{Gauge} + \mathcal{L}_{Higgs} \tag{10.1}$$

where

$$D_{L\mu} = \partial/\partial x^{\mu} + ½ig_2\boldsymbol{\sigma}\cdot\mathbf{W}_{\mu} - ig_1B_{\mu}/2 \tag{10.2}$$

$$D_{R\mu} = \sigma_3\partial/\partial x^{\mu} + ig_1B_{\mu}|Q|$$

$$\mathcal{D}_{qL\mu} = \mathcal{D}_{\mu} + ½ig_2\boldsymbol{\sigma}\cdot\mathbf{W}_{\mu} + ig_1B_{\mu}/6 \tag{10.3}$$

---

[114] As stated in earlier footnotes the complexon theory that we develop and use for quark dynamics in the Standard Model is not required. Our Standard Model could use Dirac fermion dynamics for the up-type quarks and tachyon dynamics for down-type quarks (and conventional gluons). Then the (broken) Left-handed Extended Lorentz group would be the basic space-time group rather than L_C.

$$\mathcal{D}_{qR\mu} = \sigma_3\mathcal{D}_\mu + ig_1B_\mu|Q| \qquad (10.4)$$

where $\mathcal{D}_\mu$ is given by eq. 9.20, $\Psi_q^\dagger$ is given by eq. 9.5b and $\Psi_{qL}^\dagger$ and $\Psi_{qR}^\dagger$ are its left-handed and right-handed parts respectively, and where

$$\mathcal{L}_{Gauge} = \mathcal{L}_{GaugeEW} + \mathcal{L}_{GaugeC} \qquad (10.5)$$

with

$$\mathcal{L}_{GaugeEW} = -¼\, F_W{}^{a\mu\nu}F_W{}^a{}_{\mu\nu} - ¼\, F_B{}^{\mu\nu}F_{B\mu\nu} + \mathcal{L}_{EW}{}^{ghost} \qquad (10.6)$$

and

$$\mathcal{L}_{GaugeC} = \mathcal{L}_{CG} + \mathcal{L}^{ghost} + \mathcal{L}_C{}^{ghost} \qquad (10.7)$$

The ElectroWeak gauge bosons $W_\mu{}^a$ and B_μ field tensors are:

$$F_W{}^a{}_{\mu\nu} = \partial W^a{}_\mu\partial/\partial x^\nu - \partial W^a{}_\nu/\partial x^\mu + g_2f^{abc}W^b{}_\mu W^c{}_\nu \qquad (10.8)$$

$$F_{B\mu\nu} = \partial B_\mu\partial/\partial x^\nu - \partial B_\nu/\partial x^\mu \qquad (10.9)$$

$\mathcal{L}_{EW}{}^{ghost}$ is the Faddeev-Popov ghost terms for the ElectroWeak $W_\mu{}^a$ gauge bosons. $\mathcal{L}_{CG}$ is defined by eqs. 9.15 and 9.16. $\mathcal{L}^{ghost}$ is the color SU(3) Faddeev-Popov ghost terms defined by eq. 9.42 for the complexon Lorentz gauge (eq. 9.37). $\mathcal{L}_C{}^{ghost}$ is the complexon color SU(3) constraint ghost terms defined using the Faddeev-Popov mechanism (eq. 9.51). The Higgs sector $\mathcal{L}_{Higgs}$ will be discussed in the context of the three generation model in a subsequent chapter.

The coordinates are

$$x_c = (x^0, \mathbf{x_r} + i\mathbf{x_i}) \qquad (10.10)$$

and the action is therefore

$$I_{SM1} = \int dx^0 d^3x_r d^3x_i\, \mathcal{L}_{SM1} \qquad (10.11)$$

The Lagrangian is supplemented with the following condition on all complexon fields $\Phi_{\ldots}$:

$$\nabla_r\cdot\nabla_i\Phi\ldots = 0 \qquad (10.12)$$

Non-complexon fields $\Omega\ldots$ in the left-handed formulation under consideration satisfy the subsidiary condition:

$$[\nabla_r\cdot\nabla_i - (\nabla_r{}^2\nabla_i{}^2)^{½}]\Omega\ldots = 0 \qquad (10.13)$$

as exemplified by the discussion of eqs. 3.82 – 3.86, section 3.8, and other sections.

Renormalization and Divergence Issues

The theory derived from $\mathcal{L}_{SM1}$ when calculated perturbatively in conventional quantum field theory has divergences that are not renormalizable due to its 7-dimensional nature. This issue does not appear to be curable by known renormalization methods. However if the theory is reformulated in the Two-Tier formalism (next section) developed by Blaha[115] then a finite theory results. This is also true for the three generation Standard Model to be developed later.

10.2 Two-Tier Formulation of One Generation Standard Model

The general idea of Two-Tier formulations of the Standard Model were presented in Blaha (2003) and Blaha (2005a). In this section we will develop a Two-Tier formulation of the one generation Standard Model described in the previous section. This formulation goes beyond our earlier books by using complex spatial coordinates which necessitate a slightly more complicated Two-Tier formulation. (Appendix 10-A describes the basic features of the Two-Tier formalism.)

The basic ansatz of the Two-Tier formalism is to replace every appearance of a coordinate x with a variable that is in part a quantum field:

$$x^\mu \rightarrow X^\mu = (y^0, \mathbf{y} + \mathbf{Y}(y^0, \mathbf{y})) \tag{10.14}$$

where $\mathbf{Y}(y^0, \mathbf{y})$ is the spatial part of a free massless vector field identical to the free QED field.

Then one finds that the momentum space free field Feynman propagators G(k) of all particles acquires a Gaussian factor exp(h(k)):

$$G(k) \rightarrow G(k)\exp(h(k)) \tag{10.15}$$

so that all perturbation theory diagrams are finite. The result is a finite perturbative result in all calculations to any order in perturbation theory. Blaha (2005a) shows that Two-Tier theories are finite, Poincare covariant, and unitary. (See appendix 10-A for a more detailed description and Blaha (2005a) for a complete discussion.)

Simple Two-Tier Formalism

In this subsection we will describe the basic Two-Tier formalism. Taking the Lagrangian described in appendix 10-A:

$$\mathscr{L}(y) = \mathscr{L}_F(X_\mu(y))J + \mathscr{L}_C(X^\mu(y), \partial X^\mu(y)/\partial y^\nu, y) \tag{10-A.26}$$

where

$$X_\mu(y) = y_\mu + i\, Y_\mu(y)/M_c^2 \tag{10.16}$$

[115] Blaha (2005a) and Blaha (2003). Blaha(2005a) follows this book in this volume and treats the Two-Tier formalism in detail.

with M_c being a large mass scale and $Y_\mu(y)$ a vector quantum field, and where J is the absolute value of the Jacobian of the transformation from X to y coordinates:

$$J = |\partial(X)/\partial(y)| \qquad (10\text{-A.21})$$

The lagrangian term $\mathscr{L}_C$ is

$$\mathscr{L}_C = +\tfrac{1}{4} M_c^4 F^{\mu\nu} F_{\mu\nu} \qquad (10.17)$$

with

$$F_{\mu\nu} = \partial X_\mu/\partial y^\nu - \partial X_\nu/\partial y^\mu \qquad (10.18)$$
$$\equiv i\,(\partial Y_\mu/\partial y^\nu - \partial Y_\nu/\partial y^\mu)/M_c^2$$

The sign in eq. 10.17 is not negative – superficially contrary to the conventional electromagnetic Lagrangian. The reason for this difference is that the quantum field part of X^μ is imaginary. Thus $\mathscr{L}_C$ winds up having the correct sign after taking account of the factor of i in the field strength $F_{\mu\nu}$.

Defining

$$F_{Y\mu\nu} = (\partial Y_\mu/\partial y^\nu - \partial Y_\nu/\partial y^\mu) \qquad (10.19)$$

we see the Lagrangian assumes the form of a conventional electromagnetic Lagrangian:

$$\mathscr{L}_C = -\tfrac{1}{4} F_Y^{\mu\nu} F_{Y\mu\nu} \qquad (10.20)$$

The action of this theory is

$$I = \int d^4y\, \mathscr{L}(y) \qquad (10.21)$$

The further development of the theory is described in Appendix 10-A and Blaha (2005a).

Two-Tier L_C-based Standard Model Theory

In the present case we will need two variables X_r^μ and X_i^μ for the Standard Model based on L_C covariance that we have defined since we have complex spatial 3-coordinates. We define them similarly to the previous case:

$$X_{r\mu}(y_r) = y_{r\mu} + i\, Y_{r\mu}(y_r)/M_c^2 \qquad (10.22)$$

$$X_{i\mu}(y_i) = y_{i\mu} + i\, Y_{i\mu}(y_i)/M_c^2 \qquad (10.23)$$

where we choose the same mass scale for both the "real" and "imaginary" variables. The Two-Tier, single generation, version of the Standard Model given in eq. 10.1 then has an action of the form

$$I_{SM1tt} = \int dy^0 d^3y_r d^3y_i \left(\mathscr{L}_{SM1}(X_r^{\mu}(y_r), \mathbf{X}_i^{k}(y_i))J_2\right)\Big|_{y_i^0 = 0,\ Y_r^0 = Y_i^0 = 0} +$$

$$+ \int dy_r^0 d^3y_r\, \mathscr{L}_C(X_r^{\mu}(y_r), \partial X_r^{\mu}(y_r)/\partial y_r^{\nu}, y_r) +$$

$$+ \int dy_i^0 d^3y_i \mathscr{L}_C(X_i^{\mu}(y_i), \partial X_i^{\mu}(y_i)/\partial y_i^{\nu}, y_i) \qquad (10.24)$$

with the replacements

$$x^{\mu} \equiv x_r^{\mu} \rightarrow X_r^{\mu}(y_r) \qquad (10.25)$$

$$x_i^{k} \rightarrow X_i^{k}(y_i) \qquad (10.26)$$

for $\mu = 0, 1, 2, 3$ and $k = 1, 2, 3$ are made in $\mathscr{L}_{SM1}$ in eq. 10.1 followed by defining $y_r^0 = y^0$ and making an L_C transformation to a frame where $y_i^0 = 0$, and where J_2 is the absolute value of the Jacobian of the transformation from (X_r, X_i) to (y_r, y_i) coordinates:

$$J_2 = |\partial(X_r, X_i)/\partial(y_r, y_i)| \qquad (10.25)$$

We also choose gauges where $Y_r^0 = Y_i^0 = 0$. These transformations and gauge choices are discussed in detail in the following chapter. The Lagrangian terms $\mathscr{L}_C(X_r^{\mu}(y_r), \partial X_r^{\mu}(y_r)/\partial y_r^{\nu}, y_r)$ and $\mathscr{L}_C(X_i^{\mu}(y_i), \partial X_i^{\mu}(y_i)/\partial y_i^{\nu}, y_i)$ have the form:

$$\mathscr{L}_C = +\tfrac{1}{4} M_c^4 F^{\mu\nu}F_{\mu\nu} \qquad (10.26)$$

with

$$F_{\mu\nu} = \partial X_{\mu}/\partial y^{\nu} - \partial X_{\nu}/\partial y^{\mu} \qquad (10.27)$$

$$\equiv i\,(\partial Y_{\mu}/\partial y^{\nu} - \partial Y_{\nu}/\partial y^{\mu})/M_c^2$$

or defining

$$F_{Y\mu\nu} = (\partial Y_{\mu}/\partial y^{\nu} - \partial Y_{\nu}/\partial y^{\mu}) \qquad (10.28)$$

we see each Lagrangian assumes the form of the conventional electromagnetic Lagrangian:

$$\mathscr{L}_C = -\tfrac{1}{4} F_Y^{\mu\nu}F_{Y\mu\nu} \qquad (10.29)$$

The Lagrangian is supplemented with the following condition on all complexon fields $\Phi_{\ldots}$:

$$(\partial/\partial X_r^{\ k}(y_r))\ (\partial/\partial X_i^{\ k}(y_i))\Phi\ldots = 0 \qquad (10.30)$$

summed over k = 1, 2, 3. Non-complexon fields Ω... in our left-handed formulation satisfy the subsidiary condition:

$$\{(\partial/\partial X_r^{\ k}(y_r))\ (\partial/\partial X_i^{\ k}(y_i)) - ((\partial/\partial X_r^{\ k}(y_r))^2\ (\partial/\partial X_i^{\ m}(y_i))^2)^{1/2}\}\Omega\ldots = 0 \qquad (10.31)$$

summed over k = 1, 2, 3 and over m = 1, 2, 3. (See the discussions on subsidiary conditions of eqs. 3.82 – 3.86, section 3.8, and other sections.)

Feynman Propagators

The momentum space free field Feynman propagators G...(k) of all particles and ghosts in our Two-Tier Standard Model acquires a Gaussian factor exp(h(k)):

$$G\ldots(k) \rightarrow G\ldots(k)\ \exp(h(k)) \qquad (10.32)$$

so that all perturbation theory diagrams are finite. The result is a finite perturbative result in all calculations to any order in perturbation theory. Blaha (2005a) shows that Two-Tier theories are finite, Poincare covariant, and unitary. (See appendix 10-A for a more detailed description and Blaha (2005a) for a complete discussion.)

An example of the Two-Tier effect on propagators is the case of the Two-Tier photon propagator. The Two-Tier photon propagator[116] is:

$$iD_F^{\ TT}(y_1 - y_2)_{\mu\nu} = -i \int \frac{d^4p\ e^{-ip\cdot z}\ g_{\mu\nu} R(\mathbf{p}, z)}{(2\pi)^4\ (p^2 + i\varepsilon)} \qquad (10.33)$$

(since the imaginary parts can be taken to be zero: $y_{1i}^{\ \mu} - y_{2i}^{\ \mu} = 0$) where

$$z^\mu = y_{1r}^{\ \mu} - y_{2r}^{\ \mu} \qquad (10.34)$$

$$R(\mathbf{p}, \mathbf{z}) = \exp[-p^ip^j\Delta_{Tij}(z)/M_c^{\ 4}] \qquad (10.35)$$
$$= \exp\{-\mathbf{p}^2[A(v) + B(v)\cos^2\theta] / [4\pi^2 M_c^{\ 4}|\mathbf{z}|^2]\} \qquad (10.36)$$

with i, j = 1, 2, 3, and with $\Delta_{Tij}(z)$ the commutator of the positive frequency part $Y^+_k(y)$ and the negative frequency part $Y^-_k(y)$ of $Y_k(y)$:

---

[116] Blaha (2005a).

$$\Delta_{Tij}(z) = [Y^+_j(y_{1r}), Y^-_k(y_{2r})] = \int d^3k\, e^{ik\cdot(y_{1r} - y_{2r})} (\delta_{jk} - k_jk_k/\mathbf{k}^2)/[(2\pi)^3 2\omega_k] \quad (10.37)$$

and

$$v = |z^0|/|\mathbf{z}| \quad (10.38)$$

$$A(v) = (1 - v^2)^{-1} + .5v \ln[(v - 1)/(v + 1)] \quad (10.39)$$

$$B(v) = v^2(1 - v^2)^{-1} - 1.5v \ln[(v - 1)/(v + 1)] \quad (10.40)$$

$$\mathbf{p}\cdot\mathbf{z} = |\mathbf{p}|\, |\mathbf{z}| \cos\theta \quad (10.41)$$

with $|\mathbf{p}|$ being the length of a spatial vector **p**, $|\mathbf{z}|$ being the length of a spatial vector **z**, and with $|z^0|$ being the absolute value of z^0.

The gaussian factors R(**p**, z), which appear in all Two-Tier propagators, damp the large momentum behavior of all perturbation theory integrals producing a completely finite perturbation theory while giving the usual results of perturbation theory at energies that are small compared to the mass scale M_c.

Complexon Feynman Propagator

In the case of complexons the Two-Tier Feynman propagator differs from the non-complexon case by having an integration over imaginary spatial 3-momenta, a derivative of a delta function embodying the orthogonality of the real and imaginary 3-momenta, and two factors of R(**p**, z): one factor being R($\mathbf{p}_r$, z_r) and the other factor being R($\mathbf{p}_i$, z_i) (where the time components $z_r^0 = z^0$ and $z_i^0 = 0$ since there is only one real time coordinate[117]) thus providing large momentum convergence for both real and imaginary 3-momentum integrations.

For a scalar complexon particle we earlier found the Feynman propagator:

$$i\Delta_{CTF}(x - y) = \theta(x^+ - y^+)\langle 0|\phi_{CT}(x)\, \phi_{CT}(y)|0\rangle + \theta(y^+ - x^+)\langle 0|\phi_{CT}(y)\phi_{CT}(x)|0\rangle$$

$$= i\int d^4p_rd^3p_i(2\pi)^{-7}\delta'(\mathbf{p}_r\cdot\mathbf{p}_i/m^2)e^{-ip^+(x^- - y^-) - ip^-(x^+ - y^+) + i\mathbf{p}_\perp\cdot(\mathbf{x}_\perp - \mathbf{y}_\perp) - i\mathbf{p}_i\cdot(\mathbf{x}_i - \mathbf{y}_i)}/(p^2 + m^2 + i\varepsilon) \quad (4.60)$$

in conventional quantum field theory.

In the case of Two-Tier quantum field a scalar complexon particle has the the Feynman propagator

$$i\Delta_{CTFtt}(x - y) = i\int d^4p_rd^3p_i(2\pi)^{-7}\delta'(\mathbf{p}_r\cdot\mathbf{p}_i/m^2)\, R(\mathbf{p}_r, z_r)R(\mathbf{p}_i, z_i)\cdot$$

$$\cdot\, e^{-ip^+(x^- - y^-) - ip^-(x^+ - y^+) + i\mathbf{p}_\perp\cdot(\mathbf{x}_\perp - \mathbf{y}_\perp) - i\mathbf{p}_i\cdot(\mathbf{x}_i - \mathbf{y}_i)}/(p^2 - m^2 + i\varepsilon) \quad (10.42)$$

where the time components $z_r^0 = z^0$ and $z_i^0 = 0$, since there is only one time coordinate, and where $p^2. = p^{0\,2} - p_r^2 + p_i^2$.

---

[117] We can arrange for $z_i^0 = 0$ by making a L_C transformation to an inertial frame where z is real.

Propagators for other types of particles are similarly modified in the Two-Tier formalism (See Blaha 2005a).

10.3 Reason for Parity Violating Form of the Standard Model

The issue of parity violation[118] discovered 50 years ago is now "explained" by the (broken) L_C covariance of the Standard Model.

[118] T. D. Lee and C. N. Yang, Phys. Rev. **104**, 254 (1956).

11. The Three Generation Standard Model

11.1 Epitome of Our One Generation Standard Model

At this point in the derivation of our formulation of the Standard Model we have seen how we can generate exactly four species of spin ½ fields with real energies (and thus real times) by using the complex Lorentz group that we designate L_C. L_C and its inhomogeneous extension have the same generators, Poincaré algebra, as the inhomogeneous Lorentz group. However L_C boosts, and thus spatial rotations and translations, have complex parameters in general. The inhomogeneous Lorentz group has real parameters. The one generation Standard Model that we have constructed begins with L_C covariance. L_C covariance is broken to Lorentz covariance – the covariance we observe in the experimentally determined Standard Model.

We find four species of spin ½ particles using L_C boosts that are constrained by conditions that preserve the reality of the energy of a spinor boosted from a spinor for a particle at rest. (This requirement enables us to have one real time coordinate – a condition maintained under conventional Lorentz transformations.) Two types of spin ½ particle have real momenta: one type is a Dirac particle and the other is a tachyon particle. We identify these particles as leptons with the Dirac particle being the "electron" and the tachyon being the "neutrino." The other two types of spin ½ particle have complex 3-momenta with the Dirac-like particle being the "up quark" and the tachyonic particle being the "down quark" in this one generation Standard Model.

Further we found that the one-generation form of this Standard Model has parity-violating ElectroWeak interactions and SU(3) strong color interactions of a form similar to the experimentally determined Standard Model.[119]

The only experimentally certain, remaining, major Standard Model features to be derived are the existence of three generations with generation mixing; and the coupling constants, and vector boson, lepton and quark masses. In this chapter we propose a solution for the generations problem.

The set of postulates that led to the one-generation Standard Model plus the postulates to be presented are sufficient to derive our Standard Model and thus *we can claim a derivation of our Standard Model* which is identical to the usual Standard Model with the exception that our quarks have complex spatial 3-momenta (in such a way as to make the quarks "fuzzy").[120] The set of postulates required to derive the Standard Model will be specified and discussed later. This derivation takes as its model the derivation of geometry from Euclid's five postulates. The natural next step would then be the theoretical investigation of the nature and meaning of these postulates

---

[119] Augmented versions of the Standard Model with CP violation are not part of this interim formulation. The possibility of Majorana particles is also not considered although a Majorana variant could be constructed.

[120] We will address the issue of complex spatial 3-momenta in a subsequent footnote on partons. We believe complex quark momenta play a role in parton dynamics.

together with a derivation/calculation of the values of the masses, coupling constants and mixing angles which are, at present, only determined experimentally.

11.2 The Source of the Three Generations

The common approach to the explication of the three generations of leptons and quarks is to insert them manually in the Standard Model using a Higgs mechanism formulation with generational mixing. This approach, while phenomenologically acceptable, answers one mystery with another mystery at the same level of intellectual depth.

For example, we see leptonic mass terms inserted in the Standard Model Lagrangian with the form

$$\mathscr{L}_{\text{LepMass}} = \sum_{a,b} \bar{L}_a(\varphi/\rho_0)M_{ab}R_b + \text{c.c.} \tag{11.1}$$

where L and R embody three generations of left and right handed leptons, a and b are generation indices, φ is a Higgs boson that acquires a vacuum expectation value (unitary gauge) and ρ_0 is a constant. The matrix M_{ab} is more or less "pulled out of a hat" in the sense that the values of the matrix elements do not have a clear theoretical basis although various ad hoc symmetry schemes have been proposed. The origin of the generations and the form of the mass matrices is a mystery.

In this section we propose a somewhat deeper hypothesis by extending our four dimensional space-time. We postulate the existence of an additional compactified, complex 4-dimensional space that determines the mass matrices and mixing of the three generations.

We will assume the following postulate:

Postulate 11.1: The space-time of our experience is a subspace of a larger 8 complex dimensional space-time with a U(8) symmetry group. The U(8) symmetry is broken to a U(4)⊗U(4) symmetry.

Postulate 11.2: One of the U(4) symmetries is further broken to L_C symmetry from which the four fermion species are obtained as well as the parity violating form of the ElectroWeak sector of the Standard Model and the Strong interaction: color SU(3). L_C is broken to the Lorentz group of our 4-dimensional, real space-time.

Postulate 11.3: The four complex dimensions of the other U(4) symmetry are compactified. This U(4) symmetry, which has a set of sixteen generators equivalent to the sixteen U(2)⊗U(2) generators, is the source of four generations of each fermion species as well as the form of each species' mass matrix. Three of the generations are the known fermion generations.

Postulate 11.4: The fourth generation is assumed to be decoupled from the three known generations and to constitute Dark Matter.

Postulate 11.2 leads to tachyons as well as conventional Dirac particles. Both conventional Dirac particles and tachyons come in two varieties: those with effectively real 3-momenta and those with complex 3-momenta. Thus there are four species of fermions as stated in postulate 11.2. These species were described in detail in previous chapters.

There are also three generations of fermions, leptons and quarks. Thus each of the four species of fermions occurs in three "sub-species" which we call generations. However, there is apparently only one generation of vector bosons: ElectroWeak bosons and SU(3) color gluons. Nature thus appears to distinguish between fundamental fermions and vector bosons.

In the following sections we will provide a formulation for the three fermion generations based on postulate 11.3.

We note that the subsequent development <u>for fermions</u> in terms of U(2)⊗U(2) generators is simply a convenient formalism. The set of 16 U(2)⊗U(2) generators is a convenient equivalent to the 16 U(4) generators. Thus the U(4) symmetry is not broken to U(2)⊗U(2) for fermions.

11.3 Argument Leading to the Three Generation Postulate 11.3

In this section we examine the well-known example of mixing in ElectroWeak Theory of the SU(2)⊗U(1) vector bosons. In particular we consider the mass matrix of the vector bosons which leads us to suggest an U(2)⊗U(2) form for the mass matrix sector. This similarity provides the basis of Postulate 11.3 as an analogue of the U(2)⊗U(2) vector boson mass matrix mixing to the four generation fermion mixed mass matrix. The mixing of the three known generations forces the ElectroWeak and color SU(3) interactions to be the same for the three generations.

Then, a succeeding section suggests the one dimensional fermion representation constitutes a separate one generation Standard Model (with a different set of ElectroWeak bosons and color SU(3) gluons) whose particles only interact with the three generation Standard Model particles via gravitation. *This one generation sector is identified with Dark Matter and Dark Energy.*

The vector bosons that appear in ElectroWeak theory can be written as a column vector as a preliminary to considering the mass matrix and mixing amongst the ElectroWeak vector bosons. (This analysis is not incompatible with the usual Higgs formulation.) First we define the SU(3)⊗U(1) vector bosons with

$$\mathscr{W}_{31} = \begin{bmatrix} W^+ \\ W^0 \\ W^- \\ B \end{bmatrix} \quad \text{and} \quad \mathscr{W}_{\text{phys}} = \begin{bmatrix} W^+ \\ Z \\ W^- \\ A \end{bmatrix} = R\,\mathscr{W}_{31} \tag{11.2}$$

where we suppress the Lorentz indices on the vector boson fields for clarity, and the Weinberg rotation R is given by eq. 11.8.

The field equations for the massive vector bosons can be written in a matrix form based on eq. 11.2. The effective diagonal mass matrix term in the Lagrangian (after spontaneous breakdown) is

$$\mathscr{W}_{phys}{}^{\dagger} M_r \mathscr{W}_{phys} \tag{11.3}$$

where M_r is the diagonal mass matrix (eq. 11.9 below). If we transform the representation $\mathscr{W}_{31}$ to the U(2)⊗U(2) representation $\underline{2}\otimes\underline{2}$ with

$$\mathscr{W}_{31} = E\, \mathscr{W}_{22} \tag{11.4}$$

where E is given by eq. 11.15 and its inverse by eq. 11.17, and where $\mathscr{W}_{22}$ is the U(2)⊗U(2) representation $\underline{2}\otimes\underline{2}$ equivalent, then

$$M_{vb22} = EM_{vb}E^{-1} = ER^{-1}M_r RE^{-1} \tag{11.5}$$

where M_{vb} is the non-diagonal mass matrix:

$$M_{vb} = \begin{bmatrix} m_W & 0 & 0 & 0 \\ 0 & m_Z\cos^2\theta_W + m_A\sin^2\theta_W & 0 & (m_A - m_Z)\cos\theta_W\sin\theta_W \\ 0 & 0 & m_W & 0 \\ 0 & (m_A - m_Z)\cos\theta_W\sin\theta_W & 0 & m_Z\sin^2\theta_W + m_A\cos^2\theta_W \end{bmatrix} \tag{11.6}$$

where m_W is the mass of the W^+ and W^-, m_Z is the mass of the Z, m_A is the mass of the electromagnetic field A (the photon) which is, in fact, zero, and θ_W is the Weinberg angle. The Z and A fields are related to the W^0 and B fields by a Weinberg angle rotation (eq. 11.2). M_{vb22} is thus

$$M_{vb22} = \begin{bmatrix} m_W & 0 & 0 & 0 \\ 0 & a + m_W/2 & b & a - m_W/2 \\ 0 & b & c & b \\ 0 & a - m_W/2 & b & a + m_W/2 \end{bmatrix} \tag{11.7a}$$

where

$$a = ½(m_Z\cos^2\theta_W + m_A\sin^2\theta_W)$$
$$b = 2^{-½}(m_A - m_Z)\cos\theta_W\sin\theta_W \qquad (11.7b)$$
$$c = m_Z\sin^2\theta_W + m_A\cos^2\theta_W$$

The Weinberg rotation in matrix form is

$$R = \begin{bmatrix} 1 & 0 & 0 & 0 \\ 0 & \cos\theta_W & 0 & -\sin\theta_W \\ 0 & 0 & 1 & 0 \\ 0 & \sin\theta_W & 0 & \cos\theta_W \end{bmatrix} \qquad (11.8)$$

with $R^{-1} = R^{\dagger}$, and the diagonal, physical mass matrix is

$$M_r = \begin{bmatrix} m_W & 0 & 0 & 0 \\ 0 & m_Z & 0 & 0 \\ 0 & 0 & m_W & 0 \\ 0 & 0 & 0 & m_A \end{bmatrix} \qquad (11.9)$$

is the diagonal mass matrix. (Note: $m_A = 0$.)

At this point we note that ElectroWeak theory was forced to have an SU(2)⊗U(1) symmetry to avoid the tachyon neutrino cascading decay to negative energy states. Initially an SU(2)⊗SU(2) symmetry appeared to be the appropriate symmetry choice (item D in section 7.3). In the present situation the Weinberg rotation matrix also has a constraint. The superselection rule for electric charge does not allow a state to be a superposition of states of differing electric charge. Thus the simplicity of the rotation matrix R.

However, despite these restrictions we find the "unrotated", non-diagonal mass matrix M_{vb} (eq. 11.4) to be a sum of U(2)⊗U(2) matrices showing that the fundamental symmetry of this mixed mass matrix is U(2)⊗U(2) although various physical requirements force the <u>vector fields</u> to have SU(2)⊗U(1) symmetry.

These considerations lead us to propose that the four fermion species have a further symmetry U(2)⊗U(2) that leads to the three observed generations of fermions. It gives a 4-vector to each fermion species just as a 4-vector (eq. 11.20 below) is needed for the vector bosons. Vector bosons must be in the regular representation of the group SU(2)⊗U(1). Thus three W's and one B vector meson. On the other hand fermions can be in the fundamental representations <u>2</u> of their SU(2) groups. So the U(2)⊗U(2) fermion symmetry is the product of spinor representations and can be represented as a 4-column.

U(2)⊗U(2) and the Non-Diagonal Vector Boson Mass Matrix

In this subsection we will show the SU(2)⊗U(1) vector boson mass matrix requires an expansion in U(2)⊗U(2) matrices. First we define a 4-dimensional U(2)⊗U(2) representation as a product of an "unprimed" and "primed" spinor representation.

$$\underline{2} \times \underline{2}' \rightarrow \underline{4}$$

where the 4-columns of the $\underline{4}$ representation have the form:

$$s_4 = \begin{bmatrix} s^{11} \\ s^{12} \\ s^{21} \\ s^{22} \end{bmatrix} \qquad (11.10)$$

The superscripts are spinor indices. The first index is that of the unprimed representation and the second index is that of the primed representation. Note the indices of the prime representation change more rapidly than the unprimed representation as one goes down the column. Based on this convention for the 4-dimensional representation the spinor matrices combine to produce 4×4 matrices which we will write in terms of the Pauli matrices σ_i using the functional notation:

$$\sigma_i\sigma'_k = \sigma_i(\sigma'_k) \qquad (11.11)$$

where the prime on the Pauli matrix indicates it is for the primed U(2). (Note the Pauli matrices are <u>not</u> matrix multiplied in eq. 11.11. Rather they represent the matrix elements $[\sigma_i]_{ab}[\sigma'_k]_{cd}$ where a, b, c, and d equal 1 or 2.) The functional form in eq. 11.11 implements the concept that σ'_k appears in each of the four matrix elements of σ_i multiplied by the value of the σ_i matrix element. For example

$$\sigma_2\sigma'_3 = \sigma_2(\sigma'_3)$$

$$= \begin{bmatrix} 0 & 0 & -i & 0 \\ 0 & 0 & 0 & i \\ i & 0 & 0 & 0 \\ 0 & -i & 0 & 0 \end{bmatrix} \qquad (11.12a)$$

We also define $\sigma_0 = I$, the two-dimensional identity matrix, and use the notation $\sigma_0(\sigma'_i) = \sigma_0\sigma'_i$. For example,

$$\sigma_0(\sigma'_3) = \sigma_0\sigma'_3$$

$$= \begin{bmatrix} 1 & 0 & 0 & 0 \\ 0 & -1 & 0 & 0 \\ 0 & 0 & 1 & 0 \\ 0 & 0 & 0 & -1 \end{bmatrix} \qquad (11.12b)$$

Similarly we define $\sigma_0(\sigma'_0) = \sigma_0\sigma'_0 = 4 \times 4$ identity matrix, and $\sigma_i(\sigma'_0) = \sigma_i\sigma'_0$. In this notation the non-diagonal, vector boson mass matrix expands in U(2)⊗U(2) matrices as

$$\begin{aligned} M_{vb22} = {} & \tfrac{1}{4}(2m_W + m_A + m_Z)\sigma_0(\sigma'_0) + \tfrac{1}{4}(m_A - m_Z)\cos(2\theta_W)\sigma_0(\sigma'_3) + \\ & + \tfrac{1}{4}[m_W - m_Z\sin^2\theta_W - m_A\cos^2\theta_W][\sigma_3(\sigma'_0) + \sigma_3(\sigma_3')] + \\ & + \tfrac{1}{4}2^{-\frac{1}{2}}(m_A - m_Z)\sin(2\theta_W)[\sigma_0(\sigma'_1) - \sigma_3(\sigma_1')] + \\ & + \tfrac{1}{4}[m_Z\cos^2\theta_W + m_A\sin^2\theta_W - m_W][\sigma_1(\sigma'_0) - \sigma_1(\sigma_3')] + \\ & + \tfrac{1}{4}2^{-\frac{1}{2}}(m_A - m_Z)\sin(2\theta_W)[\sigma_1(\sigma'_1) + \sigma_2(\sigma_2')] \end{aligned} \qquad (11.13)$$

The necessary appearance of factors such as $\sigma_0(\sigma'_3)$ and $\sigma_0(\sigma_0')$ in the expansion of the mass matrix indicates a U(2)⊗U(2) group structure is needed in the mass sector. (The expression for M_{vb22} is expressed in terms of U(2)⊗U(2) generators because the identity matrices appear in its expansion.) One cannot just get by with U(1) and SU(2) in this case. The doubling of the Weinberg angle in eq. 11.13 is suggestive of the doubling of the boost parameter when a spin ½ fermion is boosted (as we saw in earlier chapters).

11.4 Fermion Generations

In the previous section we established that the ElectroWeak vector boson mass matrix required an expansion in U(2)⊗U(2) matrices despite the fact that the ElectroWeak symmetry was forced to be SU(2)⊗U(1) (rather than SU(2)⊗SU(2)) to avoid difficulties with the tachyonic fermion sector.

We now turn to a consideration of fermions. A one generation Standard Model was developed in earlier chapters that captured many of the features of the conventional Standard Model developed through over fifty years of experimental and theoretical effort. In this section we will use Postulate 11.2 to transform the one generation Standard Model into a three generation Standard Model (plus possibly a hitherto unobserved one generation Standard Model that might describe dark matter and dark energy).

We now assume (Postulate 11.2) that each of the four fermion species has a further U(2)⊗U(2) symmetry similar to, but differing from, the U(2)⊗U(2) found in the ElectroWeak mass matrix expansion. We use the notation developed in eqs. 11.10 – 11.12. Thus each fermion species become a 4-column field of the form:

$$\Psi = [\psi_h] = \begin{bmatrix} \psi_{11} \\ \psi_{12} \\ \psi_{21} \\ \psi_{22} \end{bmatrix} \tag{11.14}$$

where h = 1, 2, 3, 4. Thus the fermion species are:

Ψ_E = the four charged leptons
Ψ_v = the four neutrinos leptons
Ψ_U = the four up-type quarks
Ψ_D = the four down-type quarks

We can transform this representation of U(2)⊗U(2) into a $\underline{3}$ + $\underline{1}$ pair of irreducible representations by forming symmetric combinations of the indices (the $\underline{3}$ representation) and an antisymmetric combination of the indices (the $\underline{1}$ representation). The $\underline{3}$ representation will consist of the three generations of each species.

A matrix implementing the transformation to a $\underline{3}$ + $\underline{1}$ representation is

$$E = \begin{bmatrix} 1 & 0 & 0 & 0 \\ 0 & 2^{-1/2} & 2^{-1/2} & 0 \\ 0 & 0 & 0 & 1 \\ 0 & 2^{-1/2} & -2^{-1/2} & 0 \end{bmatrix} \tag{11.15}$$

It transforms Ψ into the $\underline{3}$ + $\underline{1}$ representation:

$$E\Psi = \begin{bmatrix} \psi_{12} \\ 2^{-1/2}(\psi_{12} + \psi_{21}) \\ \psi_{22} \\ 2^{-1/2}(\psi_{12} - \psi_{21}) \end{bmatrix} \tag{11.16}$$

The top three elements in the column form a $\underline{3}$ representation that we identify with the three generations of leptons and quarks. The fourth element is a $\underline{1}$ representation that we discuss in the following section.

The inverse of E is its Hermitian conjugate:

$$E^{-1} = E^{\dagger} = \begin{bmatrix} 1 & 0 & 0 & 0 \\ 0 & 2^{-1/2} & 0 & 2^{-1/2} \\ 0 & 2^{-1/2} & 0 & -2^{-1/2} \\ 0 & 0 & 1 & 0 \end{bmatrix} \qquad (11.17)$$

We now take the fermion mass matrix of eq. 11.1 for one of the four species and expand it in terms of U(2)⊗U(2) matrices. In order to accomplish this task we must add a fourth row and column to the mass matrix M_{ab}. In the absence of significant experimental data for a fourth generation of fermions we choose the form

$$M_f = \begin{bmatrix} m_{11} & m_{12} & m_{13} & 0 \\ m_{21} & m_{22} & m_{23} & 0 \\ m_{31} & m_{32} & m_{33} & 0 \\ 0 & 0 & 0 & m_{44} \end{bmatrix} \qquad (11.18)$$

where the upper 3 × 3 submatrix is M_{ab} and m_{44} is the mass of a fourth generation fermion. The zeroes in rows and columns reflect the provisional assumption of no mixing between the three known generations and the fourth generation.

We assume the three generation mass matrix implement a $\underline{3}$ representation of U(2)⊗U(2) and the fourth generation matrix element implements a $\underline{1}$ representation of U(2)⊗U(2) (in a manner analogous to the vector bosons in eq. 11.2) Thus we must use E to transform M_f into the $\underline{2} + \underline{2}$ representation that better brings out the U(2)⊗U(2) structure.

$$M_{f22} = E^{-1} M_f E \qquad (11.19)$$

from eq. 11.2 and eq. 11.16. We find

$$M_{f22} = \begin{bmatrix} m_{11} & am_{12} & am_{12} & m_{13} \\ am_{21} & a^2(m_{22} + m_{44}) & a^2(m_{22} - m_{44}) & am_{23} \\ am_{21} & a^2(m_{22} - m_{44}) & a^2(m_{22} + m_{44}) & am_{23} \\ m_{31} & am_{32} & am_{32} & m_{33} \end{bmatrix} \qquad (11.20)$$

where $a = 2^{-1/2}$.

We can expand M_{f22} in terms of U(2)⊗U(2) matrices just as we did for the vector boson mass matrix in eq. 11.13. To that end we define the matrices of

U(2)⊗U(2) in terms of Pauli matrices and also Dirac matrices. The notation of eq. 11.11 and 11.12 extends to all sixteen U(2)⊗U(2) matrices:

$$\sigma_i\sigma'_k = \sigma_i(\sigma'_k) \tag{11.21}$$

where we define the 2 × 2 identity matrices:

$$\sigma_0 = I \qquad \text{and} \qquad \sigma'_0 = I' \tag{11.22}$$

Then we see that $\sigma_a(\sigma'_b)$ for a, b = 0, 1, 2, 3 can be expressed in terms of Dirac matrices as

$$\begin{aligned}
\sigma_0(\sigma'_k) &= ½\varepsilon_{ijk}\sigma^{ij} = ¼\, i\varepsilon_{ijk}[\gamma^i,\gamma^j] \\
\sigma_1(\sigma'_k) &= \gamma^0\gamma^k \\
\sigma_2(\sigma'_k) &= -i\gamma^k \\
\sigma_3(\sigma'_k) &= ½\varepsilon_{ijk}\gamma^0\sigma^{ij} \\
\sigma_1(\sigma'_0) &= \gamma^5 \\
\sigma_2(\sigma'_0) &= -i\gamma^5\gamma^0 \\
\sigma_3(\sigma'_0) &= \gamma^0
\end{aligned} \tag{11.23}$$

for k = 1, 2, 3. Furthermore $\sigma_i(\sigma'_k)$ is Hermitian and its square is the 4×4 identity matrix:

$$\begin{aligned}
[\sigma_i(\sigma'_k)]^\dagger &= \sigma_i(\sigma'_k) \\
[\sigma_i(\sigma'_k)]^2 &= \sigma_0(\sigma'_0)
\end{aligned} \tag{11.23a}$$

for i, k = 0,1,2,3. We expand M_{f22} in terms of U(2)⊗U(2) matrices using the below identity.

If M is a 4 × 4 matrix with row index a and column index d, then M can be expanded in terms of $\sigma_c(\sigma'_d)$ for c, d = 0, 1, 2, 3

$$M = \sum_{c,d} \alpha_{cd}\sigma_c(\sigma'_d) \tag{11.24}$$

with

$$\alpha_{cd} = ¼ \text{ Tr } \sigma_c(\sigma'_d)M \tag{11.25}$$

where Tr indicates trace. We find

$$M_{f22} = \sum_{c,d} \alpha_{cd}\sigma_c(\sigma'_d) \tag{11.26}$$

where the expansion coefficients matrix α is

$$\alpha = \tfrac{1}{4}\begin{bmatrix} (m_{11}+m_{22}+m_{33}+m_{44}) & a(m_{12+}+m_{23+}) & ia(m_{12-}+m_{23-}) & (m_{11}-m_{33}) \\ a(m_{12+}+m_{23+}) & (m_{13+}+m_{22}-m_{44}) & im_{13-} & a(m_{12+}-m_{23+}) \\ ia(m_{12-}+m_{23-}) & im_{13-} & (m_{13-}+m_{22}-m_{44}) & ia(m_{12-}-m_{23-}) \\ (m_{11}-m_{33}) & a(m_{12+}-m_{23+}) & ia(m_{12-}-m_{23-}) & (m_{11}-m_{22}+m_{33}-m_{44}) \end{bmatrix}$$

(rows indexed by c ↓, columns by d →)

(11.27)

where $a = 2^{-1/2}$, and $m_{ik\pm} = m_{ik} \pm m_{ki}$. Thus α is symmetric. The ten independent components (the upper right part) of α determine the ten independent components of the mass matrix M_f (eq. 11.18).

The symmetry of α enables us to write

$$M_{f22} = \sum_c \alpha_{cc}\Omega_{+cc} + 2\sum_{c<d} \alpha_{cd}\Omega_{+cd} \qquad (11.28)$$

in terms of the ten symmetric matrices: (The mass matrix components m_{ik} can be real or complex.)

$$\Omega_{+cd} = \tfrac{1}{2}[\sigma_c(\sigma'_d) + \sigma_d(\sigma'_c)] \qquad (11.29)$$

($c \leq d$) as well as in terms of $\sigma_i(\sigma'_k)$ (eq. 11.24). It also will be important to define the six antisymmetric matrices ($c \leq d$)

$$\Omega_{-cd} = \tfrac{1}{2}[\sigma_c(\sigma'_d) - \sigma_d(\sigma'_c)] \qquad (11.30)$$

11.5 Space-Time Group Structure Implied by the Multi-Generation Standard Model

We now show that linear combinations of the matrices defined by eqs. 11.29 and 11.30 form a spinor representation of the generators of the Lorentz group, the U(3) group and a U(1) group: sixteen matrices with six matrices serving as Lorentz group generators, one matrix generator of U(1) and nine matrices serving as U(3) generators.

Lorentz Group Generators

Six matrices defined by eqs. 11.29 and 11.30 can be used to define boost and rotation generator matrices of the Lorentz group. We define the Lorentz boost generators (in our notation):

$$K_i = i\Omega_{-0i} \qquad (11.31)$$

and the three rotation generators

$$J_i = \Omega_{+0i} \qquad (11.32)$$

(Note the Hermitian conjugate relations $K_i^\dagger = -K_i$, and $J_i^\dagger = J_i$ are the same as those of the corresponding Lorentz generators in our notation.)

Their commutation relations are

$$[K_i, K_j] = -i\varepsilon_{ijk}J_k$$
$$[J_i, J_j] = i\varepsilon_{ijk}J_k \qquad (11.33)$$
$$[J_i, K_j] = i\varepsilon_{ijk}K_k$$

U(1) Generator

The matrix $\sigma_0(\sigma'_0) = \Omega_{+00}$ serves as the generator of a U(1) symmetry.

U(3) Generators

The nine matrices Ω_{-ik} and Ω_{+ik} (for i, k = 1, 2, 3), together with linear combinations of Ω_{+0k} and Ω_{-0k}, form a spinor representation of the U(3) generators.

Gell-Mann SU(3) Generators

$$F_1 = \tfrac{1}{2}(\Omega_{+13} - \Omega_{-13} + \Omega_{+01} + \Omega_{-01})$$
$$F_2 = \tfrac{1}{2}(\Omega_{+23} - \Omega_{-23} + \Omega_{+02} + \Omega_{-02})$$
$$F_3 = \tfrac{1}{2}(\Omega_{+33} + \Omega_{+03} + \Omega_{-03})$$
$$F_4 = \tfrac{1}{2}(\Omega_{+10} + \Omega_{-10} + \Omega_{+13} + \Omega_{-13}) \qquad (11.34)$$
$$F_5 = \tfrac{1}{2}(\Omega_{+20} + \Omega_{-20} + \Omega_{+23} + \Omega_{-23})$$
$$F_6 = \tfrac{1}{2}(\Omega_{+11} + \Omega_{+22})$$
$$F_7 = \tfrac{1}{2}(\Omega_{+21} + \Omega_{-21} - \Omega_{+12} - \Omega_{-12})$$
$$F_8 = (2\sqrt{3})^{-1}(-\Omega_{+00} + 2\Omega_{+33} + \Omega_{+03} - 5\Omega_{-03})$$

The eight SU(3) generators are block diagonal with zeroes in the fourth row and fourth column. They satisfy the standard SU(3) commutation relations:

$$[F_i, F_j] = i\, f_{ijk}F_k \qquad (11.35)$$

where f_{ijk} are the totally antisymmetric SU(3) structure constants for i, j, k = 1, 2, …,8. On the other hand the Lorentz generators do not commute with the F_i.

$$[F_i, K_j] \neq 0 \qquad (11.36)$$
$$[F_i, J_j] \neq 0$$

An additional U(1) Generator needed for U(3)

$$F_9 = \tfrac{1}{4}(3\Omega_{+00} - \Omega_{+33} + 2\Omega_{+03}) \qquad (11.37)$$

This generator consists of a 3×3 identity matrix with the element in the fourth row and column having the value 0.

We note also

$$[F_i, F_9] = 0 \qquad (11.38)$$
$$[F_9, K_j] \neq 0$$
$$[F_9, J_j] \neq 0$$

for j = 1, 2, 3.

As a result of the above considerations we see that all fermion species <u>*mass matrices can be expressed as a linear combination of U(3) generator matrices*</u> *(eq. 11.24).*

11.6 U(8) Symmetry

Having shown that we can decompose the set of sixteen 4 × 4 matrices into generators for the Lorentz group, U(3) and U(1) we can now understand the nature of the overall "space-time" symmetry and its breakdown into the space-time of our experience and into internal degrees of freedom that we call generations.

<u>***Summary***</u>

Space-time has a U(8) symmetry with a coordinate space of 8 complex dimensions. This symmetry is broken to U(4)⊗U(4). One of the U(4) groups is then further broken to the complex Lorentz group L_c which is then further broken to the Lorentz group of our space-time. The other U(4) group is compactified in some manner. It creates four generations of each fermion species and generates particle mass matrices. Three of these generations have been experimentally detected. The fourth generation does not interact with the three known generations. The fourth generation has its own ElectroWeak interactions and color SU(3) interactions mediated by different vector bosons from the three generation sector. We identify fourth generation fermions with dark matter and fourth generation photons with dark energy.

Nature seems to have taken the L_c group from one U(4) subgroup and used it "to create" the fermion species. And it used the other U(4) group to create the generations and their mixings.

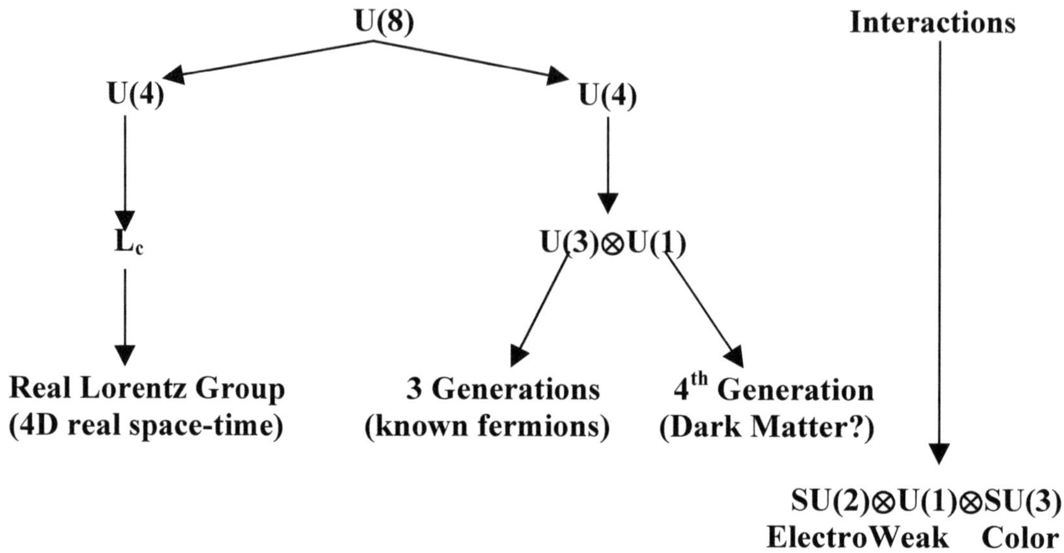

Figure 11.1. Group hierarchy of extended "space-time" leading to the Standard Model, and the Standard Model generations and interactions. Nature takes L_C from one U(4) subgroup to define the space-time structure, and U(3)⊗U(1) from the other U(4) to define the generations and their mixing. Thus the generators of L_C commute with the generators of the generations subgroup U(3)⊗U(1) to avoid mixing the "space-time group" and the generations internal group (to avoid the type of mistake made in the SU(6) theory of the 1960's).

Fig. 11.1 displays the pattern of symmetry breakdown the above summary embodies. The mechanism(s) that implements the pattern of breakdown is not evident and reflects a level of understanding that is as yet unexplored. The compactification of the generation U(4) dimensions can be "explained" in two possible ways: 1) it may take the form of compactifications used in SuperString theories; or 2) it may reflect the manner in which the universe evolved from the Big Bang. The second mechanism is described in a theory of the Big Bang in complex space-time based on a generalized complex Robertson-Walker metric developed by Blaha (2004). In this theory the real part of space-time expanded but the imaginary part of space-time remained small with a radius of the order of the Planck length. In any case the only remnants of the compactified U(4) are the generations themselves, and the mass matrices of the four fermion species with mixing between the three known generations.

The U(8) theory is not inconsistent with a form of Higgs mechanism to generate the mass matrices. However, it is also possible that another mechanism is at play here and we consider a mass term generated by an unknown mechanism from the U(3)×U(1) subgroup of U(4). We examine this possibility in the next section.

11.7 Mass Matrices

In this section we will show two ways of obtaining the mass matrices of the four fermion species. Since the existence of Higgs particles is at present still an open question experimentally we will first show a method to obtain the mass matrices using the U(3)×U(1) subgroup of U(4). Then we will examine an alternate approach based on the Higgs mechanism.

Mass Matrices from the U(3) ×U(1) Subgroup of U(4)

We will consider free dynamical equations for fermions. The result generalizes directly to the case of interacting fermions.

First consider the ordinary free Dirac equation:

$$(i\gamma^\mu \partial/\partial x^\mu - m)\psi(x) = 0 \tag{11.39}$$

which can be rewritten in terms of the Hermitian matrices Ω_{+ik} and Ω_{-ik} as

$$[(\Omega_{+30} + \Omega_{-30})i\partial/\partial x^0 + (\Omega_{+2k} + \Omega_{-2k})\partial/\partial x^k - m]\psi(x) = 0 \tag{11.40}$$

or in terms of $\sigma_i(\sigma'_k)$

$$[\sigma_3(\sigma'_0)i\partial/\partial x^0 + \sigma_2(\sigma'_k)\partial/\partial x^k - m]\psi(x) = 0 \tag{11.41}$$

(The "missing" i in the spatial derivatives term is embedded within $\Omega_{+2k} + \Omega_{-2k}$ and $\sigma_2(\sigma'_k)$.)

We now replace the mass term with a U(3)×U(1) matrix expression to yield the generalized equation (and in a corresponding way for the three other species' dynamical equations):

$$[i\gamma^\mu \partial/\partial x^\mu \Omega_{+00} + \Omega_{+00}{}^\gamma \exp(i\sum_{k=1}^{9} a_k F_k + a_0 \Omega_{+00})]\psi(x, z) = 0 \tag{11.42}$$

where Ω_{+00} is the identity matrix in "generation" space, where $\Omega_{+00}{}^\gamma$ is the identity matrix in the spinor space of the Dirac matrices γ^μ, where a_k are a set of nine real constants and the F_k matrices are given by eq. 11.34 and 11.37. where a_0 is a real constant, and where the bare mass term is absorbed within the exponential. The origin of the constants a_k and a_0 is not known.

Eq. 11.42 becomes

$$[i\gamma^\mu \partial/\partial x^\mu \Omega_{+00} - \Omega_{+00}{}^\gamma M_{f22}]\psi_0(x) = 0 \tag{11.43}$$

after evaluating the exponential expression where M_{f22} is given by eq. 11.28.

Mass Matrices Based on the Higgs Mechanism

In this section we consider a more conventional Standard Model approach based on the Higgs Mechanism. To that end we define Higgs boson second rank tensor fields $\varphi_{ab}(x)$ (symmetric in a and b) for each fermion species that has a potential of the form

$$V(\varphi) = (\varphi_{ab}\varphi_{ab} - 2\varphi_{ab}\alpha_{ab} + \alpha_{ab}\alpha_{ab})^2 \tag{11.44}$$

with implied summations over a and b. The fermion interaction term has the form

$$\overline{\psi}(x)[\sum_c \varphi_{cc}\Omega_{+cc} + 2\sum_{c<d} \varphi_{cd}\Omega_{+cd}]\psi(x) \tag{11.45}$$

which becomes the mass matrix term (plus other terms)

$$\overline{\psi}(x)M_{f22}\psi(x) \tag{11.46}$$

upon implementation of spontaneous symmetry breaking via the Higgs Mechanism. Again the implementation of the Standard Model formalism within the Two-Tier quantum field theory formalism eliminates all divergences from the theory and thus makes renormalization via t'Hooft's approach unnecessary.

Diagonalizing the Mass Matrices

After obtaining the mass matrices for each species we can proceed to diagonalize them in a two step process. First we make each species block diagonal with a "three generation" block and a fourth generation block. Then the three generation block can be diagonalized in the well-known manner. It is, of course, sensible to combine these diagonalizations as we will see. We will denote the four fermion species as $\underline{E}$ (charged leptons), $\underline{N}$ (neutrinos), $\underline{U}$ (up-type quarks) and $\underline{D}$ (down-type quarks). Then the fermion mass terms are:

$$\mathscr{L}_{\text{FermionMasses}} = \overline{\Psi}_{\underline{E}L}M_{f22\underline{E}}\Psi_{\underline{E}R} + \overline{\Psi}_{\underline{N}L}M_{f22\underline{N}}\Psi_{\underline{N}R} +$$

$$+ \overline{\Psi}_{\underline{U}L}M_{f22\underline{U}}\Psi_{\underline{U}R} + \overline{\Psi}_{\underline{D}L}M_{f22\underline{D}}\Psi_{\underline{D}R} + \text{c.c.} \tag{11.47}$$

where the generation indices are not shown and c.c. represents the complex conjugate of the displayed terms.

We now proceed to diagonalize by defining the physical mass eigenstates as

$$\begin{aligned} \Psi_{\underline{S}Lp} &= A_{\underline{S}L}E\Psi_{\underline{S}L} \\ \Psi_{\underline{S}Rp} &= A_{\underline{S}R}E\Psi_{\underline{S}L} \end{aligned} \tag{11.48}$$

for $\underline{S}$ = $\underline{E}$, $\underline{N}$, $\underline{U}$, $\underline{D}$ where the subscript p indicates the physical mass eigenstates and each field, denoted $\Psi_{\underline{S}Lp}$ and $\Psi_{\underline{S}Rp}$ is a column vector of four generations of the respective species. The matrices $A_{\underline{S}L}$ and $A_{\underline{S}R}$, and their inverses, are augmented by a fourth row and column with the only non-zero element (with value one) being the diagonal element where the fourth row and column intersect. The physical, diagonal mass matrices $M_{\underline{S}p}$ for the fermion species are then

$$M_{\underline{S}p} = A_{\underline{S}L} E M_{f22\underline{S}} E^{-1} A_{\underline{S}R}^{-1} \tag{11.49}$$

where $A_{\underline{S}L}$ and $A_{\underline{S}R}$ are non-singular for all $\underline{S}$. Implementing the transformation throughout the fermion Lagrangian causes the charge changing ElectroWeak currents to have the form:

$$\bar{\Psi}_{\underline{S}Lp} C_{\underline{ST}} \gamma^{\mu} \Psi_{\underline{T}Lp} \tag{11.50}$$

where

$$C_{\underline{ST}} = A_{\underline{S}L} E E^{-1} A_{\underline{T}L}^{-1} = A_{\underline{S}L} A_{\underline{T}L}^{-1} \tag{11.51}$$

The combinations of $\underline{S}$ and $\underline{T}$ in eq. 11.51 that terms in the ElectroWeak currents have are $\underline{S}$ = $\underline{N}$ and $\underline{T}$ = $\underline{E}$; $\underline{S}$ = $\underline{E}$ and $\underline{T}$ = $\underline{N}$; $\underline{S}$ = $\underline{D}$ and $\underline{T}$ = $\underline{U}$; $\underline{S}$ = $\underline{U}$ and $\underline{T}$ = $\underline{D}$. The strong interaction, electromagnetic and neutral currents are unchanged by these transformations. We note

$$C_{\underline{ST}} C_{\underline{ST}}^{\dagger} = I \tag{11.52}$$

$$|\det C_{\underline{ST}}| = 1 \tag{11.53}$$

Is $A_{\underline{S}L}$ Different from $A_{\underline{S}R}$?

The U(4) subgroup that is the source of fermion generations has no relation to the U(4) subgroup that breaks down to Lc and then to the Lorentz group of our space-time. Therefore *it is reasonable to assume that parity violation does not crop up in the dynamics of the U(4) generations sector* since the breakdown of U(8) to U(4)⊗U(4) is the result of a parity neutral mechanism. Consequently it appears reasonable to assume

$$A_{\underline{S}R} = A_{\underline{S}L} \tag{11.54}$$

For $\underline{S}$ = $\underline{E}$, $\underline{N}$, $\underline{U}$, $\underline{D}$. Under this assumption the physical (diagonal) mass matrix for each fermion species simplifies to

$$M_{\underline{S}p} = A_{\underline{S}L} E M_{f22\underline{S}} E^{-1} A_{\underline{S}L}^{-1} \tag{11.55}$$

Neutrino Mass Matrix

We now consider the case of the neutrino mass matrix. In this case the four generations of physical left-handed neutrinos are given by

$$\Psi_{\underline{N}Lp} = A_{\underline{N}L} E \Psi_{\underline{N}L} \tag{11.56}$$

and

$$C_{\underline{EN}} = A_{\underline{E}\,L} A_{\underline{N}L}^{-1} \tag{11.57}$$

The neutrino mass matrix is

$$M_{\underline{N}p} = A_{\underline{N}L} E M_{f22\underline{N}} E^{-1} A_{\underline{N}L}^{-1} \tag{11.58}$$

using eq. 11.55.
Experimental efforts[121] have determined the $C_{\underline{EN}}$ matrix to be approximately

$$C_{\underline{EN}} = \begin{bmatrix} -(2/3)^{1/2} & (1/3)^{1/2} & 0 & 0 \\ (1/6)^{1/2} & (1/3)^{1/2} & (1/2)^{1/2} & 0 \\ (1/6)^{1/2} & (1/3)^{1/2} & -(1/2)^{1/2} & 0 \\ 0 & 0 & 0 & 1 \end{bmatrix} \tag{11.59}$$

We note again that we assume no mixing between the three known generations and the unknown fourth generation.

In order to have a concrete example we now assume $A_{\underline{E}L} \approx I$ (the identity matrix) based on the physical notion that the charged leptons, which have vastly different masses, do not mix very much. Therefore

$$C_{\underline{EN}} \approx A_{\underline{N}L}^{-1} \tag{11.60}$$

and thus

$$A_{\underline{N}L}^{-1} \approx \begin{bmatrix} -(2/3)^{1/2} & (1/3)^{1/2} & 0 & 0 \\ (1/6)^{1/2} & (1/3)^{1/2} & (1/2)^{1/2} & 0 \\ (1/6)^{1/2} & (1/3)^{1/2} & -(1/2)^{1/2} & 0 \\ 0 & 0 & 0 & 1 \end{bmatrix} \tag{11.61}$$

Substituting eq. 11.61 and $A_{\underline{N}L}$ (its Hermitian conjugate) in

$$M_{f22\underline{N}} = E^{-1} A_{\underline{N}L}^{-1} M_{\underline{N}p} A_{\underline{N}L} E \tag{11.62}$$

[121] E. Ma, arXiv:hep-ph/0612013 (2006) and references therein; P. F. Harrison, D. H. Perkins, and W. G. Scott, Phys. Lett. **B530**, 167 (2002).

(which is obtained from eq. 11.58) we find $M_{f22\underline{N}}$ is approximately

$$\begin{bmatrix} \frac{2}{3}e + \frac{1}{3}\mu & \frac{1}{6}\sqrt{2}(\mu - e) & \frac{1}{6}\sqrt{2}(\mu - e) & \frac{1}{3}(\mu - e) \\ \frac{1}{6}\sqrt{2}(\mu - e) & \frac{1}{12}e + \frac{1}{6}\mu + \frac{1}{4}\tau + \frac{1}{2}x & \frac{1}{12}e + \frac{1}{6}\mu + \frac{1}{4}\tau - \frac{1}{2}x & \sqrt{2}(\frac{1}{12}e + \frac{1}{6}\mu - \frac{1}{4}\tau) \\ \frac{1}{6}\sqrt{2}(\mu - e) & \frac{1}{12}e + \frac{1}{6}\mu + \frac{1}{4}\tau - \frac{1}{2}x & \frac{1}{12}e + \frac{1}{6}\mu + \frac{1}{4}\tau + \frac{1}{2}x & \sqrt{2}(\frac{1}{12}e + \frac{1}{6}\mu - \frac{1}{4}\tau) \\ \frac{1}{3}(\mu - c) & \sqrt{2}(\frac{1}{12}e + \frac{1}{6}\mu - \frac{1}{4}\tau) & \sqrt{2}(\frac{1}{12}e + \frac{1}{6}\mu - \frac{1}{4}\tau) & \frac{1}{6}e + \frac{1}{3}\mu + \frac{1}{2}\tau \end{bmatrix} \tag{11.63}$$

with rows and columns numbered from zero to four, and with e representing the mass of the electron neutrino, μ representing the mass of the μ neutrino, τ representing the mass of the τ neutrino, and x represents the fourth generation neutrino.

The neutrino α matrix, $\alpha_{\underline{N}}$ (eq. 11.27), then has form (with c, d = 0, 1, 2, 3):

c d→
↓

$$\begin{bmatrix} e + \mu + \tau + x & a_+ & 0 & -7e/12 + \frac{1}{6}\mu - \frac{1}{4}\tau - \frac{1}{2}x \\ a_+ & \frac{1}{3}\sqrt{2}(\mu - e) - \frac{1}{12}e - \frac{1}{6}\mu - \frac{1}{4}\tau + \frac{1}{2}x & 0 & a_- \\ 0 & 0 & -\frac{1}{12}e - \frac{1}{6}\mu - \frac{1}{4}\tau + \frac{1}{2}x & 0 \\ -7e/12 + \frac{1}{6}\mu - \frac{1}{4}\tau - \frac{1}{2}x & a_- & 0 & \frac{1}{2}e - \frac{1}{2}\tau \end{bmatrix} \tag{11.64}$$

<u>divided by 4</u> where

$$\begin{aligned} a_\pm &= (m_{12+} \pm m_{23+})/\sqrt{2} \\ &= \tfrac{1}{3}(\mu - e) \pm (\tfrac{1}{12}e + \tfrac{1}{6}\mu + \tfrac{1}{4}\tau - \tfrac{1}{2}x)/\sqrt{2} \end{aligned} \tag{11.65}$$

By eq. 11.26

$$M_{f22\underline{N}} = \sum_{c,d} \alpha_{\underline{N}cd}\sigma_c(\sigma'_d) \tag{11.66}$$

$$= \tfrac{1}{4}(e + \mu + \tau + x)\Omega_{+00} + \tfrac{1}{2}a_+\Omega_{+01} + \tfrac{1}{2}(-7e/12 + \tfrac{1}{6}\mu - \tfrac{1}{4}\tau - \tfrac{1}{2}x)\Omega_{+03} +$$
$$+ \tfrac{1}{2}a_-\Omega_{+13} + \tfrac{1}{4}[\tfrac{1}{3}\sqrt{2}(\mu - e) - \tfrac{1}{12}e - \tfrac{1}{6}\mu - \tfrac{1}{4}\tau + \tfrac{1}{2}x]\Omega_{+11} +$$

$$+ \tfrac{1}{4}(-\tfrac{1}{12}e - \tfrac{1}{6}\mu - \tfrac{1}{4}\tau + \tfrac{1}{2}x)\Omega_{+22} + \tfrac{1}{8}(e - \tau)\Omega_{+33}$$

The four neutrino masses, e, μ, τ, x, (and the mass mixing parameters) set seven distinct coefficients of the Ω_+ matrices, or equivalently ten coefficients of the $\sigma_c(\sigma'_d)$ matrices.

11.8 Determination of the Values of the Coupling Constants and Mass Matrices

While the previous developments have determined the structure of the Standard Model Lagrangian (modulo CP violation terms and with the existence of Higgs bosons still an open question experimentally) the values of the Electroweak coupling constants and Weinberg angle, the strong interaction coupling constant, (and the gravitational coupling constant upon unification with quantum gravity), the masses of ElectroWeak vector bosons, and the mass matrices of the four fermion species remain to be determined.

There are two general methods to pin down the values of constants within the framework of Quantum Field Theory: 1) imposition of a (broken) symmetry group structure that sets the values of constants; and 2) the existence of consistency conditions that set the values of some or all of the theory's constants. There are many model theories illustrating approach 1.

The only major 4-dimensional quantum field theory illustrating approach 2 is the Johnson-Baker-Willey model[122] of massless Quantum Electrodynamics within which this author[123] developed an approximate solution of the Schwinger-Dyson equations to obtain a solution of the eigenvalue function for the physical fine structure constant α. This theory has been superseded by ElectroWeak theory. Nevertheless it illustrates the possibility of using consistency conditions to determine the value of constants appearing in a quantum field theory.

11.9 Renormalization of the Standard Model

The Standard Model that we have developed has the same renormalizability as the conventional Standard Model except that we have assigned complex 3-momenta to quarks. This was an arbitrary choice. However it made a point of difference between quarks and leptons, and we see that they differ very much in experiments. In addition it made the introduction of the SU(3) color interaction seem natural.

However, the extra three imaginary spatial dimensions would make our theory non-renormalizable in the quark sector <u>if</u> we followed the standard renormalization approach.

Therefore, we suggest that a different form of quantum field theory[124] be used in which our Standard Model, and a unified theory of the Standard Model and Quantum Gravity, are both fully renormalizable. In fact all diagrams calculated in perturbation theory are finite in this new approach. The reader is directed to Blaha (2005) for a detailed account of this new form of quantum field theory.

---

[122] M. Baker and K. Johnson, Phys. Rev. **D8**, 1110 (1973) and references therein.

[123] S. Blaha, Phys. Rev. **D9**, 2246 (1974).

[124] Blaha (2005a). This book is the second book in this volume and the second edition of Blaha (2003). Blaha (2005a) discusses this type of quantum field theory in detail including issues such as anomalies and unitarity.

Thus our form of the Standard Model is fully renormalizable in the new form of quantum field theory.

11.10 Unification with Quantum Gravity

The unification of our Standard Model with Quantum gravity has a Lagrangian of the form[125]

$$\mathrm{L}_{\mathrm{Unified}} = \int d^4y\ \mathscr{L}_{\mathrm{Unified}} \tag{11.67}$$

$$\mathscr{L}_{\mathrm{Unified}} = \mathrm{J}\sqrt{g(\mathrm{X})}\big(\mathscr{L}_{\mathrm{F}}^{\mathrm{Grav}}(\mathrm{X}^\mu) + \mathscr{L}_{\mathrm{SM3}}(\mathrm{X}^\mu)\big) + \mathscr{L}_{\mathrm{C}} \tag{11.68}$$

with

$$\mathscr{L}_{\mathrm{F}}^{\mathrm{Grav}}(\mathrm{X}^\mu) = (2\kappa^2)^{-1}{}_{\mathrm{X}}\mathrm{R} \tag{11.69}$$

$$\mathfrak{L}_{\mathrm{SM3}} = \Psi_{\mathrm{L}}{}^{\dagger}\gamma^0 i\gamma^\mu \mathrm{D}_{\mathrm{L}\mu}\Psi_{\mathrm{L}} - \Psi_{\mathrm{R}}{}^{\dagger}\gamma^0 i\gamma^\mu \mathrm{D}_{\mathrm{R}\mu}\Psi_{\mathrm{R}} + \tag{11.70}$$

$$+ \Psi_{\mathrm{qL}}{}^{\dagger}\gamma^0 i\gamma^\mu \mathfrak{D}_{\mathrm{qL}\mu}\Psi_{\mathrm{qL}} + \Psi_{\mathrm{qR}}{}^{\dagger}\gamma^0 i\gamma^\mu \mathfrak{D}_{\mathrm{qR}\mu}\Psi_{\mathrm{qR}} + \mathfrak{L}_{\mathrm{Gauge}} + \mathfrak{L}_{\mathrm{Higgs}}$$

where $\mathscr{L}_{\mathrm{SM3}}$ is the complete "normal" Quantum Field theory Lagrangian for our three plus one generation Standard Model (see eq. 10.1 for the one generation version), with the generations indices suppressed, and written in a general covariant form; g(X) is the absolute value of the determinant of $g_{\mu\nu}$; and J is the Jacobian of eq. A.21 of Blaha (2005a). *All particle fields in* $\mathscr{L}_{SM3}$ *are assumed to be functions of the* X^μ *coordinate only:*

$$\mathrm{X}^\mu(\mathrm{y}) = \mathrm{y}^\mu + i\ \mathrm{Y}^\mu(\mathrm{y})/\mathrm{M_c}^2 \tag{11.71}$$

where $Y^\mu(y)$ *is a free vector quantum field. The dependence of the particle fields on the "underlying" coordinates* y^μ *is assumed to be solely through* X^μ. The Lagrangian $\mathscr{L}_{\mathrm{Unified}}$ is a separable Lagrangian of the type of eq. A.26 of Blaha (2005a) embodying the composition of extrema described in Appendix A of that book. The generational dependence of $\mathscr{L}_{\mathrm{SM3}}$ is in its mass/Higgs terms.

The coordinate part of the Lagrangian $\mathscr{L}_{\mathrm{C}}$ is

$$\mathscr{L}_{\mathrm{C}} = -\tfrac{1}{4}\, \mathrm{F_Y}^{\mu\nu}\mathrm{F}_{\mathrm{Y}\mu\nu} \tag{11.72}$$

with

$$\mathrm{F}_{\mathrm{Y}\mu\nu} = \partial \mathrm{Y}_\mu/\partial \mathrm{y}^\nu - \partial \mathrm{Y}_\nu/\partial \mathrm{y}^\mu \tag{11.73}$$

---

[125] See chapter 9 of Blaha (2005a) for a detailed discussion of Two-Tier quantum field theory, Two-Tier Quantum Gravity and the form of this Lagrangian.

and

$$F_Y^{\mu\nu} = \eta^{\mu\alpha}\eta^{\nu\beta}F_{Y\alpha\beta} \tag{11.74}$$

The path integral formulation of this type of quantum field theory has been shown to be fully renormalizable and unitary in Blaha (2005a) following in this volume. We state again that all diagrams in its perturbation theory are finite and do not require regularization, cutoffs, etc. Anomalies resulting from the problems of dimensional regularization due to γ_5 dependence, such as the triangle anomaly, are absent from this theory.

11.11 Fermion Feynman Propagators of our MultiGeneration Standard Model

We now turn to consider the consideration of the perturbation theory of our Standard Model. We will begin by calculating the free field Feynman propagators in Two-Tier quantum field theory for a Dirac fermion such a charged lepton, for a tachyonic fermion such a neutrino, for a complexon fermion such as a u-type quark, and for a tachyonic complexon fermion such as a d-type quark.

Free Two-Tier quantum field theories are the same as the corresponding quantum field theory propagator except each integrand also has a Gaussian exponential damping factor denoted R(**p**, y).

The Two-Tier Feynman propagator for a Dirac fermion has the form[126]

$$iS_F^{TT}(y_1 - y_2) = i\int \frac{d^4p\, e^{-ip\cdot(y_1 - y_2)}\,(\not{p} + m)\,R(\mathbf{p}, y_1 - y_2)}{(2\pi)^4\,(p^2 - m^2 + i\varepsilon)} \tag{11.75}$$

using eq. 10.35 where[127]

$$R(\mathbf{p}, y_1 - y_2) = \exp[-p^i p^j \Delta_{Tij}(y_1 - y_2)/M_c^4] \tag{11.76}$$

(summations are over spatial indices only) and

$$\Delta_{Tij}(z) = \int d^3k\, e^{-ik\cdot z}(\delta_{ij} - k_i k_j/\mathbf{k}^2)/[(2\pi)^3 2\omega_k] \tag{11.77}$$

See the discussion of section 10.2 for tachyon and complexon propagators which also applies to our three plus one generation theory with the addition of generational indices.

[126] A generalization of eq. 7.45 Blaha (2005a).
[127] Eqs. 6.53 – 6.62 of Blaha (2005a), the second book in this volume.

11.12 Fermion Self-Mass as a Function of Momentum

In our Two-Tier models there is a mass scale M_c which is large or it would have been detected experimentally. M_c might be as large as the Planck mass. For energies that are small compared to M_c the lowest order fermion self-mass contribution (Fig. 11.2) yields the usual result found in conventional quantum field theory. However for energies comparable to, or larger than, M_c the self-mass is proportional to M_c

$$\Sigma(p) \propto M_c \tag{11.78}$$

So the "small" physical masses that are observed at low energy ($\ll M_c$) become large due to self-mass effects at very large energies. Physically this growth in the effective mass of a fermion (or massive vector boson) is due the growth in the cloud of virtual Y-quanta surrounding the particle, which generate the Gaussian factor $R(\mathbf{p}, y_1 - y_2)$.

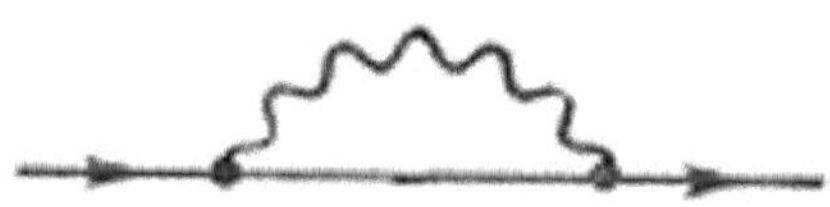

Figure 11.2. Lowest order self-mass diagram of a fermion.

Fig. 11.3 graphically depicts the cloud of Y-quanta surrounding a fermion. (The Y vector field appears in the coordinates eq. 11.71.)

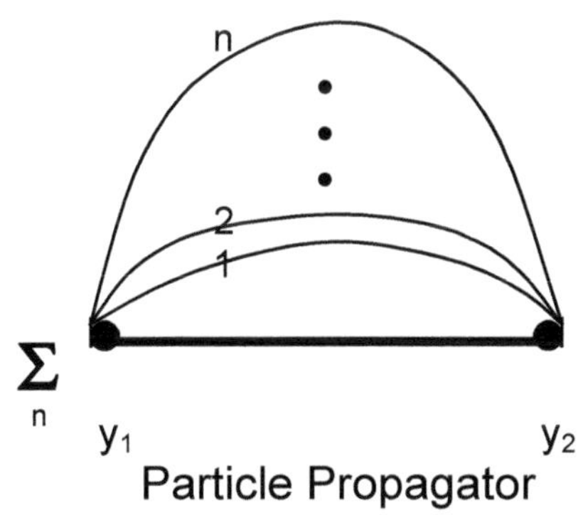

Figure 11.3. Propagating fermion surrounded by Y quanta.

The manner of growth in the self-mass of particles at high energy is the opposite of the change in effective electric charge with energy. As the energy of a probe becomes "infinite" the "infinite" bare charge of a particle manifests itself while at low energies the electric charge of a particle is the familiar physical charge. Physically we attribute the change of electric charge with energy to vacuum polarization effects. At large distances (low energy) the charge of a particle is masked by the polarization effects of positrons in the vacuum that are "drawn" by the charge of the (negative charge) particle. At very small distances (large energy) the polarization effects are diminished and the large, less shielded bare charge shows its large size.

In this respect we see the contrasting behavior of charge shielding and self-mass enhancement.

Appendix 11-A. Higgs Mechanism for Tachyons

The Higgs mechanism is currently the favored mechanism for spontaneous symmetry breaking and to give masses to fermions and bosons. The nature of the free tachyon Lagrangian terms,

$$\mathcal{L}_{\text{free}} = \psi_T^\dagger i\gamma^0\gamma^5(\gamma^\mu\partial/\partial x^\mu + m_0)\psi_T(x) \quad (11\text{-A.1})$$

where m_0 is a possible bare mass, requires a Higgs sector that contributes to a tachyon mass through spontaneous symmetry breaking having the general form:

$$\mathcal{L}_{\text{Higgs}} = \tfrac{1}{2}\partial\phi/\partial x^\mu\partial\phi/\partial x_\mu - \psi_T^\dagger i\gamma^0\gamma^5\psi_T\phi - V(\phi) \quad (11\text{-A.2})$$

where

$$V(\phi) = g^2(\phi^2 - (\delta m)^2)^2 \quad (11\text{-A.3})$$

Note the quadratic term in ϕ – the "mass" term has the negative sign of a tachyon – again showing that tachyons are a feature of modern physics. In the present case the quartic term "stabilizes" the tachyon field which then can be shifted to the minimum of the potential.

The spontaneous symmetry breaking resulting from the potential $V(\phi)$ causes the mass of the tachyon to change to

$$m = m_0 \pm \delta m \quad (11\text{-A.4})$$

A choice of vacuum state corresponding to the positive sign in eq. 11-A.4 causes the tachyon mass to increase. A choice of the vacuum state corresponding to the negative sign causes the tachyon mass to decrease, and could cause m to become negative. However this event would not make the tachyon into a normal particle. Rather it would essentially transform the tachyon into its antiparticle.

REFERENCES

Akhiezer, N. I., Frink, A. H. (tr), 1962, *The Calculus of Variations* (Blaisdell Publishing, New York, 1962).

Bjorken, J. D., Drell, S. D., 1964, *Relativistic Quantum Mechanics* (McGraw-Hill, New York, 1965).

Bjorken, J. D., Drell, S. D., 1965, *Relativistic Quantum Fields* (McGraw-Hill, New York, 1965).

Blaha, S., 1995, *C++ for Professional Programming* (International Thomson Publishing, Boston, 1995).

_______, 1998, *Cosmos and Consciousness* (Pingree-Hill Publishing, Auburn, NH, 1998 and 2002).

_______, 2002, *A Finite Unified Quantum Field Theory of the Elementary Particle Standard Model and Quantum Gravity Based on New Quantum Dimensions™ & a New Paradigm in the Calculus of Variations* (Pingree-Hill Publishing, Auburn, NH, 2002).

_______, 2004, *Quantum Big Bang Cosmology: Complex Space-time General Relativity, Quantum Coordinates,™ Dodecahedral Universe, Inflation, and New Spin 0, ½, 1 & 2 Tachyons & Imagyons* (Pingree-Hill Publishing, Auburn, NH, 2004).

______, *2005a, Quantum Theory of the Third Kind: A New Type of Divergence-free Quantum Field Theory Supporting a Unified Standard Model of Elementary Particles and Quantum Gravity based on a New Method in the Calculus of Variations* (Pingree-Hill Publishing, Auburn, NH, 2005).

______, 2005b, *The Metatheory of Physics Theories, and the Theory of Everything as a Quantum Computer Language* (Pingree-Hill Publishing, Auburn, NH, 2005).

______, 2005c, *The Equivalence of Elementary Particle Theories and Computer Languages: Quantum Computers, Turing Machines, Standard Model, Superstring Theory, and a Proof that Gödel's Theorem Implies Nature Must Be Quantum* (Pingree-Hill Publishing, Auburn, NH, 2005).

______, 2006a, *The Foundation of the Forces of Nature* (Pingree-Hill Publishing, Auburn, NH, 2006).

______, 2006b, *A Derivation of ElectroWeak Theory based on an Extension of Special Relativity; Black Hole Tachyons; & Tachyons of Any Spin.* (Pingree-Hill Publishing, Auburn, NH, 2006).

_______, 2007a, *Physics Beyond the Light Barrier: The Source of Parity Violation, Tachyons, and A Derivation of Standard Model Features* (Pingree-Hill Publishing, Auburn, NH, 2007).

______, 2007b, *The Origin of the Standard Model: The Genesis of Four Quark and Lepton Species, Parity Violation, the ElectroWeak Sector, Color SU(3), Three Visible Generations of Fermions, and One Generation of Dark Matter with Dark Energy* (Pingree-Hill Publishing, Auburn, NH, 2007).

______, *2008a, A Direct Derivation of the Form of the Standard Model From GL(16) (Pingree-Hill Publishing, Auburn, NH, 2008).*

______, 2008b, *A Complete Derivation of the Form of the Standard Model With a New Method to Generate Particle Masses Second Edition* (Pingree-Hill Publishing, Auburn, NH, 2008)

______, 2009, *The Algebra of Thought & Reality: The Mathematical Basis for Plato's Theory of Ideas, and Reality Extended to Include A Priori Observers and Space-Time Second Edition* (Pingree-Hill Publishing, Auburn, NH, 2009).

______, 2010a, *Operator Metaphysics: A New Metaphysics Based on a New Operator Logic and a New Quantum Operator Logic that Lead to a Mathematical Basis for Plato's Theory of Ideas and Reality* (Pingree-Hill Publishing, Auburn, NH, 2010).

______, 2010b, *The Standard Model's Form Derived from Operator Logic, Superluminal Transformations and GL(16)* (Pingree-Hill Publishing, Auburn, NH, 2010).

______, 2010c, *SuperCivilizations: Civilizations as Superorganisms* (McMann-Fisher Publishing, Auburn, NH, 2010).

______, 2011a, *21st Century Natural Philosophy Of Ultimate Physical Reality* (McMann-Fisher Publishing, Auburn, NH, 2011).

______, 2011b, *All the Universe! Faster Than Light Tachyon Quark Starships & Particle Accelerators with the LHC as a Prototype Starship Drive Scientific Edition* (Pingree-Hill Publishing, Auburn, NH, 2011).

______, 2011c, *From Asynchronous Logic to The Standard Model to Superflight to the Stars* (Blaha Research, Auburn, NH, 2011).

______, 2012a, *From Asynchronous Logic to The Standard Model to Superflight to the Stars volume 2: Superluminal CP and CPT, U(4) Complex General Relativity and The Standard Model, Complex Vierbein General Relativity, Kinetic Theory, Thermodynamics* (Blaha Research, Auburn, NH, 2012).

______, 2012b, *Standard Model Symmetries, And Four And Sixteen Dimension Complex Relativity; The Origin Of Higgs Mass Terms* (Blaha Reasearch, Auburn, NH, 2012).

______, 2013a, *Multi-Stage Space Guns, Micro-Pulse Nuclear Rockets, and Faster-Than-Light Quark-Gluon Ion Drive Starships* (Blaha Research, Auburn, NH, 2013).

______, 2013b, *The Bridge to Dark Matter; A New Sister Universe; Dark Energy; Inflatons; Quantum Big Bang; Superluminal Physics; An Extended Standard Model Based on Geometry* (Blaha Reasearch, Auburn, NH, 2013).

______, 2014a, *Universes and Megaverses: From a New Standard Model to a Physical Megaverse; The Big Bang; Our Sister Universe's Wormhole; Origin of the Cosmological Constant, Spatial Asymmetry of the Universe, and its Web of Galaxies; A Baryonic Field between Universes and Particles; Megaverse Extended Wheeler-DeWitt Equation* (Blaha Reasearch, Auburn, NH, 2014).

______, 2014b, *All the Megaverse! Starships Exploring the Endless Universes of the Cosmos Using the Baryonic Force* (Blaha Research, Auburn, NH, 2014).

______, 2014c, *All the Megaverse! II Between Megaverse Universes: Quantum Entanglement Explained by the Megaverse Coherent Baryonic Radiation Devices – PHASERs Neutron Star Megaverse Slingshot Dynamics Spiritual and UFO Events, and the Megaverse Microscopic Entry into the Megaverse* (Blaha Research, Auburn, NH, 2014).

______, 2015a, *PHYSICS IS LOGIC PAINTED ON THE VOID: Origin of Bare Masses and The Standard Model in Logic, U(4) Origin of the Generations, Normal and Dark Baryonic Forces, Dark Matter, Dark Energy, The Big Bang, Complex General Relativity, A Megaverse of Universe Particles* (Blaha Research, Auburn, NH, 2015).

______, 2015b, *PHYSICS IS LOGIC Part II: The Theory of Everything, The Megaverse Theory of Everything, U(4)⊗U(4) Grand Unified Theory (GUT), Inertial Mass = Gravitational Mass, Unified Extended Standard Model and a New Complex General Relativity with Higgs Particles, Generation Group Higgs Particles* (Blaha Research, Auburn, NH, 2015).

______, 2015c, *The Origin of Higgs ("God") Particles and the Higgs Mechanism: Physics is Logic III, Beyond Higgs – A Revamped Theory With a Local Arrow of Time, The Theory of Everything Enhanced, Why Inertial Frames are Special, Universes of the Mind* (Blaha Research, Auburn, NH, 2015).

______, 2015d, *The Origin of the Eight Coupling Constants of The Theory of Everything: U(8) Grand Unified Theory of Everything (GUTE), S^8 Coupling Constant Symmetry, Space-Time Dependent Coupling Constants, Big Bang Vacuum Coupling Constants, Physics is Logic IV* (Blaha Research, Auburn, NH, 2015).

______, 2016a, *New Types of Dark Matter, Big Bang Equipartition, and A New U(4) Symmetry in the Theory of Everything: Equipartition Principle for Fermions, Matter is 83.33% Dark, Penetrating the Veil of the Big Bang, Explicit QFT Quark Confinement and Charmonium, Physics is Logic V* (Blaha Research, Auburn, NH, 2016).

______, 2016b, *The Periodic Table of the 192 Quarks and Leptons in The Theory of Everything: The U(4) Layer Group, Physics is Logic VI* (Blaha Research, Auburn, NH, 2016).

______, 2016c, *New Boson Quantum Field Theory, Dark Matter Dynamics, Dark Matter Fermion Layer Mixing, Genesis of Higgs Particles, New Layer Higgs Masses, Higgs Coupling Constants, Non-Abelian Higgs Gauge Fields, Physics is Logic VII* (Blaha Research, Auburn, NH, 2016).

______, 2016d, *Unification of the Strong Interactions and Gravitation: Quark Confinement Linked to Modified Short-Distance Gravity; Physics is Logic VIII* (Blaha Research, Auburn, NH, 2016).

______, 2016e, *MoND: Unification of the Strong Interactions and Gravitation II, Quark Confinement Linked to Large-Scale Gravity, Physics is Logic IX* (Blaha Research, Auburn, NH, 2016).

______, 2016f, *CQ Mechanics: A Unification of Quantum & Classical Mechanics, Quantum/Semi-Classical Entanglement, Quantum/Classical Path Integrals, Quantum/Classical Chaos* (Blaha Research, Auburn, NH, 2016).

______, 2016g, *GEMS: Unified Gravity, ElectroMagnetic and Strong Interactions: Manifest Quark Confinement, A Solution for the Proton Spin Puzzle, Modified Gravity on the Galactic Scale* (Pingree Hill Publishing, Auburn, NH, 2016).

______, 2016h, *Unification of the Seven Boson Interactions based on the Riemann-Christoffel Curvature Tensor* (Pingree Hill Publishing, Auburn, NH, 2016).

______, 2017a, *Unification of the Eleven Boson Interactions based on 'Rotations of Interactions'* (Pingree Hill Publishing, Auburn, NH, 2017).

______, 2017b, *The Origin of Fermions and Bosons, and Their Unification* (Pingree Hill Publishing, Auburn, NH, 2017).

______, 2017c, *Megaverse: The Universe of Universes* (Pingree Hill Publishing, Auburn, NH, 2017).

______, 2017d, *SuperSymmetry and the Unified SuperStandard Model* (Pingree Hill Publishing, Auburn, NH, 2017).

______, 2017e, *From Qubits to the Unified SuperStandard Model with Embedded SuperStrings: A Derivation* (Pingree Hill Publishing, Auburn, NH, 2017).

______, 2017f, *The Unified SuperStandard Model in Our Universe and the Megaverse: Quarks, ... ,* (Pingree Hill Publishing, Auburn, NH, 2017).

______, 2018a, *The Unified SuperStandard Model and the Megaverse SECOND EDITION A Deeper Theory based on a New Particle Functional Space that Explicates Quantum Entanglement Spookiness (Volume 1)* (Pingree Hill Publishing, Auburn, NH, 2018).

______, 2018b, *Cosmos Creation: The Unified SuperStandard Model, Volume 2, SECOND EDITION* (Pingree Hill Publishing, Auburn, NH, 2018).

______, 2018c, *God Theory (*Pingree Hill Publishing, Auburn, NH, 2018).

______, 2018d, *Immortal Eye: God Theory: Second Edition* (Pingree Hill Publishing, Auburn, NH, 2018).

______, 2018e, *Unification of God Theory and Unified SuperStandard Model THIRD EDITION* (Pingree Hill Publishing, Auburn, NH, 2018).

______, 2019a, *Calculation of: QED α = 1/137, and Other Coupling Constants of the Unified SuperStandard Theory* (Pingree Hill Publishing, Auburn, NH, 2019).

______, 2019b, *Coupling Constants of the Unified SuperStandard Theory SECOND EDITION* (Pingree Hill Publishing, Auburn, NH, 2019).

______, 2019c, *New Hybrid Quantum Big_Bang–Megaverse_Driven Universe with a Finite Big Bang and an Increasing Hubble Constant* (Pingree Hill Publishing, Auburn, NH, 2019).

______, 2019d, *The Universe, The Electron and The Vacuum* (Pingree Hill Publishing, Auburn, NH, 2019).

______, 2019e, *Quantum Big Bang – Quantum Vacuum Universes (Particles)* (Pingree Hill Publishing, Auburn, NH, 2019).

______, 2019f, *The Exact QED Calculation of the Fine Structure Constant Implies ALL 4D Universes have the Same Physics/Life Prospects* (Pingree Hill Publishing, Auburn, NH, 2019).

______, 2019g, *Unified SuperStandard Theory and the SuperUniverse Model: The Foundation of Science* (Pingree Hill Publishing, Auburn, NH, 2019).

______, 2020a, *Quaternion Unified SuperStandard Theory (The QUeST) and Megaverse Octonion SuperStandard Theory (MOST)* (Pingree Hill Publishing, Auburn, NH, 2020).

______, 2020b, *United Universes Quaternion Universe - Octonion Megaverse* (Pingree Hill Publishing, Auburn, NH, 2020).

______, 2020c, *Unified SuperStandard Theories for Quaternion Universes & The Octonion Megaverse* (Pingree Hill Publishing, Auburn, NH, 2020).

______, 2020d, *The Essence of Eternity: Quaternion & Octonion SuperStandard Theories* (Pingree Hill Publishing, Auburn, NH, 2020).

______, 2020e, *The Essence of Eternity II* (Pingree Hill Publishing, Auburn, NH, 2020).

______, 2020f, *A Very Conscious Universe* (Pingree Hill Publishing, Auburn, NH, 2020).

______, 2020g, *Hypercomplex Universe* (Pingree Hill Publishing, Auburn, NH, 2020).

______, 2020h, *Beneath the Quaternion Universe* (Pingree Hill Publishing, Auburn, NH, 2020).

______, 2020i, *Why is the Universe Real? From Quaternion & Octonion to Real Coordinates* (Pingree Hill Publishing, Auburn, NH, 2020).

______, 2020j, *The Origin of Universes: of Quaternion Unified SuperStandard Theory (QUeST); and of the Octonion Megaverse (UTMOST)* (Pingree Hill Publishing, Auburn, NH, 2020).

______, 2020k, *The Seven Spaces of Creation: Octonion Cosmology* (Pingree Hill Publishing, Auburn, NH, 2020).

______, 2020l, *From Octonion Cosmology to the Unified SuperStandard Theory of Particles* (Pingree Hill Publishing, Auburn, NH, 2020).

______, 2021a, *Pioneering the Cosmos* (Pingree Hill Publishing, Auburn, NH, 2021).

______, 2021b, *Pioneering the Cosmos II* (Pingree Hill Publishing, Auburn, NH, 2021).

______, 2021c, *Beyond Octonion Cosmology* (Pingree Hill Publishing, Auburn, NH, 2021).

______, 2021d, *Universes are Particles* (Pingree Hill Publishing, Auburn, NH, 2021).

______, 2021e, *Octonion-like dna-based life, Universe expansion is decay, Emerging New Physics* (Pingree Hill Publishing, Auburn, NH, 2021).

______, 2021f, *The Science of Creation New Quantum Field Theory of Spaces* (Pingree Hill Publishing, Auburn, NH, 2021).

______, 2021g, *Quantum Space Theory With Application to Octonion Cosmology & Possibly To Fermionic Condensed Matter* (Pingree Hill Publishing, Auburn, NH, 2021).

______, 2021h, *21st Century Natural Philosophy of Octonion Cosmology , and Predestination, Fate, and Free Will* (Pingree Hill Publishing, Auburn, NH, 2021).

______, 2021i, *Beyond Octonion Cosmology II : Origin of the Quantum; A New Generalized Field Theory (GiFT); A Proof of the Spectrum of Universes; Atoms in Higher Universes* (Pingree Hill Publishing, Auburn, NH, 2021).

Eddington, A. S., 1952, *The Mathematical Theory of Relativity* (Cambridge University Press, Cambridge, U.K., 1952).

Fant, Karl M., 2005, *Logically Determined Design: Clockless System Design With NULL Convention Logic* (John Wiley and Sons, Hoboken, NJ, 2005).

Feinberg, G. and Shapiro, R., 1980, *Life Beyond Earth: The Intelligent Earthlings Guide to Life in the Universe* (William Morrow and Company, New York, 1980).

Gelfand, I. M., Fomin, S. V., Silverman, R. A. (tr), 2000, *Calculus of Variations* (Dover Publications, Mineola, NY, 2000).

Giaquinta, M., Modica, G., Souchek, J., 1998, *Cartesian Coordinates in the Calculus of Variations* Volumes I and II (Springer-Verlag, New York, 1998).

Giaquinta, M., Hildebrandt, S., 1996, *Calculus of Variations* Volumes I and II (Springer-Verlag, New York, 1996).

Gradshteyn, I. S. and Ryzhik, I. M., 1965, *Table of Integrals, Series, and Products* (Academic Press, New York, 1965).

Heitler, W., 1954, *The Quantum Theory of Radiation* (Claendon Press, Oxford, UK, 1954).

Huang, Kerson, 1992, *Quarks, Leptons & Gauge Fields 2nd Edition* (World Scientific Publishing Company, Singapore, 1992).

Jost, J., Li-Jost, X., 1998, *Calculus of Variations* (Cambridge University Press, New York, 1998).

Kaku, Michio, 1993, *Quantum Field Theory*, (Oxford University Press, New York, 1993).

Kirk, G. S. and Raven, J. E., 1962, *The Presocratic Philosophers* (Cambridge University Press, New York, 1962).

Landau, L. D. and Lifshitz, E. M., 1987, *Fluid Mechanics 2nd Edition*, (Pergamon Press, Elmsford, NY, 1987).

Misner, C. W., Thorne, K. S., and Wheeler, J. A., 1973, *Gravitation* (W. H. Freeman, New York, 1973).

Rescher, N., 1967, *The Philosophy of Leibniz* (Prentice-Hall, Englewood Cliffs, NJ, 1967).

Rieffel, Eleanor and Polak, Wolfgang, 2014, *Quantum Computing* (MIT Press, Cambridge, MA, 2014).

Riesz, Frigyes and Sz.-Nagy, Béla, 1990, *Functional Analysis* (Dover Publications, New York, 1990).

Sagan, H., 1993, *Introduction to the Calculus of Variations* (Dover Publications, Mineola, NY, 1993).

Sakurai, J. J., 1964, *Invariance Principles and Elementary Particles* (Princeton University Press, Princeton, NJ, 1964).

Weinberg, S., 1972, *Gravitation and Cosmology* (John Wiley and Sons, New York, 1972).

Weinberg, S., 1995, *The Quantum Theory of Fields Volume I* (Cambridge University Press, New York, 1995).

INDEX

About the Author

Stephen Blaha is a well-known Physicist and Man of Letters with interests in Science, Society and civilization, the Arts, and Technology. He had an Alfred P. Sloan Foundation scholarship in college. He received his Ph.D. in Physics from Rockefeller University. He has served on the faculties of several major universities. He was also a Member of the Technical Staff at Bell Laboratories, a manager at the Boston Globe Newspaper, a Director at Wang Laboratories, and President of Blaha Software Inc. and of Janus Associates Inc. (NH).

Among other achievements he was a co-discoverer of the "r potential" for heavy quark binding developing the first (and still the only demonstrable) non-Aeolian gauge theory with an "r" potential; first suggested the existence of topological structures in superfluid He-3; first proposed Yang-Mills theories would appear in condensed matter phenomena with non-scalar order parameters; first developed a grammar-based formalism for quantum computers and applied it to elementary particle theories; first developed a new form of quantum field theory without divergences (thus solving a major 60 year old problem that enabled a unified theory of the Standard Model and Quantum Gravity without divergences to be developed); first developed a formulation of complex General Relativity based on analytic continuation from real space-time; first developed a generalized non-homogeneous Robertson-Walker metric that enabled a quantum theory of the Big Bang to be developed without singularities at $t = 0$; first generalized Cauchy's theorem and Gauss' theorem to complex, curved multi-dimensional spaces; received Honorable Mention in the Gravity Research Foundation Essay Competition in 1978; first developed a physically acceptable theory of faster-than-light particles; first derived a composition of extremums method in the Calculus of Variations; first quantitatively suggested that inflationary periods in the history of the universe were not needed; first proved Gödel's Theorem implies Nature must be quantum; provided a new alternative to the Higgs Mechanism, and Higgs particles, to generate masses; first showed how to resolve logical paradoxes including Gödel's Undecidability Theorem by developing Operator Logic and Quantum Operator Logic; first developed a quantitative harmonic oscillator-like model of the life cycle, and interactions, of civilizations; first showed how equations describing superorganisms also apply to civilizations. A recent book shows his theory applies successfully to the past 14 years of history and to *new* archaeological data on Andean and Mayan civilizations as well as Early Anatolian and Egyptian civilizations.

He first developed an axiomatic derivation of the form of The Standard Model from geometry – space-time properties – The Unified SuperStandard Model. It unifies all the known forces of Nature. It also has a Dark Matter sector that includes a Dark ElectroWeak sector with Dark doublets and Dark gauge interactions. It uses quantum coordinates to remove infinities that crop up in most

interacting quantum field theories and additionally to remove the infinities that appear in the Big Bang and generate inflationary growth of the universe. It shows gravity has a MOND-like form without sacrificing Newton's Laws. It relates the interactions of the MOND-like sector of gravity with the r-potential of Quark Confinement. The axioms of the theory lead to the question of their origin. We suggest in the preceding edition of this book it can be attributed to an entity with God-like properties. We explore these properties in "God Theory" and show they predict that the Cosmos exists forever although individual universes (or incarnations of our universe) "come and go." Several other important results emerge from God Theory such a functionally triune God. The Unified SuperStandard Theory has many other important parts described in the Current Edition of *The Unified SuperStandard Theory* and expanded in subsequent volumes.

Blaha has had a major impact on a succession of elementary particle theories: his Ph.D. thesis (1970), and papers, showed that quantum field theory calculations to all orders in ladder approximations could not give scaling deep inelastic electron-nucleon scattering. He later showed the eigenvalue equation for the fine structure constant α in Johnson-Baker-Willey QED had a zero at α = 1 not 1/137 by solving the Schwinger-Dyson equations to all orders in an approximation that agreed with exact results to 4th order in α thus ending interest in this theory. In 1979 at Prof. Ken Johnson's (MIT) suggestion he calculated the proton-neutron mass difference in the MIT bag model and found the result had the wrong sign reducing interest in the bag model. These results all appear in Physical Review papers. In the 2000's he repeatedly pointed out the shortcomings of SuperString theory and showed that The Standard Model's form could be derived from space-time geometry by an extension of Lorentz transformations to faster than light transformations. This deeper space-time basis greatly increases the possibility that it is part of THE fundamental theory. Recently, Blaha showed that the Weak interactions differed significantly from the Strong, electromagnetic and gravitation interactions in important respects while these interactions had similar features, and suggested that ElectroWeak theory, which is essentially a glued union of the Weak interactions and Electromagnetism, possibly modulo unknown Higgs particle features, be replaced by a unified theory of the other interactions combined with a stand-alone Weak interaction theory. Blaha also showed that, if Charmonium calculations are taken seriously, the Strong interaction coupling constant is only a factor of five larger than the electromagnetic coupling constant, and thus Strong interaction perturbation theory would make sense and yield physically meaningful results.

In graduate school (1965-71) he wrote substantial papers in elementary particles and group theory: The Inelastic E- P Structure Functions in a Gluon Model. Phys. Lett. B40:501-502,1972; Deep-Inelastic E-P Structure Functions In A Ladder Model With Spin 1/2 Nucleons, Phys.Rev. D3:510-523,1971; Continuum Contributions To The Pion Radius, Phys. Rev. 178:2167-2169,1969; Character Analysis of U(N) and SU(N), J. Math. Phys. 10, 2156 (1969); and The Calculation of the Irreducible Characters of the Symmetric Group in Terms of the

Compound Characters, (Published as Blaha's Lemma in D. E. Knuth's book: *The Art of Computer Programming Vols. 1 – 4*).

In the early 1980's Blaha was also a pioneer in the development of UNIX for financial, scientific and Internet applications: benchmarked UNIX versions showing that block size was critical for UNIX performance, developing financial modeling software, starting database benchmarking comparison studies, developing Internet-like UNIX networking (1982) and developing a hybrid shell programming technique (1982) that was a precursor to the PERL programming language. He was also the manager of the AT&T ten-year future products development database. His work helped lead to commercial UNIX on computers such as Sun Micros, IBM AIX minis, and Apple computers.

In the 1980's he pioneered the development of PC Desktop Publishing on laser printers and was nominated for three "Awards for Technical Excellence" in 1987 by PC Magazine for PC software products that he designed and developed.

Recently he has developed a theory of Megaverses – actual universes of which our universe is one – with quantum particle-like properties based on the Wheeler-DeWitt equation of Quantum Gravity. He has developed a theory of a baryonic force, which had been conjectured many years ago, and estimated the strength of the force based on discrepancies in measurements of the gravitational constant G. This force, operative in D-dimensional space, can be used to escape from our universe in "uniships" which are the equivalent of the faster-than-light starships proposed in the author's earlier books. Thus travel to other universes, as well as to other stars is possible.

Blaha also considered the complexified Wheeler-DeWitt equation and showed that its limitation to real-valued coordinates and metrics generated a Cosmological Constant in the Einstein equations.

The author has also recently written a series of books on the serious problems of the United States and their solution as well as a book on the decline of Mankind that will follow from current social and genetic trends in Mankind.

In the past twenty years Dr. Blaha has written over 80 books on a wide range of topics. Some recent major works are: *From Asynchronous Logic to The Standard Model to Superflight to the Stars*, *All the Universe!*, *SuperCivilizations: Civilizations as Superorganisms, America's Future: an Islamic Surge, ISIS, al Qaeda, World Epidemics, Ukraine, Russia-China Pact, US Leadership Crisis, The Rises and Falls of Man – Destiny – 3000 AD: New Support for a Superorganism MACRO-THEORY of CIVILIZATIONS From CURRENT WORLD TRENDS and NEW Peruvian, Pre-Mayan, Mayan, Anatolian, and Early Egyptian Data, with a Projection to 3000 AD,* and *Mankind in Decline: Genetic Disasters, Human-Animal Hybrids, Overpopulation, Pollution, Global Warming, Food and Water Shortages, Desertification, Poverty, Rising Violence, Genocide, Epidemics, Wars, Leadership Failure.*

He has taught approximately 4,000 students in undergraduate, graduate, and postgraduate corporate education courses primarily in major universities, and large companies and government agencies.

Recently he developed a quantum theory, The Unified SuperStandard Theory (UST), which describes elementary particles in detail without the difficulties of conventional quantum field theory. He found that the internal symmetries of this theory could be exactly derived from an octonion theory called QUeST. He further found that another octonion theory (UTMOST) describes the Megaverse. It can hold QUeST universes such as our own universe. It has an internal symmetry structure which is a superset of the QUeST internal symmetries.

www.ingramcontent.com/pod-product-compliance
Ingram Content Group UK Ltd.
Pitfield, Milton Keynes, MK11 3LW, UK
UKHW052228270726
14060UKWH00004B/664

9 781737 264088